WIRELESS SENSOR NETWORKS

Edited by **Suraiya Tarannum**

Wireless Sensor Networks
Edited by Suraiya Tarannum

Published by InTech
Janeza Trdine 9, 51000 Rijeka, Croatia

Publishing Process Manager Sonja Mujacic

Technical Editor Martina Sirotic

Cover Designer InTech Design Team

Additional hard copies can be obtained from orders@intechopen.com

Wireless Sensor Networks, Edited by Suraiya Tarannum
p. cm.
ISBN 978-953-307-325-5

For EU product safety concerns:
IN TECH d.o.o., Prolaz Marije Krucifikse Kozulić 3, 51000 Rijeka, Croatia,
info@intechopen.com or visit our website at intechopen.com.

Contents

1

Literature Review of MAC, Routing and Cross Layer Design Protocols for WSN

Tayseer AL-Khdour, Uthman Baroudi
King Fahd University of Petroleum and Minerals
Saudi Arabi

1. Introduction

A WSN is composed of a large number of sensor nodes that are communicating using a wireless medium. The sensor nodes are deployed in the environment to be monitored in ad hoc structure. In WSN, there is sink node that collects data from all sensors, and usually not all nodes hear all other nodes. WSN is considered a multi-hop network.
Although a WSN is a wireless multi-hop network, the ease of deployment of sensor nodes, the system lifetime, the data latency, and the quality of the network distinguish WSN from traditional multi-hop wireless networks. These features must be taken into account when designing different protocols that control the operation of WSN such as MAC protocols and routing protocols. Therefore, Many MAC and Routing protocols are proposed for WSN. These protocols take into account the distinguished features of WSN. Moreover, Cross layer design protocols are proposed for WSN. In cross layer design protocols, different layers interact to optimize the performance of the WSN protocol.
In this chapter, we will present a survey of the most well known protocols for WSN. A survey of the most well-known MAC protocols is presented in section 0. Section 0 presents discussion of routing protocols of WSN and classification of these protocols according to data traffic models. The routing protocols are also classified as: data centric protocols, hierarchical protocols, location-based protocols and QoS-aware protocols. In section 0, we will present some cross layer design protocols for WSN. A summery of the cross layer design protocols is presented at the end of the section.

2. MAC protocols for WSN

In designing a MAC protocol for a Wireless Sensor Network (WSN), some of the unique features of WSN must be taken into consideration. Low-power consumption must be the main goal of the protocol. The coordination and synchronization between nodes must be minimized in the protocol. The MAC protocol must be able to support a large number of nodes. It must have a high degree of scalability. The MAC protocol must take into account the limited bandwidth availability. Since sensor nodes of a WSN are deployed randomly without a predefined infrastructure, the first objective of the MAC protocol for a WSN is the creation of the network infrastructure. The second objective is to share the medium communication between the sensor nodes (Ian et al. 2002).

IEEE 802.11 is a well-known MAC protocol for Ad hoc network (IEEE working group 1999). The energy constraints in the sensor nodes make it is unpractical to apply the IEEE 802.11 protocol directly in WSN. IEEE 802.11 has a power save mode. The power save mode in IEEE 802.11 is designed for a single hop network, where all nodes can hear each other. This is not the case in WSN. A set of MAC protocols for the WSN were proposed. Most of the existing protocols aimed to save power consumption in the sensor nodes. In the following subsections, we will discuss most of MAC protocols for WSN

2.1 S-MAC protocol

The main goal of S-MAC is to reduce energy consumption while supporting good scalability and collision avoidance. (Wei et al. 2004) extend PAMAS (Sureh S. and Cauligi 1998) by using a single channel for transmitting data packets and control packets. In designing S-MAC protocol they assume that WSN composed of many small nodes deployed in an Ad Hoc fashion. Moreover they assume that most communication will be between nodes as peers rather than one base station. It is assumed that the sensor nodes are self configured and the sensor network is dedicated to a single application or a few collaborative applications. The sensor network has the ability of in-network processing.

Ye et al identify four sources for energy wasting. The first source is collisions which will cause retransmission the packet. Transmission will consume power. The second source is overhearing; picking a packet intended to another node. The third source of energy consumption is transmission of control packets. The final source of energy consumption is idle listening. S-MAC reduces the energy waste due to these reasons. The basic idea of S-MAC is to let the node sleep and listen periodically. In sleeping mode, the node turns its radio off. The listening period is fixed according to physical layer and MAC layer parameters. The complete cycle of listening and sleeping periods is called a frame. The duty cycle is defined as the ratio of the listening interval to the frame length. Neighboring nodes can be scheduled to listen and sleep at the same time. Two neighboring nodes may have different schedules if they are synchronized by different two nodes. Nodes exchange their schedule by broadcasting a SYNC packet to their immediate neighbors. The period to send a SYNC packet is called the synchronization period. If a node wishes to transmit a packet to its neighbor it must wait until its neighbor becomes in its listening period. Fig. 1 shows 4 neighboring nodes *A*, *B*, *C*, and *D*. Nodes *A* and *C* are synchronized together (they have the same schedule , they listen and they sleep at the same time) while nodes *B* and *D* are synchronized together.

Fig. 1. S-MAC: Neighboring nodes *A* and *B* have different schedules. They synchronize with nodes *C* and *D* respectively

S-MAC forms nodes into a flat, peer-to-peer topology. To choose a schedule the node firstly listens for a fixed amount of time (at least the synchronization period). If the node does not receive a schedule within the synchronization period, the node chooses its own schedule and starts to follow it, and then it announces its schedule to its neighbors by broadcasting the SYNC packet. If it hears a schedule from one of its neighbors before it chooses or announces its own schedule, it follows that schedule. If a node receives a different schedule after it announces its own schedule, then there will be two cases, in the first case, the node has not other neighbors, then it discard its own schedule and it will follow the new schedule. In the second case, the node already follows a schedule with one of its neighbors; therefore it will adopt both schedules by waking up at the listening intervals of the two schedules. To maintain the schedule, each node maintains a schedule table that stores the schedules of all its known neighbors. To prevent case two in which neighbors miss each other forever when they follow two different schedules, a periodic neighbor discovery is introduced. Each node periodically listens for the whole synchronization period. If multiple nodes wish to talk to the same node that is in listening period, then all of them must contend for the medium. IEEE 802.11 scheme with RTS and CTS is used to avoid collision, which will save energy consumption due to the packets collision and retransmissions.

To avoid overhearing which is one of the sources of energy consumptions, each interfering nodes must go to sleep after they hear RTS and CTS. All immediate neighbors of both sender and receiver should sleep after they hear RTS or CTS. To reduce the delay due to sleeping, a technique called adaptive listening is integrated in S-MAC. Each node will wake up for a short period at the end of the transmission. In this way, if the node is the next-hop node, its neighbor is able to pass the data immediately to it instead of waiting for its scheduled listening time.

To reduce energy consumed due to control packet overhead, a message passing technique is included in S-MAC. If a node wishes to transmit a long message, the long message is fragmented into fragments and the node will transmit them in burst; one RTS and one CTS are used for all the fragments. When a node sends data, it waits for ACK. The ACK is useful to solve the hidden terminal problem. Data fragment and ACK packets have a duration field. If a node wakes up or joins the network and it receives a data or ACK packet, it will go to sleep for the period in the duration field in data or ACK packet.

Synchronization among neighboring nodes is required to remedy their clock drift. Synchronization is achieved by making all nodes exchange a relative timestamps and letting the listening period is longer than clock drift.

A disadvantage of S-MAC is that the listening interval is fixed regardless whether the node has data to send or there are data intended to it. a Traffic Aware, Energy Efficient MAC protocol is proposed for WSN (TEEM) (Chansu & Young-Bae 2005). They extend the S-MAC protocol by reducing the listening interval.

2.2 A Traffic Aware, Energy Efficient MAC protocol for Wireless Sensor Networks (TEEM)

The TEEM protocol is an extension to S-MAC In S-MAC protocol the listening interval is fixed while in TEEM protocol the listening interval depends on the traffic. In TEEM protocol; all nodes will turn their radio off much earlier when no data packet transfer exists. Furthermore, the transmission of a separate RTS is eliminated. In TEEM protocol; each listening interval is divided into two parts instead of three parts as in S-MAC protocol. In the first part of the listening interval, the node sends a SYNC packet when it has any data message ($SYNC_{data}$). If the node has no data message, it will send a SYNC packet ($SYNC_{nodata}$) in the second part of its listening interval. $SYNC_{data}$ is combined with RTS packet to form $SYNC_{rts}$. If a node does not receive $SYNC_{data}$ in the first part of its listening interval and it has no data to send it will send $SYNC_{nodata}$ in the second part of its listening interval. If a node receives a $SYNC_{rts}$ that is intended to another node, it will turn its radio off and goes to sleep until its successive listening interval starts. The intended receiver will send CTS in the second part of its listening interval. The performance evaluation of TEEM protocol shows that the percentage of sleeping time in TEEM is greater than the percentage of sleeping time in S-MAC. The number of control packets in TEEM is less than the number of control packets in S-MAC. Energy consumption in TEEM is the least compared with S-MAC and IEEE 802.11. Although the power consumption is reduced in the TEEM by decreasing the listening interval, the latency will increase since decreasing the listening interval depends only on the local traffic, traffic in the node itself and in the neighboring node, and does not take into account the traffic in the whole network. To take into account the delay in the whole network, Lin et al propose a sensor medium access control protocol with a dynamic duty cycle, DSMAC (Peng et al. 2004). DSMAC intend to achieve a good tradeoff between power consumption and latency.

2.3 Medium ACCES Control with a Dynamic duty cycle for sensor network (DSMAC)

In S-MAC the duty cycle is fixed. In DSMAC the duty cycle is changed based on average delay of the data packet and the power consumption (Peng et al. 2004). The duty cycle is defined as the ratio of the listening interval to the frame length; the frame length is the sleeping interval plus the listening interval. Duty cycle can be changed by changing the sleeping interval while fixing listening interval. As in S-MAC, the nodes in DSMAC form groups of peers. Each set of neighbors follow a common schedule. In DSMAC, one- hop packet latency is proposed which is the time since a packet gets into the queue until it is successfully sent out. The packet latency is recorded in the packet header and sent to the receiver. The receiver calculates the average packet latency. The average packet latency is an estimation of the current traffic. If the average packet latency is larger than a threshold delay (D_{max}), and if the energy consumption level greater than a threshold energy (E_{max}), then the duty cycle will be doubled by decreasing the sleeping interval such that the new frame length is half of the original frame length. Otherwise the duty cycle will be halved by doubling the sleeping interval, doubling the sleeping interval will double frame length. The purpose of changing the duty cycle by two (or half) is to maintain the old schedule, which enables neighboring nodes to communicate using the old schedule.

2.4 Timeout-MAC (T-MAC)

In T-MAC, the node will keep listening and transmitting as long as it is in an active period (Tijs & Koen 2003). An active period ends when no activation event has occurred for a specific time TA. An activation event may be firing of a periodic frame timer, reception of any data on the radio, sensing of communication on the radio, end-of-transmission of a node's own data packet or acknowledgement, or the knowledge that a data exchange of a neighbor has ended. Communications between nodes in T-MAC is performed using RTS/CTS mechanism. The node that wishes to transmit data must send an RTS and wait for the CTS. If it does not receive CTS within the TA period the node will go to sleep. The node does not receive CTS in three cases; the receiver has not received the RTS, the receiver receives RTS but it is prohibited from replying, or the receiver is sleeping. It is accepted and recommended for the node to go to sleeping in the third case. But it is not an optimal decision to go to sleeping in the first two cases. To take into account all the three cases; when the node does not receive CTS to the first RTS it will resend another RTS and if it does not receive a response to the second RTS then it will go to sleeping. Sending two RTS packets without getting a CTS indicates that the receiver cannot reply now so it is convenient for the sender to go to sleeping. TA must be long enough to receive at least the start of the CTS packet. Overhearing avoidance is achieved by the same technique used in S-MAC. One problem of the T-MAC is the early sleeping problem, which occurs in case of asymmetric communication where there are four consecutive nodes: *A*, *B*, *C*, and *D*. node *A* sends data to *B* which its final destination is *C*, at the same time *C* wishes to send data to node *D* but it cannot transmit data since a collision will occur at node *B* with the transmission form *A* to *B*, so node *C* will go to sleeping. Moreover, node *D* will go to sleeping. Later when node *B* wishes to forward the data to node *C*, it will find that node *C* is sleeping which will make node *B* to go to sleeping and transmit its data later which will increase the delay and decrease the throughput. Two solutions are proposed: future request-to-send and taking priority on full buffers (Tijs & Koen 2003).

2.5 GANGS Protocol

There are some applications, in which most of the traffic in the nodes is a forwarding traffic. For these network models, Biaz et al propose a MAC protocol (GANGS) in which the nodes are organized into clusters 0(Saad & Yawen 2004). The communication within the cluster is contention based and the communication between cluster heads is TDMA based. GANGS is an energy efficient MAC protocol. As the other protocols, the nodes in GANGS are organized into clusters. Each cluster has a head. The heads form the backbone of the sensor network. The communication between nodes within cluster is contention based while the communication between heads is TDMA based. The frame is divided into multiple contention free TDMA slots and one contention slot. Number of TDMA slots depends on the number of neighboring clusters heads. The radios of all normal nodes will be turned OFF through TDMA slots while the radios of all heads are turned ON through the entire frame.
Establishing the cluster consists of three stages: local maximum stage, inter-cluster stage and reconfiguration stage. In the local maximum stage, the nodes communicate with their neighbors and exchange their energy information. The node that has the local maximum energy claims that it is the head and sends this claim to its neighbors. In the Inter-cluster phase, new heads are added to construct the backbone. Any node that it is not a head may be in the range of one head and accepts it as a head, in the range of multiple heads and it needs to choose one of them, or it is not in the range of any head. If it is in the range of multiple heads and if it has a maximum energy, then it will be the new head, otherwise the node will select the head with the maximum power. If it is not in the range of any head, then it sends a message to a node with local maximum power to demand head service. The node with local maximum power will be the new head. Since the head consumes more energy, eventually it will no longer have the maximum energy and reconfiguration must be performed to select new heads.
As any TDMA based protocol, Synchronization between the cluster heads is needed. To arrange the TDMA schedule each head knows number of its neighbors, each head randomly choose a number in the range [1, number of neighbors+1]. Each head sends the chosen number to its neighbors. If the chosen number is the same, the head with less number of

neighbors will change its schedule. All the nodes will synchronize themselves with the head to which they belong to it.

3. Routing Protocols for WSN

WSN has distinguished characteristics over traditional wireless network that makes routing in WSN is very challenging. First; it is not possible to build a global addressing scheme due to the deployment of huge number of sensor nodes, therefore the classical IP-based routing protocols cannot be applied to sensor networks. Second, Most applications of the sensor networks require the data flow from multiple sources to a particular sink. Third, the generated data has significant traffic redundancy in it. Furthermore, sensor nodes have limited power resource and processing capacity. Due to such differences many routing protocols for WSN are proposed. The routing protocols are classified as data centric, hierarchical, or location based (Kemal & Mohamed 2005). Data-centric protocols are query-based and depend on naming of desired data. Hierarchical protocols aim at clustering the nodes so that cluster heads can do some aggregation and reduction of data to reduce energy. Location based protocols utilize the position information to relay data to the desired region rather than the whole network.

Flooding is a classical mechanism to relay data in sensor network without using any routing protocol. In flooding, each sensor node receives a data packet; it will broadcast data to all its neighbors (Sandra & Stephen 1988). Eventually the data packet will reach its destination. To reduce the data traffic in the network, gossiping is implemented in which a receiving node send packet to a randomly selected neighbors. In flooding and gossiping, a lot of energy is wasted due to unnecessary transmissions. In addition to energy loss, flooding and gossiping have many drawbacks such as implosion where duplicated message sent to the same node, and overlap where many nodes sense the same region and send similar packets to the same neighbors.

3.1 Data-Centric protocols

In data-centric routing protocol, the sink sends queries to specific regions and the sensor nodes located in the selected region will send the corresponding data to the sink (Kemal & Mohamed 2005)0. To specify the properties of the requested data, attribute-based naming is usually used. Many data centric routing protocols are proposed.

Directed Diffusion: In Directed Diffusion, a naming scheme for the data is used; attribute-value pairs for the data are used (Chalermek C. et al. 2000). The sensor nodes are queried on demand using attribute-value pairs. To create a query, an interest is defined using a list of attribute-value pairs such as name of objects, interval, duration and geographical area. The interest is broadcasted by the sink. Each node receives the interest will cache it along with the reply link to a neighbor from which the interest is received. The reply link which is called a gradient is characterized by data rate, duration and expiration time. To establish the path between the sink and source, each node will compare the attribute of received data with the values in the cached interest. Using the gradients, the receiving node will specify the outgoing link. Path repairs are possible in Directed Diffusion, when a path between a source and sink fails, a new path should be identified. Multiple paths are identified in advances so that when a path fails one of the alternative paths is chosen without any cost of searching for another path. Directed Diffusion has many advantages; since all communication is neighbor-to-neighbor there is no need for addressing mechanism. Using caching will reduce processing delay. Moreover, Direct Diffusion is energy efficient since the transmission is on demand and there is no need for maintaining global network topology. On the other hand, directed diffusion can not be applied to all sensor network-application since it is based on query-driven data delivery model. It can not be used for applications that require continues data delivery such as environmental monitoring. In addition, the data naming scheme used in Directed Diffusion is application dependent, it must be defined in advance.

Rumor Routing: Rumor Routing (David & Deborah 2002) is another variation of the Directed Diffusion. It is based on a query-driven data delivery model. In Rumor Routing, the queries are routed only to the nodes that have observed a particular event instead of querying the entire network as in Directed Diffusion. In Rumor Routing, each node maintains a list of neighbors and events table with forwarding information to all the events it knows. When a node senses an event, it adds it to its event table with a distance of zero to the event, and it generates an agent. An agent is a long-lived packet that travels the network in order to propagate information about local events to all the nodes. The agent contains an events table similar to the table in the nodes. Any node may generate a query for an event; if the node has a route to the event, it will transmit the query. If it does not, it will forward the query in a random direction. This continues until the query TTL expires, or until the query reaches a node that has observed the target event. If the node that originated the query determines that the query did not reach a destination it can retransmit or flood the query.

A New Gradient Based Routing Protocol: (Li et al. 2005) proposes a new gradient-based routing protocol. The proposed protocol takes into account the minimum hop count and remaining energy of each node while relaying data from source node to the sink. The optimal routes can be established autonomously with the proposed protocol. A simple acknowledgement scheme, which is implemented without extra overheads, is proposed. Data aggregation is performed to save transmission energy. To handle the frequent change of the topology of the network, a scheme for frequent change of the topology of the network is provided.

O(1)-Reception Routing Protocol: (Abdelmalik et al. 2007) proposes a technique that enables the best route selection based on exactly one message reception. It is called O(1)-reception . In O(1)-reception, each node delays forwarding of routing messages (RREQs) for an interval inversely proportional to its residual energy. This energy-delay mapping technique makes it possible to enhance an existing min-delay routing protocol into an energy-aware routing that maximizes the lifetime of sensor networks. They also identify comparative elements that help to perform a thorough posteriori comparison of the mapping functions in terms of the route selection precision. The O(1)-reception routing enhances the basic diffusion routing scheme by delaying the interests forwarding for an interval inversely proportional to the residual energy: nodes compute a forwarding delay based on their residual energy and defer the forwarding of interest messages for this period of time. As maximum lifetime routing should combine the min and the max–min metrics, in the energy-delay mapping function, nodes with high residual-energy forward interests without delay to make diffusion equivalent to the min energy routing, and nodes with low residual-energy delay forwarding of interests for a time interval to make diffusion equivalent to the max–min residual energy routing.

Energy-Balancing Multipath Routing (EMPR): The basic idea of EMBR is that the base station finds multipath to the source of the data and selects one of them for data transmission (Yunfeng & Nidal 2006). The base station dynamically updates the available energy of each node along the path based on the amount of packets being sent and received. The base station then uses the updated energy condition to periodically select a new path from multiple paths. The base station takes the role of the server and all sensor nodes work as clients. Base station does every thing from querying specific sensing data, broadcasting control packets, routing path selection and maintenance to work as the interface to the outside networks. Sensor nodes are only responsible for sensing data and forwarding packets to the base station. Topology construction is initiated by the base station at any time. The base station broadcasts Neighbor Discovery (ND) packet to the whole network. Upon receiving this packet, every node records the address of the last hop from which it receives and stores it in the neighbors list in ascending order of receiving time. The node changes the source address of the packet to itself. Then it broadcasts the packet. If the new packet is already received the node drops the ND packet and does not rebroadcast. After the completion of Neighbors discovery, the base station broadcasts another packet, Neighbors collection (NC) to collect the neighbor information of each node. Upon receiving the NC packet, the node replies a NCR (Neighbors Collection Reply) packet by flooding. The base station now has a vision of the topology of the networks through the neighbor's information of all nodes. After the topology construction, the base station constructs a weighted directed graph. The weight of each edge is the available energy of the head node. In the data transmission phase, the base station broadcasts enquiry (DE) for sensing data with specific features. Then the sensor nodes satisfying an enquiry will reply with Data Enquiry Reply (DER) packet. On the other hand, the sensor node does not satisfy the enquiry will rebroadcast DE. The base station calculates the shortest path to the desired node in the weighted node.

3.2 Hierarchical Protocols

In hierarchical routing protocols, clusters are formed. For each cluster, a head node is assigned dynamically, a set of nodes will attach the head node, and the head nodes can communicate with the sink either directly or through upper level of heads. Data aggregation is usually performed at each head.

Low-Energy Adaptive Clustering Hierarchy LEACH: (Wendi et al 2002) propose a LEACH. In LEACH, the nodes organize themselves into clusters. In designing the LEACH, it is assumed that all the nodes in the network can transmit with enough power to reach the base station (BS) of the network and each node has sufficient computational power to support different MAC protocols and perform signal processing functions. Regarding the network model it is assumed that the network consists of nodes that always have data to send to the end user and the nodes which are located close to each other have correlated data.

In LEACH, the nodes organize themselves into local clusters. One of the nodes is identified as a cluster head and all other nodes in the cluster send their data to the cluster head. The cluster head is responsible for processing the data received from the nodes and transmit the resulted data to the base station. Since the cluster head performs data processing and transmission, it will consume more power than normal nodes. The cluster head must be changed through the system life time. Each node must take its turn to act as a cluster head. Operation of LEACH is divided into rounds. Each round begins with a set-up phase followed by a steady-state phase. In set-up phase, the clusters are formed and the cluster head is assigned. In the steady state phase, the nodes will transmit their data. The algorithm to select a cluster head is a distributed algorithm. Each node makes autonomous decision to be a cluster head. During each round, there are k clusters so there must be k heads. At round $r+1$ which starts at time t, each node selects itself to be a cluster head with probability $P_i(t)$. $P_i(t)$ is chosen such that the expected value of the cluster head must be k. To ensure that all nodes will act as cluster head equal number of times, each node must be a cluster head once in N/k rounds. In (Windy et al 2002) a new probability is proposed to take into account the energy in each node

After identifying the clusters heads, each node must determine the cluster to which it belongs. Each cluster head broadcasts advertisement message containing the head's id using non-persistent CSMA scheme. Each node determines its cluster by selecting the head whose advertise signal is the strongest signal. This head is the closest head to the node. The node will transmit a joint request message to the chosen cluster head using CSMA. Upon receiving all the joint request messages the cluster head sets up the TDMA schedule and transmit this schedule to the nodes in the cluster. Each node will turn OFF its radio all the time slots except their assigned slots. This will end up the set-up phase and start the steady state phase.

The steady state phase is divided into frames; each node sends its data to the cluster head once per frame during its assigned slot. All nodes must be synchronized and start their set-up phase at the same time. This can be done by transmitting a synchronization pulse by the base station to all nodes. To reduce energy dissipation each non head node use power control to set the least amount of energy in the transmitted signal to the base station based on the received strength of the cluster head advertisement. When a cluster head receives the data from all nodes, it performs data aggregation and the resultant data will be sent to the base station. Processing the data locally within the cluster reduces the data to be sent to the base station; therefore the consumed energy will reduced. This is an advantage of the LEACH. To reduce inter-cluster interference, each cluster communicates using direct sequence spread spectrum DSSS. Each cluster uses a unique spreading code.

The distributed cluster formulation algorithm does not offer guarantee about placement and number of cluster head nodes. An alternative algorithm is a central cluster formation; base station (BS) cluster formation. The central cluster formation produce better clusters by dispersing the cluster head nodes throughout the network. In the central algorithm, each node sends information about its current location and its energy level to the BS. The BS computes the average energy level. Any node has energy level less than the average cannot be a cluster head, other nodes can be clusters heads. The BS use simulated annealing to find the cluster heads. The solution must minimize the amount of energy for non-cluster head and find k the optimal number of clusters k_{opt}. When the cluster heads and associated clusters are found the BS broadcasts a message that contains the cluster head ID for each node. (Windy et al 2002) propose a formula to find the optimum number of clusters that minimize the total consumed energy

The frame size in LEACH is fixed regardless of the active nodes in the cluster since it is assumed that all nodes have data to send. This is not the real case all the time, sometimes some of the nodes are active and other nodes are not active.

Energy-Aware Data-Centric Routing Algorithm (EAD): (Azziddine et al. 2005) propose EAD. EAD is designed for event driven application. In EAD, a tree rooted at the base station is constructed. The tree consists of leaf and non-leaf nodes. A non-leaf node is a node that has at least one child. On the other hand, a leaf node is a node that has no child. All the leaf nodes of the tree will turn their radio OFF most of the time. On the other hand, all the non-leaf nodes will turn their radio ON all the time. When an event occurs, the leaf nodes will collect the related data and turn its radio ON to transmit the data to its parent. When a non-leaf node receives data from all its children, it will aggregate the data and send it to its parent. All the nodes use CSMA/CA for transmitting the data. Since the radio of the non-leaf sensor nodes will always be ON, they will lose much power than the leaf nodes. The tree will be reconstructed from time to time. (Azziddine et al. 2005) proposes an energy aware algorithm to build the tree. One of the disadvantages of EAD is that the non-leaf nodes will be awake all the time even though there are not events to detect. This makes EAD unsuitable for applications with periodic data traffic.

To build a tree rooted at the sink, the sink initiates the process of building the tree. Building the tree is performed by broadcasting control messages. Each control message consists of four fields: *type, level, parent, power*. For the sender node v , $type_v$ represents its status; 0: undefined; 1: leaf node; 2: non-leaf node. $level_v$ refers to the number of hops from v to the sink. $parent_v$ is the next hop of v in the path to the sink; $power_v$ is the residual power E_v. Initially each node has status 0. The sink broadcasts $msg(2,0,NULL,\infty)$. When a node v receives $msg(2 , level_u , parent_u , E_u)$ from node u , it becomes a leaf node, sense the channel until it is idle, then waits for T_2^v time , if the channel is still idle, v broadcasts $msg(1 , level_u +1 , u , E_v)$. If v receives $msg(1 , level_u , parent_u , E_u)$ from u , it senses the channel until it is idle, waits for T_1^v if the channel is still idle , v broadcasts $msg(2 , level_u +1 , u , E_v)$. And it becomes non-leaf node. If node v receives more than one message from different nodes before it broadcasts its message, it will select the node with larger energy as its parent. If both nodes have the same energy, it will select one of them randomly. The waiting node will go back to sensing state, if another node occupies the common channel before it times out. If a node v with status 1 receives $msg(2 , level_w , v , E_w)$ from node w indicating that v is its parent, v broadcasts $msg(2 , level_v , parent_v , E_v)$ immediately after the channel is idle. The process will continue until each node becomes leaf or non-leaf node. A sensor with status 2 becomes a leaf node if it detects that it has no children. Both T_1^v and T_2^v are chosen such that no two neighboring broadcasts are scheduled at the same time. On the other hand, to force the neighboring sensors with higher energy to broadcast earlier than those nodes with a lower residual power, both T_1^v and T_2^v must be monotonically decreasing functions of E_v. One of the disadvantages of EAD is that all the nodes are connected to the sink through few nodes that are close to the sink. These nodes are considered as gateways. These nodes will be non-leaf nodes for most of time; they will consume a lot of energy. Therefore, they will die early. When they die, the rest of the nodes will be isolated. However, those isolated nodes still have non-consumed energy. Therefore, energy utilization is not so efficient in EAD. (Tayseer & Baroudi 2007) generalize EAD such that any node can act as a gateway.

A Generalized Energy-Aware Data Centric Routing For Wireless Sensor Network ($EAD_{General}$): (Tayseer & Uthman 2007) generalize EAD such that any node can act as a gateway. To generalize EAD, they assume that each node has the ability to transmit its data for long distance, i.e. its transmission can reach the sink. Each node has power control capability such that the transmission energy depends on the distance to the destination node. When a node sends data to its nearest neighbor, the transmission energy will be small compared with the transmission energy required to transmit data to the sink. In $EAD_{General}$, a new phase; Selecting Gateways (SG), is added. In this phase, gateway nodes are selected. It is assumed that the network is virtually divided into tiers. Each tier includes all nodes that can hear a signal transmitted with specific energy from the sink. For example, $tier_0$ includes all nodes that can hear the signal transmitted from sink with transmission energy equals to E_0. $Tier_1$ includes all nodes that can hear the signal transmitted from sink with transmission energy equals to E_1, where $E_1>E_0$ and so on. Initially, the nodes of $tier_0$ will be considered as potential candidate gateways. Based on their energy level, some of these nodes will advertise themselves as gateways. They will act as gateways until their residual energy drops below a threshold value E_{th}. Then new gateways will be selected from the nodes of $tier_1$. The selected nodes will act as gateways until their residual energy drops below E_{th} and so on. When all tiers are considered and no more nodes can be selected as gateways based on the current E_{th}, a new cycle will start, in this cycle new gateways will be selected from $tier_0$ using smaller value of E_{th} and so on. To select the gateways, the sink broadcasts an *ADV* message. The *ADV* message contains a field for E_{th}. Initially *ADV* message is broadcasted with energy E_0 such that it reaches the nodes of $tier_0$ only. When a node receives the *ADV* message, it compares its residual energy with E_{th}, and then it responds with a *JOIN* message. A *JOIN* message contains a confirmation field. *Confirmation* is set to 1, if the node's residual energy is greater than E_{th}, i.e. the node can be a gateway and it selects the sink as its parent, otherwise *confirmation* is set to 0. After the node sends its *JOIN* message, it will act as gateway in the current round. Assuming reliable channel, it does not need a confirmation from the sink to be a gateway. All nodes send JOIN message with *confirmation field=1* will be considered gateways. If the sink receives *JOIN* messages from all nodes in the target tier and the *confirmation field =0* in all the received *JOIN* messages, then no node from the target tier can be a gateway, since we assume that all nodes can reach the sink, the sink will broadcast a new *ADV* message with higher transmission energy E_1 using the same E_{th} to select a gateway from the next tier. The nodes of the next tier will respond with *JOIN* messages according to their energy. The process will continue until all tiers are considered and no node has energy greater than E_{th}; no node can be a gateway. A new cycle will start from $tier_0$ with new E_{th}, $E_{th}(new)=eE_{th}(current)$, where $0<e<1$. Following the same procedure as above, new gateway nodes will be selected from $tier_0$. For each cycle, a fixed E_{th} will be used, and at the beginning of each new cycle, E_{th} will be reduced by the factor e. The sink and nodes will exchange messages using the CSMA mechanism. The node has to be ON until it receives the *ADV* message from the sink and then it sends the *JOIN* message. Since the node does not need confirmation from the sink, it will go to sleep immediately after sending the *JOIN* message.

A Generalized Energy-Efficient Time-Based Communication Protocol For Wireless Sensor Networks (GET): GET is proposed by (Tayseer & Uthman 2009). In designing GET, they assume that each node has the ability to transmit its data for long distance, i.e. its transmission can reach the sink. Each node has power control capability such that the transmission energy depends on the distance to the destination node. When a node sends data to its nearest neighbor, the transmission energy will be small compared to the transmission energy required to transmit data to the sink. they assumed that all nodes are synchronized. Regarding the application of the network, they assume that the event that is being monitored is periodic, so data transmission from sensor nodes to the sink will start at specific time, and it will be repeated periodically. They assume also that all the nodes that are located close to each other and have correlated data. Hence, data aggregation will be used and it will reduce data redundancy. In GET, time is divided into rounds. Each round consists of four phases: Selecting the Gateways (SG), Building the Tree (BT), Building the Schedule (BS), and Data Transmission (DT). In the first phase, gateways are selected; the gateway is selected using the algorithm proposed in (Tayseer and Uthman 2007). In the second phase, a tree rooted at the sink is built. The tree is built using building tree algorthim proposed by (Azziddine et al. 2005). They modify the buiding tree algorithm such that building tree process will be initiated by the gatewyas not by the sink Based on this tree, a TDMA schedule is built in a distributed manner in phase-3. The schedule will be built assuming that in the data transmission period, all nodes connected to the sink through the same gateway will use the same frequency to transmit their data.. For each node, they identify two time constants: Time Ready to Receive (*TRR*) and Time Ready to Transmit (*TRT*). For a node v, TRR_v represents the time slot when the node is ready to receive data from its children, while TRT_v represents the time slot when a node can transmit data to its parent. Assuming t_0 represents the time at which the periodic sensing event occurred and the data is already collected from the monitored environment. For a leaf node, $TRT_v = t_0$. TRR_v is not valid since it does not have children. On the other hand, for a non-leaf node v:

$$\begin{aligned} TRR_v &= Max(TRT_i) \qquad i = 1,2,3,...n_v^c \\ TRT_v &= TRR_v + n_v^c T_t \end{aligned} \tag{1}$$

Where i represent an index for the child of node v, , $n_v{}^c$ represents the count of v's children, and T_t represents the time needed to transmit one data packet. To build the schedule, initially, each leaf node will transmit its *TRT* value to its parent. When a parent receives *TRT* values from all its children, it calculates its *TRR* and *TRT* using (1) and builds the schedule for its children. Then it transmits its *TRT* to its parent and broadcasts the schedule to its children. The process will continue until all nodes receive their assigned time slot from their parents. Both leaf and non-leaf nodes use *CSMA/CA* protocol to exchange data (*TRT* and the Schedule). Eventually, we have a TDMA schedule for the whole sensor network.

In the fourth phase, data is transmitted from sensor nodes to the sink following the schedule prepared in phase-3. Data transmission period represents the time needed to forward all data packets in a single round. Data transmission period may be repeated many times in a single round

TinyDB: Another alternative in the same direction is the work presented in (Samuel et al. 2005). A distributed query processor for smart sensor devices (TinyDB) is proposed. In TinyDB, to disseminate queries and collecting results, a routing tree rooted at the base station is built. The routing tree is formed by forwarding a routing request (a query in TinyDB) from every node in the network. The root sends a request then all child nodes that hear this request process it and forward it on to their children, and so on, until the entire network has heard the request. Each node picks a parent node that is one level closer to the root. This parent will be responsible for forwarding the node's query results to the base station. To limit the scope of queries, a Semantic Rooting Tree (SRT) is built. This tree is built based on the routing tree. If a node knows that none of its children currently satisfies the query, it will not forward the query down the routing tree. Therefore, each node must have information about child attribute values.

Unequal Cluster Based routing (UCR): In UCR protocol, clusters with different size are constructed (Guihai et al. 2007). Cluster heads closer to the sink will have smaller cluster sizes than those farther from the sink. Thus they can preserve some energy for the inter-cluster data forwarding. A greedy geographic and energy-aware routing protocol is designed for the inter cluster communication which considers the tradeoff between the energy cost of relay paths and the residual energy of relay nodes. The UCR protocol consists of two parts: an energy-efficient unequal clustering algorithm called EEUC and an intercluster greedy geographic and energy-aware routing protocol. Initially, the base station broadcasts a beacon signal to all sensors at a fixed power level. Based on the received signal strength, each sensor node can compute the approximate distance to the base station. It not only helps nodes to select the proper power level to communicate with the base station, but also helps us to produce clusters of unequal sizes. In EEUC algorithm, heads will be identified randomly. As in LEACH protocol, the task of being a cluster head is rotated among sensors in each round to distribute the energy consumption across the network. After cluster heads have been selected, each cluster head broadcasts a CH_ADV_MSG across the network field. Each ordinary node chooses its closest cluster head, the head with the largest received signal strength, and then informs it by sending a JOIN_CLUSTER_MSG. After forming clusters, data will be transmitted from the cluster heads to the base station. Each cluster head first aggregates the data from its cluster members, and then sends the packet to the base station via a multi-hop path through other intermediate cluster heads. Before selecting the next hop node, each cluster head broadcasts a short beacon message across the network at a fixed power which consists of its node ID, residual energy, and distance to the base station. A threshold TD_MAX in the multi-hop routing protocol is proposed. If a node's distance to the base station is smaller than TD_MAX, it transmits its data to the base station directly; otherwise, it is better to find a relay node that can forward its data to the base station.

Energy-aware routing for cluster-based sensor networks: (Younis et al. 2002) proposed a hierarchical routing algorithm based on a three-tier architecture. In the proposed protocol, sensors are grouped into clusters. The cluster heads (gateways) are less energy constrained than normal sensors. It is assumed that cluster heads knows the location of the sensor nodes. Gateways maintain the states of the sensors and sets up multi-hop routes for collecting sensors data. Each gateway informs each node within its clusters the time slots in which it can transmit and in which it have to listen to other nodes transmission. The sensor nodes in the cluster can be in one of four states: sensing only, relaying only, sensing-relaying and inactive. In sensing state the sensor node senses the environment and generates the corresponding data. In the relaying only state, the node does not sense the environment but it forwards data from other active nodes. In sensing-relaying state, the node not only senses the environment but also forwards the data from other active nodes. In inactive state, the node neither senses the environment nor forwards data. The link cost is defined as the energy consumption to transmit data between two nodes, the delay optimization and the other performance cost. A least-cost path is found between sensor nodes and the gateway. The gateway monitors the available energy level at every sensor that is active. Rerouting is triggered by an application-related event requiring different set of sensors to probe the environment or the depletion of the battery of an active node.

Base-Station Controlled Dynamic Clustering Protocol (BCDCP): (Muruganathan et al. 2005) proposes a clustering-based routing protocol called Base Station Controlled dynamic

Clustering protocol (BCDCP). In BCDCP, the base station sets up clusters and routing paths, performs randomized rotation of cluster heads, and carries other energy intensive tasks. The key ideas in BCDCP are: formulation of balanced clusters where each cluster head serves an approximately equal number of member nodes, uniform placement of cluster heads throughout the entire sensor field, and the utilization of cluster-head-to-cluster-head(CH-to-CH) routing to transfer the data to the base station. Class-based addressing of the form <Location ID, Node Type ID> is used in BCDCP. The Location ID identifies the location of a node. It is assumed that the base station keeps up-to-date information on the location of all the nodes in the network. A Node Type ID describes the functionality of the sensor such as seismic sensing, and thermal sensing. BCDCP operates in two major phases: setup and data communication. In setup phase, clusters are formed, clusters' heads are selected, CH-to-CH routing paths are formed, and schedule is created for each cluster. During each setup phase, the base station receives information on the current energy status from all the nodes in the network. Based on this information, the base station computes the average energy level and then chooses a set of nodes, denoted S, whose energy levels are above the average value. Cluster heads for the current round will be chosen from the set S. To identify the cluster heads from the set and to from clusters, iterative cluster splitting algorithm is used. This simple algorithm first splits the network into two sub-clusters, and proceeds further by splitting the sub-clusters into smaller clusters. The base station repeats the cluster splitting process until the desired number of clusters is attained. Once the clusters and the cluster head nodes have been identified, the base station chooses the lowest-energy routing path and forwards this information to the sensor nodes along with the details on cluster groupings and selected cluster heads. The routing paths are selected by connecting all the cluster head nodes using the minimum spanning tree approach that minimizes the energy consumption and then a head is randomly selected to transmit data to the base station. The last step in this phase is building a TDMA Schedule for each cluster. In The data communication phase, Data gathering, Data fusion, and Data routing is performed using the TDMA schedule created in setup phase.

3.3 Location-Based Protocols

Information Location can be utilized to forward data with minimum energy consumption. If the region to be monitored is known, the query can be forwarded to that region. Many location-based routing protocols for WSN were proposed. In the successive subsections, I will survey many of these protocols.

Geographic Adaptive Fidelity (GAF): GAF is energy-aware location-based routing protocol designed for mobile ad hoc protocols, but it can be applicable to sensor networks (Ya et al. 2001) . In GAF a virtual grid for the monitored area is formed. Each node uses its GPS-indicated location to associate itself with a point in the virtual grid. Nodes associated with same point in the grid are equivalent. Some of them can be in the sleeping state to save energy while others will be in active state. Therefore, the network lifetime will increase. To balance load among nodes, equivalent nodes change their state from active to sleeping in turn. Three states are defined in GAF, discovery, sleep, and active. In the discovery state a node will determine its neighbors. While it is in sleep state, a node will turn OFF its radio. The active node will participate in data routing. A node will be in each state for particular time period which is application dependent. On the other hand, determining which nodes that will be in sleep state is application dependent. GAF is implemented for non-mobility

(GAF-basic) and mobility (GAF-mobility adaptation) of nodes. To keep the network connected, a representative node must be always active for each region on its virtual grid.

Geographic and Energy Aware Routing (GEAR): In GEAR protocol, energy aware and geographical-informed neighbor selection heuristic is used to route packets towards the destination region (Yan et al. 2001). The key idea is to restrict the number of interests in Directed Diffusion to certain regions rather than sending interest to the whole network. Each node keeps an estimated cost and a learning cost of reaching the destination through its neighbors. The estimated cost is a combination of residual energy and distance to destination. The learned cost is a refinement of the estimated cost. A hole exists in the network when a node does not have any closer neighbor to the target region. With no holes in the network, the estimated cost is equal to the learned cost. When a packet reaches the destination, the learned cost is propagated one hop back so that route setup for next packet will be adjusted. The GEAR protocol consists of two phases; in the first phase, the packets are forwarded towards the target region, when a node receives a packet, it checks its neighbors to see if there is a neighbor that is closer to the target region. The closest neighbor to the target region is selected as the next hop. When all neighbors are further than node itself, a hole exists; one of them will be selected based on the learned cost function. This selection will be updated according to the convergence of the learned cost. In the second phase, packets will be forwarded within the region; the packets are forwarded in the region by either recursive geographic forwarding or restricted flooding.

A Mesh-Based Routing Protocol for Wireless Ad-Hoc Sensor Network (MBR): In MBR protocol, the area of the sensor network is portioned into regions; mesh topology (Foad & r, Hadi 2006). The nodes can communicate to their neighbor nodes through virtual channels. Forming the mesh topology is performed in three phases. In the first phase, the base node for zoning is selected. Two setup sensors are determined. One of them is located at the largest diameter and in the boundary of the area and the second sensor is located on the boundary of other orthogonal diameter of the region. In phase two, the network is divided into regions. In phase three, each sensor nodes is assigned ID. Each sensor will be known with two features: its region coordinate (X,Y) and its ID. To transmit data between source nodes and sink a path is reserved between them firstly. To reserve a path, the source node sends a reserve message, called RAP, to the sensors in its target (X,Y). Upon receiving the RAP message, each node generates a priority number and returns it to the source node using ACK message. Sensors have higher energy will have higher priority. The source sensor will select sensors to form the path among the sensors that sends ACK message. Then data will be sent based on the path determined. After transmitting data, path must be released. This is done by sending a CRP message.

Energy-efficient geographic multicast routing: (Juan et al. 2007) proposes a novel energy-efficient multicast routing protocol called GMREE. It aims to preserve energy and network bandwidth. GMREE protocol builds multicast trees based on a greedy algorithm using local information. GMREE protocol is based in the concept of cost over progress metric and it is specially designed to minimize the total energy used by the multicast tree. The cost is defined as the energy needed to reach the furthest neighbor in the selected set of relays plus the energy that such amount of nodes will need to process the message. GMREE incorporates a relay selection function which selects nodes from a node's neighborhood taking into account not only the minimization of the energy but also the number of relays selected. Nodes only select relays based on a locally built and energy-efficient underlying graph reduction such as Gabriel graph, enclosure graph or a local shortest path tree. Thus, the topology of the resulting multicast trees really takes advantage of the benefit of sending a single message to multiple destinations through the relays which provide best energy paths.

Energy-Aware Geographic Routing for Sensor Networks with Randomly Shifted Anchors: Anchor-based geographic routing aims at finding a small number of intermediate nodes acting as anchors so that the path length (i.e. number of hops) between the source and destination can be reduced. However, some nodes (e.g., nodes near the boundary of the network) tend to be used as anchors repeatedly by multiple flows. As a result, their energy drains quickly and the lifetime of the network is reduced. Moreover, the intermediate nodes between source and destination change very little once the anchor list is set. This also contributes to the quick depletion of the energy for some nodes. To overcome these shortcomings, (Gang et al. 2007) introduces a random shift to the location of each anchor in the routing process. Each new packet will then be routed to a different anchor determined by the location of the original anchor plus the random shift. Because the shift is generated randomly, different packets will likely be routed through a different list of anchors. This allows more nodes to be involved in the routing process and the energy consumption is better distributed among nodes in the network.

On Optimal Geographic Routing in Wireless Networks with Holes and Non-Uniform Traffic: Subramanian et al. propose a randomized geographic routing scheme that can achieve a throughput capacity of $\Theta(1/\sqrt{n})$ (within a poly-logarithmic factor) even in networks with routing holes (Sundar et al. 2007). They show that the proposed scheme is throughput optimal (up to a poly-logarithmic factor) while preserving the inherent advantages of geographic routing. They also show that the routing delay incurred by the proposed scheme is within a poly-logarithmic factor of the optimal throughput-delay trade-off curve. On the other hand, Subramanian et al. construct a geographic forwarding based routing scheme that can support wide variations in the traffic requirements as much as $\Theta(1)$ rates for some nodes, while supporting $\Theta(1/\sqrt{n})$ for others. They show that the above two schemes can be combined to support non-uniform traffic demands in networks with holes.

The randomized algorithm takes as input the number of nodes in the network, the packet to be sent, as well as the number of holes. Considering the first packet in all the source nodes, The source node for every traffic flow creates Rlog(n) copies of its packet to send. It chooses Rlog(n) independent and uniformly distributed points from the unit region and sets the NEXT-DEST field in the packet to the randomly generated location in each of these copies. The Rlog(n) packets are routed from the source in a greedy geographic manner to the location in NEXTDEST. Upon receiving a packet, a node checks if it is the NEXTDEST location. If it is not the NEXT-DEST location, it searches within its neighboring nodes for the node that is closest to the NEXT-DEST location, and forwards the packet to that node. If none of its neighbor nodes is closer to the NEXT-DEST than itself, the node drops the packet. If it is the NEXT-DEST location, it checks whether it is the final destination or not. If it is the final destination, then the packet is received. Otherwise, If the final destination is one hop away from the current node, the node forwards the packet greedily to the final destination. If the final destination is more than one hop a way from the current node, the current node makes Rlog(n) copies of the packet and again generates uniform and randomly chosen locations for the NEXT-DEST in each of the packet copies, and forwards them greedily.

3.4 QoS-aware Protocols

QoS-aware protocols consider end-to-end QOS requirement while setting up the paths in the sensor network. Many QoS-aware routing protocols for WSN were proposed. In the successive subsections, I will survey many of these protocols.

Maximum Lifetime Energy Routing: (Jae-Hwan et al. 2000) presents a routing protocol for sensor networks based on a network flow approach. The protocol aims to maximize the network lifetime by defining link cost as a function of node remaining energy and the required transmission energy using that link. Finding traffic distribution is a possible solution to the routing problem. The solution to this problem maximize the network lifetime. Two maximum residual energy path algorithms were proposed to find the best link metric for the maximization problem. The two algorithms differ in their definition of link costs and the incorporation of nodes' residual energy. The least cost paths to destination are found using Bellman-Ford shortest path algorithm. The least cost path is the path whose residual energy is largest among all paths.

Maximum Life Time Data Gathering: (Konstantinos et al. 2002) models the data routes setup in sensor network as the maximum lifetime data-gathering problem. A polynomial time algorithm to solve this problem is proposed. The data-gathering schedule specifies for each round how to get and route data to sink. For each round, a schedule has one tree rooted at the sink and spans all the nodes of the network. The network lifetime depends on the duration for which the schedule remains valid. The Maximum Lifetime Data Aggregation (MLDA) protocol is proposed to set up maximum lifetime routes taking into account data aggregation. If a schedule "S" with "T" rounds is considered, it induces a flow network G. the flow network with maximum lifetime subject to the energy constraints of sensor nodes is called an optimal admissible flow network. A schedule will be constructed by using this admissible flow network. For application with no data aggregation such as video sensors, a new scenario is presented, which is called Maximum Lifetime Data Routing (MLDR). It is modeled as a network flow problem with energy constraints on sensors.

SPEED: SPEED is a real-time communication protocol for sensor networks (Tian et al. 2003). It provides three types of real-time communication services; real-time unicast, real-time area-multicast and real-time area-anycast. SPEED is a stateless, localized algorithm with minimal control overhead. End-to-end soft real-time communication is achieved by maintaining a desired delivery speed across the sensor network through a novel combination of feedback control and non-deterministic geographic forwarding. SPEED is a highly efficient and scalable protocol for sensor networks where the resources of each node are scarce. In SPEED protocol, each node should maintain information about its neighbors. Geographic forwarding is used to find the paths. SPEED protocol strives to ensure end-to-end delay for the packets in the network such that each application can estimate the end-to-end delay for the packets. SPEED protocol consists of the following components: A neighbor beacon exchange scheme, a delay estimation scheme, The Stateless Non-deterministic Geographic Forwarding algorithm (SNGF), A Neighborhood Feedback Loop (NFL), Backpressure Rerouting, and Last mile processing. SNGF is the routing module responsible for choosing the next hop candidate that can support the desired delivery speed. NFL and Backpressure Rerouting are two modules to reduce or divert traffic when congestion occurs, so that SNGF has available candidates to choose from. The last mile process is provided to support the three communication semantics mentioned before. Delay estimation is the mechanism by which a node determines whether or not congestion has occurred. And beacon exchange provides geographic location of the neighbors so that SNGF can do geographic based routing. Table 1 shows a classification of routing protocols based on the application.

	Application		
Protocol	Query Based	Event Driven	Periodic
SPIN	√		
Directed Diffusion	√		
Shah et al.			√
Rumor Routing	√		
CADR	√		
COUGAR	√		
ACQIRE	√		
GBR	√		
O(1)-Reception Routing Protocol		√	
EMPR	√		
LEACH		√	
EAD		√	
TinyDB	√		
PEGASIS		√	
TEEN		√	
APTEEN			√
UCR		√	
BCDCP		√	
GAF		√	√
MECN		√	
GEAR	√		
GOAFR		√	
MBR		√	
GMREE		√	
Zhao et al. Randomly Shifted Anchors:		√	
Chang et al			√
Kalpakis et al.		√	√
Minimum Cost Forwarding		√	
SAR		√	√
Energy-Aware QoS Routing Protocol			√
$EAD_{General}$		√	
SPEED	√		
GET			√

Table 1. Classification of Routing Protocols based on the Applications

4. Literature Review of Cross Layer design in WSN

Many researchers studied the necessity and possibility of taking advantages of cross layer design to improve the power efficiency and system throughput of Wireless sensor network. (Safwat et al. 2003) proposed Optimal Cross-Layer Designs for Energy-efficient Wireless Ad hoc and Sensor Networks . They propose Energy-Constrained Path Selection (ECPS) scheme and Energy-Efficient Load Assignment (E2LA). ECPS is a novel energy-efficient scheme for wireless ad hoc and sensor networks. it utilizes cross-layer interactions between the network layer and MAC sublayer. The main objective of the ECPS is to maximize the probability of sending a packet to its destination in at most n transmissions. To achieve this objective, ECPS employs probabilistic dynamic programming (PDP) techniques assigning a unit reward if the favorable event (reaching the destination in n or less transmissions) occurs, and assigns no reward otherwise. Maximizing the expected reward is equivalent to maximizing the probability that the packet reaches the destination in at most n transmissions. Ahmed Safwat et. al, find the probability of success at an intermediate node i right before the tth transmission $f_t(i)$:

$$f_t(i) = \begin{cases} 1 & i = D \\ \max\limits_{j} \sum\limits_{k} p_k f_{t+k}(j) & otherwise \end{cases} \quad (2)$$

where D is the destination node and j is the next hop towards the destination D. Any energy-aware route that contains D and the distance between D and the source node is less or equal to n can be used as input to ECPS. The MAC sub-layer provides the network layer with the information pertaining to successfully receiving CTS or an ACK frame, or failure to receive one. Then ECPS chooses the route that will minimize the probability of error

The objective of the E2LA scheme is to distribute the routing load among a set Z of Energy-aware routes. Packets are allotted to routes based on their willing to save energy. Similar to ECPS, E2LA employs probabilistic dynamic programming techniques and utilize cross-layer interactions between the network and MAC layers. At the MAC layer, each node computes the probability of successfully transmitting packets in α attempt. E2LA assign loads according to four distinct reward schemes (Safwat et al. 2003).

(Venkitasubramaniam et al. 2003) propose a novel distribution medium access control scheme called opportunistic ALOHA (O-ALOHA) for reachback in sensor network with mobile agent. The proposed scheme based on the principle of cross layer design that integrates physical layer characteristics with medium access control. In the O-ALOHA scheme, each sensor node transmits its information with a probability that is a function of its channel state (propagation channel gain). This function called transmission control is then designed assuming that orthogonal CDMA is employed to transmit information. In designing the O-ALOHA scheme they consider a network with n sensors communicate with a mobile agent over a common channel. It is assumed that all the sensor nodes have data to transmit when the mobile agent is in the vicinity of the network. Time is slotted into intervals with equal length equal to the time required to transmit a packet. The network is assumed to operate in time division duplex (TDD) mode. At the beginning of each slot, the collection agent transmits a beacon. The beacon is used by each sensor to estimate the propagation channel gain from the collection agent to it which is the same as the channel gain from the sensor to the collection agent. It is assumed that the channel estimation is perfect. The propagation channel gain from sensor i to the collection agent during slot t which is

$$\gamma_i^{(t)} = \frac{P_T R_{it}^2}{r_i^2 + d^2} \tag{3}$$

Where R^2_{it} : is Rayleigh Distribution, and P_T is the transmission power of each sensor, and r_i is the radial distance of sensor i , and d is the distance from collecting agent and sensor node. During the data transmission period, each sensor transmits its information with a probability $S(\gamma_i^{(t)})$ where $S(.)$ is a function that maps the channel state to a probability. Two transmission controls are proposed to map from the channel gain to the probability; Location independent transmission control (LIT) and Location aware transmission control (LAT). In LIT, the decision to transmit a packet is made by observing channel state γ alone, while in LAT, every sensor makes an estimate of its radial distance and the decision to transmit is a function of both the channel state γ and the location of sensor.
(Sichitiu 2004) proposed a deterministic schedule based energy conservation scheme. In the proposed approach, time synchronized sensors form on-off schedules that enable the sensors to be awake only when necessary. The energy conservation is achieved by making the sensor node go to sleeping mode. The proposed approach is suitable for periodic applications only, where data are generated periodically at deterministic time. The proposed approach requires the cooperation of both the routing and MAC layers. The on-off schedule is built according to the route determined by routing protocol. The proposed approach consists of two phases; the Setup and reconfiguration phase and the steady state phase. In the setup and reconfiguration phase, a route is selected from the node originating the flow to the base station then the schedules are setup along the chosen route. In the steady phase, the nodes use the schedule established in the setup and configuration phase to forward the data to the base station. In this phase, there will be three types of actions at each node; Sample action which is taking data sample from environment, Transmit action to transmit data, and Receive action to receive data. The actions at each node along with the time when each action will take place are stored in the schedule table of each node. The node can be awake ate the time of each action and go to sleep otherwise.
(Li-Chun & Chung-Wei 2004) proposed Cross layer Design of Clustering architecture for wireless Sensor Networks. The proposed scheme is called Power On With Elected Rotation (POWER). The objective of the POWER is to determine the optimal number of clusters from the cross-layer aspects of power saving and coverage performance simultaneously. The basic concept of the POWER is to select a representation sensor node in each cluster to transmit the sensing information in the coverage area of the sensor node. The representative sensor node in a cluster rotated from all the sensor nodes in each cluster. In the POWER scheme, the scheduling procedure is rotated many rounds. In each round, there are two phases; the construction table phase (CTP), to construct the rotation table and the rotational representative phase (RRP) to transmit data. In CTP, all sensor nodes employ the MAC protocol and the first sensor node accessing the channel become the initiator node, then the initiator node detects other neighboring node and form s the cluster. RRP starts after constructing the rotation table. RRP is divided into many sRPs (Sub-Rotated Period). In each sRP, one node will be a representative node and all other nodes in the cluster will be in sleeping mode.

Protocol	Layers	Approach	Evaluation method	Applica-tion	Network Topology	Cross layer Objective	Performance metrics
ECPS	MAC, Network	Mathematical Model: probabilistic dynamic programming	Experiment		Random (Static)	Maximization of probability of sending packet to its D at n transmission	Energy
E2LA	MAC, Network	Mathematical Model: probabilistic dynamic programming	Experiment		Random (Static)	Minimize Energy:- Multiple simultaneous routes Load distribution	Energy
MAC CROSS	MAC, Network	Heuristic	Simulation Hardware Implemen-tation (MICAZ)		Random (Static)	Maximize Sleep Duration	Energy
O-Aloha	Physical, MAC	Heuristic	Simulation	SENMA	Random	Maximize throughput	Throughput
POWER	Physical, MAC, Network	Heuristic			Uniform (Static)	Optimize number of cluster	Energy
Weilian Su	ALL layers	Framework (optimization Agent)	Experiment al (MICAZ)		Random	Optimize performance of WSN	Link Quality Packet Received
Shunguang Cui	Routing, MAC, Link layer	Modeling as optimization problem	Analytical		Random	Maximize network lifetime	Network lifetime
Sense-Sleep Trees (SS-Trees)	MAC, Network	Heuristic	Simulation	Surveill-ance	Mesh-based	Maximizing Network lifetime, and monitoring coverage	Network lifetime Energy consumed
Game Theoretic Approach	Applicat ion, Physical	Game Theory	Analytical		Random	Minimize total distortion	Distortion coverage
In Yeup Kong	Physical, MAC, Network	Mathematical	Analytical		Random	Maximize Network lifetime	
Cross Layer Scheduling	MAC, Network	Heuristic	Simulation	Periodic	Random	Maximize network lifetime	Network lifetime
Cross Layer design for cluster formulate-on	MAC, Physical, Network	Heuristic	Simulation	Periodic	Uniform distribution	Maximize network lifetime	Network lifetime

Table 2. Summary of Cross layer Protocols for W

(Rick et al. 2005) proposes a cross-layer sleep-scheduling-based organization approach, called Sense-Sleep Trees (SS-trees). The proposed approach aims to harmonize the various engineering issues and provides a method of increasing monitoring coverage and operational lifetime of mesh-based WSNs engaged in wide-area surveillance applications. An iterative algorithm is suggested to determine the feasible SS-tree structure. All the SS trees are rooted at the sink. Based on the computed SS-trees, optimal sleep schedules and traffic engineering measures can be devised to balance sensing requirements, network communication constraints, and energy efficiency. For channel access a simple single-channel CSMA MAC with implicit acknowledgements (IACKs) is selected. In SS-trees approach, the WSN's life cycle goes through many stages. After the initial deployment of nodes, the WSN will enter the network initialization stage, in which the sink gathers network connectivity information from sensor nodes, compute the SS-trees, and disseminate the sleep schedules to every sensor node. Then the WSN will enter the operation stage, in which the nodes will alternate between Active and sleep stages. During long periods when sensing services are not needed the entire WSN will enter the Hibernation mode to conserve energy. The SS-trees must be computed with minimizing number of shared nodes (nodes belonging to multiple SS-trees), minimizing co-SS tree neighbors of each node, and minimizing the cost of forwarding messages between the data sink and each node. Rick W. Ha et al proposes a greedy algorithm to compute the SS-trees. The proposed algorithm follows a greedy depth-first approach that constructs the SS-trees from the bottom up on a branch-by-branch basis. After computing the SS-trees, an optimal sleep schedule that maximizes energy efficiency must be determined. The length of the active and sleep period will increase the data delay. The proposed SS-Tree design streamlines the routing procedures by restricting individual sensor nodes to only maintain local connectivity information of its immediate 1-hop neighbors.

(Shuguang et al. 2005) emphasize that the energy efficiency must be supported across all layers of the protocol stack through a cross-layer design. They analyze energy-efficient joint routing, scheduling, and link adaptation strategies that maximize the network lifetime. They propose variable-length TDMA schemes where the slot length is optimally assigned according to the routing requirement while minimizing the energy consumption across the network. They show that the optimization problems can be transferred into or approximated by convex problems that can be solved using known techniques. They show that link adaptation be able to further improve the energy efficiency when jointly designed with MAC and routing. In addition to reduce energy consumption, Link adaptation may reduce transmission time in relay nodes by using higher constellation sizes such as the extra circuit energy consumption is reduced.

(Weilian and Tat 2006) propose a cross layer design and optimization framework, and the concept of using an optimization agent (OA) to provide the exchange and control of information between the various protocol layers to improve performance in wireless sensor network. The architecture of the proposed framework consists of a proposed optimization agent (OA) which facilitates interaction between various protocol layers by serving as a database where essential information such as node identification number, hop count, energy level, and link status are maintained. (Weilian and Tat 2006) conduct the performance measurements to study the effects of interference and transmission range for a group of wireless sensors. The results of their performance measurements help to facilitate the design and development of the OA. The OA can be used to trigger an increase in transmit power to overcome the effects of mobility or channel impairments due to fading when it detects a degradation due in BER. Alternatively, it can reduce the transmit power to conserve energy to prolong its lifetime operations in the absence of mobility or channel fading. The OA can also be used to provide QoS provisioning for different types of traffic. This can be done by tagging different priority traffic with different transmit power levels.

(Changsu et al. 2006) proposed an energy efficient cross-layer MAC protocol for WSN. It is named MAC-CROSS. In the proposed protocol, the routing information at the network layer is utilized for the MAC layer such that it can maximize sleep duration of each node. in MAC-CROSS protocol the nodes are categorized into three types: Communicating Parties (CP) which refers to any node currently participating in the actual data transmission, Upcoming Communicating Parties (UP) which refers to any node to be involved in the actual data transmission, and Third Parties (TP) which refers to any node are not included on a routing path. The UP nodes are asked to wake up while other TP nodes can remain in their sleep modes. The RTS/CTS control frames are modified in the MAC-CROSS protocol. The modification is needed to inform a node that its state is changed to UP or TP in the corresponding listen/sleep period. a new field; Final_destination_Addr, is added to the RTS. On the other hand, a new field; UP_Addr is added to the CTS and it informs which node is UP to its neighbors. When a node B receives an RTS from another node A including the final destination address of the sink, B's routing agent refers to the routing table for getting the UP (node C) and informs back to its own MAC. The MAC agent of Node B then transmits CTS packet including the UP information. After receiving the CTS packets from node B, C changes its state to UP and another neighbor nodes change their states to TP and will go to sleep.

Table 2 shows summary of cross-layer design protocols for WSN.

5. Conclusion

In this chapter, we present a summary for MAC, Routing, and Cross layer Design protocols for WSN. In section 0, a survey of MAC protocols for WSN is presented. The routing protocols for WSN are discussed in section 0. A classification of the routing protocols according to the application is presented in section 3. Section 0 presents a summary of cross layer design protocols for WSN. A summary of cross layer design protocols at the end of section 4.

6. References

Ian F. Akylidiz, W. Su, Y. Sankarasubramaniam, and E. Cayirci (2002). A survey on sensor networks. *IEEE Personal Communications Magazine*, August.

The working group for WLAN standards 1999). IEEE 802.11 standards, Part 11: Wireless Medium Access Control (MAC) and physical layer (PHY) specifications. *Technical report*, IEEE

Sureh S. and Cauligi S. Raghavendra 1998), “PAMAS: Power aware multi-access protocol with signalling for ad hoc networks,” *ACM Comput. Commun. Rev.*, vol. 28, no. 3, July 1998, pp. 5–26,.

Wei Ye, John Heidemann, and Deborah Estrin, Fellow 2004), “Medium Access Control With Coordinated Adaptive Sleeping for Wireless Sensor Networks”, *IEEE/ACM Transactions on Networking*, Vol 12, No. 3, June 2004, pp. 493-506,.

Chansu Suh, Young-Bae Ko. (2005), "A traffic Aware, Energy Efficient MAC Protocol for Wireless Sensor Networks", *IEEE International Symposium on Circuits and Systems, 2005. ISCAS 2005.* pp.2975 - 2978 Vol. 3 , 23-26 May 2005

Peng Lin, Chunming Qiao and Xin Wang (2004) “Medium Access Control With A Dynamic Duty Cycle For Sensor Networks,” in WCNC, Mar 2004.

Tijs van Dam and Koen Langendoen (2003), "An adaptive Energy-Efficient MAC Protocol for Wireless Sensor Networks," in ACM Sensys’03, Nov. 2003.

Saad Biaz, Yawen Dai Barowski (2004), "GANGS: an Energy Efficient MAC Protocol for Sensor Networks", ACMSE'04 April 2-3.

Kemal Akkaya and Mohamed Younis (2005), "A Survey of Routing Protocols in Wireless Sensor Networks, " in *the Elsevier Ad Hoc Network Journal*, 2005 vol. 3/3 pp. 325-349.

Sandra. M. Hedetniemi and Stephen. T. Hedetniemi (1988), "A survey of gossiping and broadcasting in communication networks," Networks, Vol. 18, No. 4, 1988, pp. 319-349,

Chalermek Intanagonwiwat, Ramesh Govindan and Deborah Estrin (2000), "Directed diffusion: A scalable and robust communication paradigm for sensor networks", in the Proceedings of the 6th Annual ACM/IEEE International Conference on Mobile Computing and Networking (MobiCom'00), Boston, MA, August 2000.

David Braginsky and Deborah Estrin (2002), "Rumor Routing Algorithm for Sensor Networks," in the Proceedings of the First Workshop on Sensor Networks and Applications (WSNA), Atlanta, GA, October 2002.

Li Xia, Xi Chen, and Xiaohong Guan Xiac (2005), A New Gradient-Based Routing Protocol in *Wireless Sensor Networks Embedded Software and Systems*, Springer Berlin, Heidelberg, 2005.

Abdelmalik Bachir, Dominique Barthel, Martin Heusse, and Andrzej Duda (2007), "O(1)-Reception routing for sensor networks," *Computer Communications* Volume 30 , Issue 13, (2007), pp. 2603-2614.

Yunfeng Chen, and Nidal Nasser (2006), "Energy-Balancing Multipath Routing Protocol for Wireless Sensor Networks," in the Proc. of the third International Conference on Quality of Service in Heterogeneous Wired/Wireless Network, Waterloo, Ontario, Canada, August 7-9, 2006.

Wendi Heinzelman, Anantha Chandrakasan, and Hari Balakrishnan (2002), "An Application-Specific Protocol Architecture for Wireless Microsensor Networks," *IEEE On Wireless Communications Trans.*, vol. 1, No. 4, Oct. 2002, pp. 660-670.

Azzedine Boukerche, Xuzhen. Cheng, Joseh. Linus (2005), "A Performance Evaluation of a Novel Energy-Aware Data-Centric Routing Algorithm in Wireless Sensor Networks", *Wireless Networks* 11, 2005, pp.619–635,

T. AL-khdour, U. Baroudi (2007), " A Generalized Energy-Aware Data Centric Routing For Wireless Sensor Network", in the Proc. of The 2007 IEEE International Conference on Signal Processing and Communications (ICSPC 2007) , Dubai, United Arab of Emirates (UAE), Nov. 24–27.

T. AL-khdour, U. Baroudi (2009), "A Generalized Energy-Efficient Time-Based Communication Protocol for Wireless Sensor Networks", *Special issue of International Journal of Internet Protocols (IJIPT)*, Vol. 4, No. 2-2009.

Samuel R. Madden Madden, Michael J. Franklin And Joseph M. Hellerstein, And Wei Hong (2005) , "TinyDB: An Acquisitional Query Processing System for Sensor Networks", *ACM Transaction on Database Systems*, Vol. 30, No 1, March 2005, Pages 122-173.

Guihai Chen, Chengfa Li , Mao Ye, and Jie Wu, (2007) "An Unequal Cluster-Based Routing Strategy in Wireless Sensor Networks ," *Wireless Networks* (JS) , April 2007.

Younis M., Youssef M. and Arisha K. (2002), "Energy-Aware Routing in Cluster-Based Sensor Networks", in the Proceedings of the 10th IEEE/ACM International Symposium on Modeling, Analysis and Simulation of Computer and Telecommunication Systems (MASCOTS2002), Fort Worth, TX, October 2002.

Muruganathan, S.D.; Ma, D.C.F.; Bhasin, R.I.; Fapojuwo, A.O. (2005), "A Centralized Energy-Efficient Routing Protocol for Wireless Sensor Networks," *IEEE Radio Communication,* March 2005, pp. S8-S13.

Ya Xu, John Heidemann, and Deborah Estrin (2001), "Geography-informed energy conservation for ad hoc routing," in the Proceedings of the 7th Annual ACM/IEEE International Conference on Mobile Computing and Networking (MobiCom'01), Rome, Italy, July 2001.

Yan Yu, Ramesh Govindan, and Deborah Estrin (2001), "Geographical and Energy-Aware Routing: A Recursive Data Dissemination Protocol for Wireless Sensor Networks," UCLA Computer Science Department Technical Report, UCLA-CSD TR-01-0023, May 2001.

Foad. Lotfifar, Hadi. Shahhoseini (2006), "A mesh-Based Routing Protocol for Wireless Ad-Hoc Sensor Networks," in the Proc. of International Wireless Communication and Mobile Computing Conference (IWCMC'06), Vancouver, British Columbia, Canda, July 3-6, 2006.

Juan A. Sanchez, Pwdro M. Ruiz, and Ivan Stojmenovic (2007), "Energy-efficient geographic multicast routing for Sensor and Actuator Networks," *Computer Communications* 30 (2007) pp. 2519–2531

Gang Zhao, Xianggian Liu, and Min-Tue Sun (2007), "Energy-Aware Geographic Routing for Sensor Networks with Randomly Shifted Anchors," in the Proc. of Wireless Communications and Networking Conference WCNC 2007, 11-15 March 2007, pp. 3454-3459

Sundar Subramanian, Sanjay Shakkottai and Piyush Gupta (2007), "On Optimal Geographic Routing in Wireless Networks with Holes and Non-Uniform Traffic," in the Proc. of 26th IEEE International Conference on Computer Communications. INFOCOM 2007, May 2007, pp. 1019-1027

Jae-Hwan. Chang, Lendros and Tassiulas (2004), "Maximum Lifetime Routing in Wireless Sensor Networks," *IEEE/ACM Transactions on Networking (TON) archive* Volume 12 , Issue 4 (August 2004) ,pages: 609 - 619

Konstantinos Kalpakis, Koustuv Dasgupta and Parag Namjoshi (2004) , "Maximum Lifetime Data Gathering and Aggregation in Wireless Sensor Networks," in the Proceedings of IEEE International Conference on Networking (NETWORKS '02), Atlanta, GA, August 2002.

Tian He, John A Stankovic, Chenyang Lu, and Tarek Abdelzaher (2003), "SPEED: A stateless protocol for real-time communication in sensor networks," in the Proceedings of International Conference on Distributed Computing Systems, Providence, RI, May 2003.

Safwati. A., Hassanein. H., Mouftah. H. (2003)," Optimal Cross-Layer Designs for Energy-Efficient Wireless Ad hoc and Sensor Networks", in the Proceedings of the IEEE International Conference of Performance, Computing, and Communications 9-11 April 2003 Page(s):123 - 128

Venkitasubramaniam P., Adireddy S., Lang Tong (2003), "Opportunistic ALOHA and cross layer design for sensor networks" , Military Communications Conference, 2003. MILCOM 2003. IEEE Volume 1, 13-16 Oct. 2003 Page(s):705 - 710

Sichitiu M.L. (2004), "Cross-Layer Scheduling for Power Efficiency in Wireless Sensor Networks" ,INFOCOM 2004. Twenty-third Annual Joint Conference of the IEEE Computer and Communications Societies , Volume 3, 2004 Page(s):1740 - 1750

Li-Chun Wang, Chung-Wei Wang (2004), "A Cross-layer Design of Clustering Architecture for Wireless Sensor Networks", in the Proceedings of the IEEE International Conference on Networking, Sensing & Control Tapel, Taiwan, March 21-23, 2004, Page(s): 547-552

Rick W. Ha, Pin-Han Ho and X. Sherman Shen (2005), "Cross-Layer Application-Specific Wireless Sensor Network Design with Single-Channel CSMA MAC over Sense-Sleep Trees", *Elsevier Journal: Computer Communications Special Issue on Energy Efficient Scheduling and MAC for Sensor Networks*, WPANs,WLANs, and WMANs, 2005

Shuguang Cui, Madan R. , Goldsmith A. , Lall S. (2005), "Joint routing, MAC, and link layer optimization in sensor networks with energy constraints " IEEE International Conference on Communications, ICC 2005 ,Volume 2, 16-20 May 2005 Page(s):725 - 729

Su. W., T.L. Lim (2006), "Cross-Layer Design and Optimization for Wireless Sensor Networks," Proceedings of the Seventh ACIS International Conference on Software Engineering, Artificial Intelligence, Networking, and Parallel/Distributed Computing, SNPD June 2006 Page(s):278 - 284

Changsu Suh, Young-Bae Ko, and Dong-Min Son (2006) , "An Energy Efficient Cross-Layer MAC Protocol for Wireless Sensor Networks", APWeb 2006, LNCS 3842, pp. 410–419, 2006.

2

Low-power Sensor Interfacing and MEMS for Wireless Sensor Networks

J.A. Michaelsen, J.E. Ramstad, D.T. Wisland and O. Søråsen
Nanoelectronic Systems Group, Department of Informatics, University of Oslo
Norway

1. Introduction

The need for low-power and miniaturized electronics is prominent in wireless sensor network (WSN) nodes—small sensor nodes containing sensors, signal processing electronics, and a radio link. The demand for long battery life of such systems, especially if used in biomedical implants or in autonomous installations, forces the development of new circuit topologies optimized for this application area. Through a combination of efficient circuit topologies and intelligent control systems, keeping the radio idle when signal transmission is not needed, the radio link budget may be dramatically reduced. However, due to the demands for continuously monitoring of the sensor in many critical applications, the sensor front-end, analog-to-digital converter (ADC), and the control logic handling the radio up/down-link may not be turned off, and for systems with long intervals between transmissions, the energy consumed by these parts will have a large impact on battery life. In this chapter, we focus on Frequency $\Delta\Sigma$ Modulator (FDSM) based ADCs because of their suitability in WSN applications. Using FDSM based converters, both sensors with analog and frequency modulated outputs may be conveniently interfaced and converted to a digital representation with very modest energy requirements.
Microelectromechanical systems (MEMS) integrated on-die with CMOS circuitry enables very compact WSN nodes. MEMS structures are used for realizing a wide range of sensors, and form vital components in radio circuits, such as mixers, filters, mixer-filters, delay lines, varactors, inductors, and oscillators. In this chapter a MEMS oscillator will be used to replace Voltage Controlled Oscillators (VCOs). The MEMS oscillator is made using a post-CMOS process. Before the die is packaged, the CMOS die is etched in order to release the MEMS structures. The top metal layers in the CMOS process acts as a mask to prevent CMOS circuitry from being etched in addition to be used as a mask to define the MEMS structures. The resulting MEMS structure consists of a metal-dielectric stack where its thickness is determined by the number of metal layers available in the CMOS process. In this chapter, we will use a deep sub-micron CMOS process to illustrate the possibility for combining MEMS and CMOS in a small die area. The MEMS oscillator is to be used as a frontend for the FDSM.
FDSM and MEMS integrated in CMOS is a versatile platform for miniaturized low-power WSN nodes. In this chapter we illustrate the benefits of this approach using simulation, showing the potential for efficient miniaturized solutions.

2. Background

Within the international research community and industry, large research and development efforts are taking place within the area of Wireless Sensor Networks (WSN) (Raghunathan et al., 2006). Wireless sensor nodes are desirable in a wide range of applications. From a research perspective, power consumption and size are main parameters where improvements are needed. In this chapter we will focus on methods and concepts for low-voltage and low-power circuits for sensor interfacing in applications where the power budget is constrained, along with MEMS structures suitable for on-die CMOS integration. These technologies enable wireless sensor network nodes (WSNNs) with a very compact size capable of being powered with a depletable energy source due to its potential for low voltage and low power consumption.

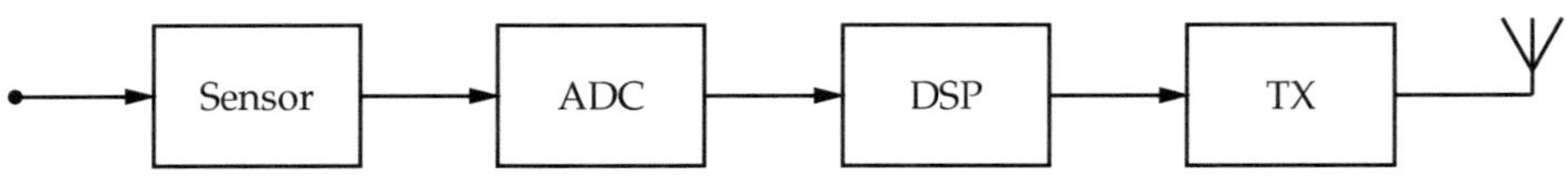

Fig. 1. Wireless sensor network node

The key components of a wireless sensor node are: 1) The sensor performing the actual measurement (pressure, light, sound, etc.), producing a small analog voltage or current. 2) An analog-to-digital (A/D) converter (ADC) converting and amplifying the weak analog sensor output to a digital representation. 3) A digital signal processing system, performing local computations on the aquired data to ready it for transmission, and for deciding when to transmit. 4) A radio transceiver for communicating the measurements. This is depicted in figure 1. The sensor readout circuitry, namely the ADC and processing logic, must continuously monitor the sensor readings in order to detect changes of interest and activate the transceiver only when needed to conserve power. For digital CMOS circuitry, an efficient way of saving power is to reduce the supply voltage, resulting in subthreshold operation of MOSFET devices, as their conductive channel will only be weakly inverted (Chen et al., 2002). In standard nanometer CMOS technology, safe operation is possible with supply voltages down to approximately 200mV (Wang & Chadrakasan, 2005). Conventional analog circuit topologies are not able to operate on these ultra low supply voltages, especially with the additional constraint of a scarce power budget (Annema et al., 2005). As a result, the ADCs currently represents a critical bottleneck in low-voltage and low-power systems, accentuating the need for new design methodologies and circuit topologies.

The sensor readout circuit must satisfy certain specifications like sufficient gain, low distortion and sufficient signal-to-quantization-noise ratio (SQNR). When studying existing Nyquist-rate ADCs, it is obvious that the analog precision is reduced as the power supply voltage is lowered (Chatterjee et al., 2005). This is mainly due to non-ideal properties of the active and passive elements, and process variations. In order to increase the SQNR, oversampled converters employing noise shaping $\Delta\Sigma$ modulators are used, trading bandwidth for higher SQNR (Norsworthy et al., 1996). ADCs are implemented either using continuous-time (CT) or Switched Capacitor (SC) components for realizing the necessary analog filter functions. SC realizations have generally been preferred for CMOS implementations as the method does not rely on absolute component values which are difficult to achieve without post-fabrication trimming. During the last few years, the power supply has moved down to 1 V in state-of-the art technologies making it hard to implement switches with sufficient conduction required for SC-filters. As a result, current SC realizations switch the opamp, eliminating the need for CMOS switches in the signal path. This method is referred to as the Switched Opamp technique (Sauerbrey et al., 2002). As a result, the most important building block for both CT and SC based $\Delta\Sigma$ modulators are the opamp, which is also the limiting component with respect to conversion speed and signal-to-noise and distortion ratio (SINAD). As mentioned earlier, the sensor readout circuitry in a battery operated wireless sensor node should allow for operation far below 1V to facilitate low power consumption. This requirement eliminates both conventional CT and SC $\Delta\Sigma$ modulators as these approaches require large amounts of power at low supply voltages to attain reasonable performance.

Several low-power ADC topologies adapted for sensor interfacing have been reported in the last few years (Yang & Sarpeshkar, 2005; Kim & Cho, 2006; Wismar et al., 2007; Taillefer & Roberts, 2007). Among them, some are utilizing the time-domain instead of the amplitude-domain to reduce the sensitivity to technology and power supply scaling (Kim & Cho, 2006; Wismar et al., 2007; Taillefer & Roberts, 2007).

The non-feedback modulator for A/D conversion was introduced in Høvin et al. (1995); Høvin et al. (1997). In contrast to earlier published $\Delta\Sigma$ based ADCs, this approach does not require a global feedback to achieve noise shaping giving new and additional freedom in practical applications. This property is particularly useful when the converter is interfacing a sensor (Øysted & Wisland, 2005). The non-feedback $\Delta\Sigma$ modulator has two important properties which make it very suitable for low-voltage sensor interfacing. First, the topology has no global feedback which opens up for increasing the speed and resolution compared to conventional methods. Second, and most important, the analog input voltage is converted to an accumulated phase representing the integral of the input signal, thus moving the accuracy requirements from the strictly limited voltage domain, to the time domain, which is unaffected by the supply voltage. The conversion from analog input voltage to accumulated phase is performed using a Voltage Controlled Oscillator (VCO). As this solution uses frequency as an intermediate value, the non-feedback ADC using a VCO for integration is normally referred to as a Frequency Delta Sigma Modulator (FDSM).

Until recently, the FDSM has mainly been used for converting frequency modulated sensor signals with no particular focus on low supply voltage. In Wismar et al. (2006), an FDSM based ADC, fabricated in 90 nm CMOS technology, is reported to operate properly down to a supply voltage of 200 mV with a SINAD of 44.2 dB in the bandwidth from 20 Hz to 20 kHz (the audio band). The measured power consumption is 0.44 µW. The implementation is based on subthreshold MOSFET devices with the bulk-node exploited as input terminal for the signal to be converted.

At the RF front-end in WSN nodes, bulky off-chip components are usually used to meet the RF performance requirements. Such components are typically external inductors, crystals, SAW filters, oscillators, and ceramic filters (Nguyen, 2005). Micromachined components have been shown to potentially replace many of these bulky off-chip components with better performance, smaller size and lower power consumption. The topic of combining MEMS directly with CMOS has been of great interest in the past years (Fedder et al., 2008). The direct integration of MEMS with CMOS reduces parasitics, reduces the packaging complexity and the need for external components becomes less prominent. It turns out that integrating MEMS after the CMOS die has been produced has been most successful which is proven by Carnegie Mellon University (Chen et al., 2005; Fedder & Mukherjee, 2008), National Tsing Hua University (Dai et al., 2005), University of Florida (Qu & Xie, 2007) and University of Oslo (Soeraasen & Ramstad, 2008; Ramstad et al., 2009). The concept of CMOS-MEMS is maturing and seems to be versatile and offer the flexibility of possibly replacing RF-front end components or sensors, both relevant in the context of WSNN.

3. Frequency Delta-Sigma Modulators

An FDSM based converter (Høvin et al., 1997) can conveniently be used in WSNNs for converting frequency modulated signals to a quantized and discrete bitstream, where the quantization noise is shaped away from the signal band. Overall, this results in frequency-to-digital (F/D) conversion with equivalent $\Delta\Sigma$ noise shaping.

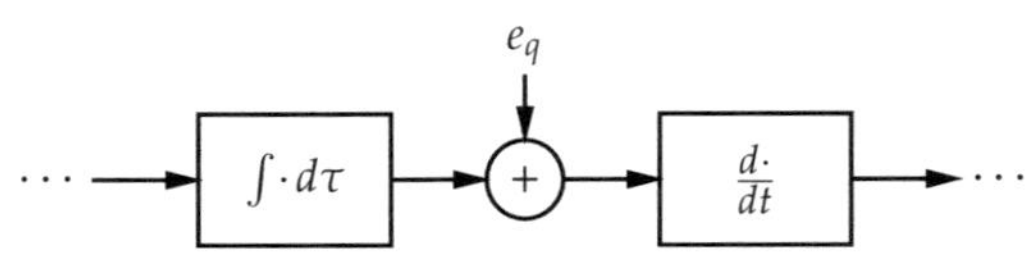

Fig. 2. FDSM overview

In the time domain, the input to the modulator, a frequency modulated (FM) signal, is $x_{fm}(t) = \cos[\theta(t)]$, where the instantaneous phase is,

$$\theta(t) = 2\pi \int_0^t f_c + f_d \cdot x(\tau)\, d\tau \tag{1}$$

f_d is the maximal deviation from the carrier frequency, f_c, while $x(\tau)$ represents the physical quantity we are measuring; assumed to be limited to ± 1. The integral of the input signal and a constant bias is now represented by the phase, $\theta(t)$. The cosine function wraps the phase every 2π, effectively performing modulo integration. By using a counter, triggered by the zero-crossings of the x_{fm} signal, the integral of the input signal is quantized to a digital value which in turn is sampled at regular intervals, $T_s = f_s^{-1}$. A digital representation of the input, x, is recovered by differentiating the quantized phase signal. This is depicted in figure 3(a).

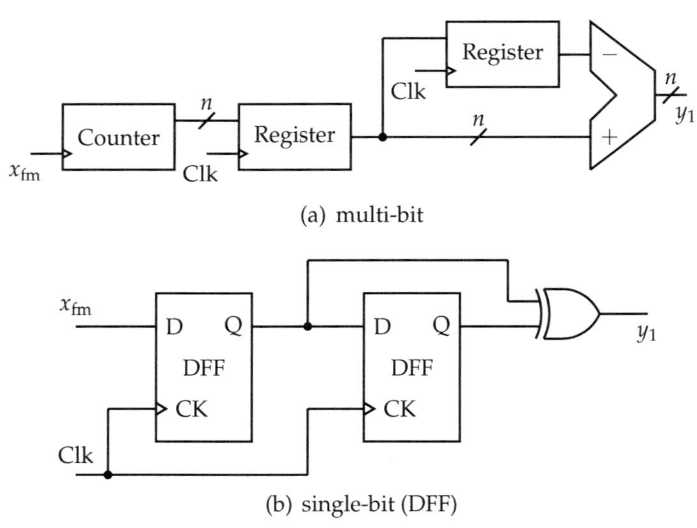

Fig. 3. First order FDSM topologies

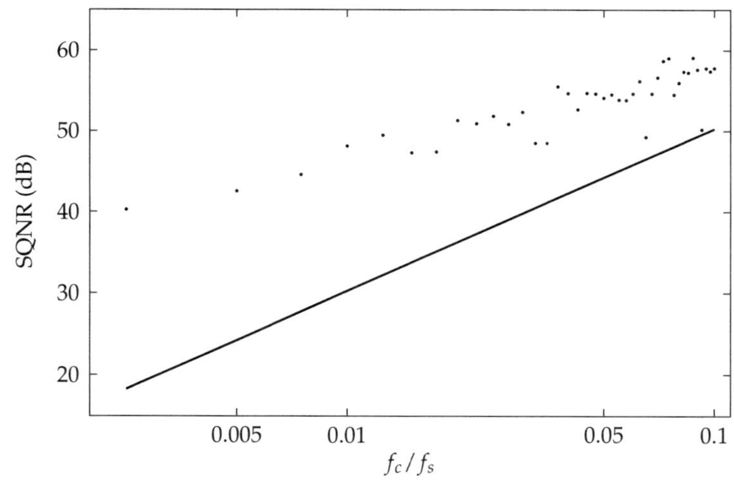

Fig. 4. Theoretical performance (solid line) and time-domain simulated performance (dots) as a function of carrier frequency and sampling frequency ratio

An important property is that quantization noise occurs after the integration, resulting in first order noise shaping of the quantization noise sequence, while the input signal is not altered. This is illustrated in figure 2, where e_q represents the quantization noise. Second order noise shaping can be obtained by integrating the quantization error from the first order FDSM. While the second order system requires a higher circuit complexity which incurs an increase in power consumption, it can be shown that the increase in performance in some cases outweighs the additional requirements (Michaelsen & Wisland, 2008).

The FDSM is inherently an oversampled system, meaning that the output bitrate, f_s, is much higher than the bandwidth of the input signal, f_b. Quantization noise is suppressed in the signal band through noise shaping. In the case of first order converters, the quantization noise will be shaped with a slope of 20 dB/decade.

If the number of zero-crossings of the FM signal during T_s is less than two, it is possible to realize the structure in figure 3(a) with only two D-flipflops (DFFs), and an XOR-gate used for subtraction, as illustrated in figure 3(b). Due to its simple implementation, the first order single-bit FDSM is a viable choice for WSNN applications because of its potential for low power consumption and low voltage operating requirements (Wismar et al., 2007). In this case, the resolution of the converter is given by (Høvin et al., 1997)

$$\mathrm{SQNR_{dB}} = 20\log_{10}\left(\frac{\sqrt{2}f_d}{f_s}\right) - 10\log_{10}\left(\frac{\pi^2}{36}\left(\frac{2f_b}{f_s}\right)^3\right) \tag{2}$$

However, in cases where $f_s/f_c \gg 1$, the actual performance may be better than predicted by equation 2. As illustrated in figure 4, this discrepancy can be significant. In this plot, f_s was held constant at 20 MHz, with $f_d = f_c \cdot 10\%$, and $f_b = 19$ kHz. The solid line represents the performance predicted by equation 2 while the dots indicate the performance from a difference equation simulation of the converter. The underlying assumption in equation 2 is that the quantization noise sequence is a white noise sequence. However, this assumption in not accurate, and it is possible to exploit pattern noise valleys for significantly improving performance (Høvin et al., 2001).

Before further processing of the digital sensor signal in the WSNN, it is usually desirable to have an output frequency that is equal to, or slightly higher than, $2f_b$. To achieve this the output bitstream is decimated by first bandlimiting the signal using a low-pass filter. This removes the out-of-band noise to avoid aliasing. After low-pass filtering, only every N-th sample is kept, where $N = {}^{f_s}/_{2f_b}$. During and after decimation, each sample must be represented by more bits to avoid quantization noise being a limiting factor. The decimation usually requires a significant amount of computation. This task is therefore done in stages, where computationally efficient filters run at the input frequency, while more accurate filters run at lower frequencies. The first stage is usually a sinc^m-filter, where m is the order of the filter, named after its $(\sin(x)/x)^m$ shaped frequency response. This class of filter has a straight forward hardware implementation (Hogenauer, 1981; Gerosa & Neviani, 2004) capable of high frequency operation. It can be shown that a sinc^{L+1} filter is sufficient for an order L $\Delta\Sigma$ modulator (Schreier & Temes, 2004). At later stages, more complex filters can be used to correct for the non-ideal features of the sinc filter such as passband droop (Altera Corporation, 2007).

The frontend of the FDSM—be it a VCO in the case of an ADC, or a device which directly converts some physical quantity to a frequency modulated signal—will to some extent have a non-linear transfer function. A non-linear FM source will in turn give rise to harmonic distortion present in the output signal. Although quantization noise is shaped away from the signal band, harmonic distortion will not be suppressed as it is impossible for the F/D converter to distinguish between what is the actual signal and what is noise and distortion. This non-linearity deteriorates the effective resolution of the measurement system. However, several digital post-processing schemes and error correction systems have been devised that are able to recover linearity to some extent (Balestrieri et al., 2005). Care must be taken when designing the post-processing system so that aliasing of values and missing output codes does not present a problem. Another issue with the FDSM frontend is phase noise, also referred to as jitter. This noise will directly add to the input signal and therefore not undergo noise shaping; raising the noise floor at the output. $1/f$ noise has shown to be particularly problematic, and careful attention to issues related to noise is critical when designing the oscillator circuit. This is especially challenging in deep sub-micron CMOS technologies.

4. Using a MEMS resonator as a VCO

4.1 The micromechanical resonator

A resonator is a component which is able to mimic full circuit functions such as filtering, mixing, line delays, and frequency locking. The resonator is a mechanical element that vibrates back and forth where the displacement of the micromechanical element generates a time varying capacitance which in turn results in an ac current at the output node. The maximum output current occurs when stimulating the resonator with an input ac voltage with a frequency equal to the resonance frequency of the resonator. The micromechanical resonator can be represented as an LCR circuit (see figure 5) where the equations describing these passive components are related to physical parameters such as mass, damping, and stiffness (Senturia, 2001; Bannon et al., 2000).

Figure 5 is a simple LRC circuit which can be described as,

$$V_i = \ddot{q}(t)L_x + \dot{q}(t)R_x + q(t)\frac{1}{C_x} \tag{3}$$

where L_x, R_x and C_x are the passive element values for a maximum displacement x of the resonator. V_i and V_o are the input and output voltages as shown in figure 5. $q(t)$ is the charge

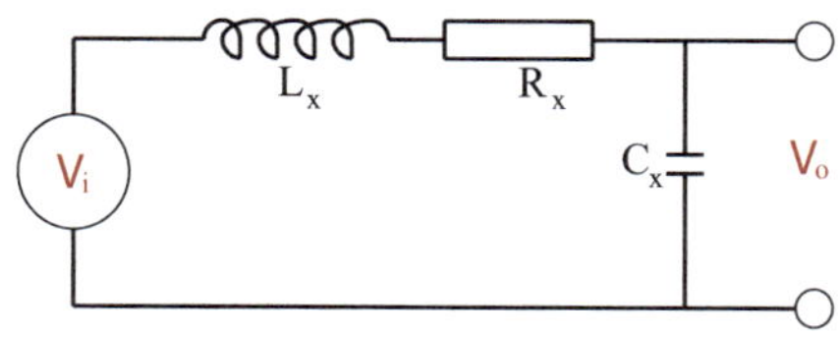

Fig. 5. A simple LCR circuit

on the capacitor which depends on the time t. By using the relationship between the output and the input ($H(t) = V_o/V_i$) from the circuit of figure 5 and by using $q = C_x V$ results in the derivation of the resonance frequency of this system:

$$f_0 = \frac{1}{2\pi}\sqrt{\frac{1}{L_x C_x}} \tag{4}$$

From the transfer function, the maximum throughput exists when the reactances of the inductor and the capacitor is equal to each other and opposite, thus this defines the resonance frequency for this micromechanical system. For RF front-end components and oscillators, it is desirable to have a good transfer of the signal through the component. A good throughput is possible by having a good Q-factor which is described by,

$$Q = \frac{\omega_0 L_x}{R_x} \tag{5}$$

where equation 5 is derived from the transfer function of figure 5 and ω_0 is the resonance frequency of the resonator ($\omega_0 = 2\pi f_0$). A large Q-factor is usually desirable to get good resonator performance. As explained in section 4.5, the resulting MEMS structures consists of a laminate of metal and dielectric, so the resulting Q-factor will be limited mostly by intrinsic material loss and gas damping which will be discussed later. A top view of a micromechanical resonator is shown in figure 6.

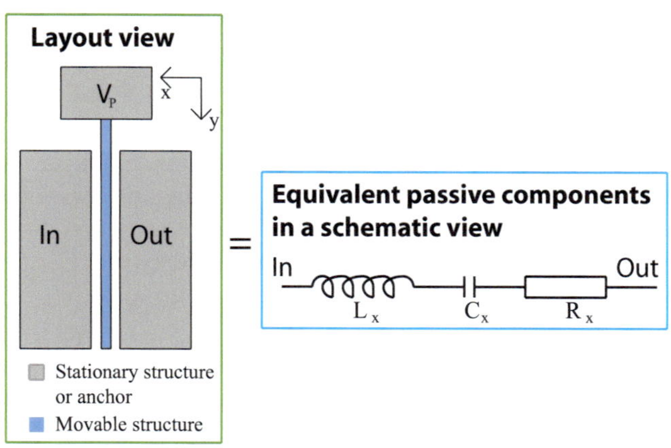

Fig. 6. The resonator analogy

Figure 6 shows a long and thin cantilever beam (fixed at one end, free to move at the other end) with two electrodes next to it. The left electrode is the input electrode while the right electrode is the output electrode. The gray areas indicate stationary elements (the anchor and the electrodes) while the blue area indicates a part which is able to move freely (the resonator). The thin and long cantilever beam moves back and forth laterally above the silicon substrate towards the two electrodes in the x-direction. At the resonance frequency of this resonator, the maximum vibration towards the electrode is x. The thickness of the beam is not shown here as this is a top view. The V_P signal applied to the beam itself is a high DC voltage which is used to cancel unwanted frequency terms and to amplify the signal of the resonator. By separating the V_P signal from the input and output ac signals, the V_P signal will not be superimposed on either of the two signals. The gap g between the resonator and the electrodes is an important parameter which will decide vital aspects of the resonator as will be shown later.

4.2 The electromechanical analogy

4.2.1 The electromechanical coupling coefficient

The micromechanical resonator is attracted due to electrostatic forces creating a capacitive coupling between the resonator and the input electrode (Kaajakari et al., 2005). A large electrode area that covers the resonator is desirable where the capacitance C is described as,

$$C = \frac{\varepsilon_0 W_r W_e}{g} \tag{6}$$

where ε_0 is the permittivity in air, W_r is the resonator width (vertical thickness, not visible in figure 6), W_e is the electrode length, and g is the gap between the resonator and the electrode. The capacitance equation is related to the electrostatic force equation (F). The electrostatic force F is derived from the potential energy equation $U = {}^1\!/_2 CV^2$ which results in:

$$F = \frac{dU}{dx} = \frac{1}{2}\frac{dC}{dx}V^2 \tag{7}$$

where V is the signal voltage. ${}^{dC}\!/_{dx}$ is the capacitance change due to a small change in the gap size g because the resonator bends towards the electrode with a displacement x. The force is proportional to the square of the voltage V which will introduce a $\cos(2\omega t)$ term (the derivation of this is not shown here). The $\cos(2\omega t)$ term will introduce oscillation at $\omega = {}^{\omega_0}\!/_2$, half the resonance frequency. In order to avoid this nonlinear relationship, a polarization voltage V_P is applied to the beam. When splitting V into $V_P + v \cdot \cos(\omega t)$ the resulting electrostatic force becomes,

$$f = V_P \frac{dC}{dx} v \tag{8}$$

Equation 8 describes the relationship between the force f and the voltage v (small signal values) that now has a linear relationship. It is now possible to derive the coefficient known as η:

$$\eta = V_P \frac{dC}{dx} \approx V_P \frac{\varepsilon_0 W_r W_e}{g^2} \tag{9}$$

η is a coefficient which describes how well the signal from the electrode is transferred to the resonator. It is an equation that is a result of the electrostatic force equation so that the force f has a linear relationship to the voltage v. A larger η results in a larger signal of the resonator. It is desirable with a large electrode area ($A_{el} = W_r W_e$) and a small gap g. Because η is inversely proportional to the square of the gap between the electrode and resonator, it is desirable to have an extremely small gap size. Both the electrode area A_{el} and the gap size g are limited by process constraints. Notice that equation 9 is a simplified equation of η as the derivation of the capacitance C with respect on the gap g is done by assuming that the gap is the same throughout the y-axis of the resonator (throughout the resonator length L).

4.2.2 Resonator output current

The output current due to the capacitive coupling explained in section 4.2.1 can be written as:

$$i_o = V_P \frac{dC}{dt} + C\frac{dv}{dt} \approx V_P \frac{dC}{dt} \tag{10}$$

The output current in equation 10 consists of two parts: One part which is amplified with the polarization voltage V_P, and one part which consists of the (small) sinusoidal voltage v. Equation 10 was derived by using $i_o = {}^{d}/_{dt}(C \cdot V)$ (Bannon et al., 2000). It is possible to further simplify this equation by neglecting the $C^{dv}/_{dt}$ part because the voltage V_P is much larger than v:

$$i_o = V_P \frac{dC}{dx}\frac{dx}{dt} \approx \eta\omega_0 x \tag{11}$$

Equation 11 was derived by using the relationship ${}^{dx}/_{dt} = \omega_0 x$. By using V_P, the output current i_o can be amplified as shown in equation 11. However, when increasing V_P, ω_0 will be reduced while the displacement x increases. This means that the current will have an exponential-like increase as V_P is increased and not a linear increase of i_o which could be expected. The fact that the operational (resonance) frequency of the resonator decreases when V_P is increased is due to an effect known as "spring-softening" which will be discussed later (Bannon et al., 2000). This spring-softening effect will be utilized in order to use the micromechanical resonator as a voltage-controlled oscillator (VCO).

4.2.3 The LCR equivalents

By using the principle of electromechanical conversion as explained in section 4.2.1, it is possible to derive formulas for L_x, C_x and R_x.

$$L_x = \frac{m_{eff}}{\eta^2} \tag{12}$$

$$C_x = \frac{\eta^2}{k_r} \tag{13}$$

$$R_x = \frac{\sqrt{k_r m_{eff}}}{Q\eta^2} \tag{14}$$

where k_r is the effective spring stiffness and m_{eff} is the effective mass of the resonator. Q is the Q-factor of the resonator which is inverse proportional to the total damping of the micromechanical resonator. All three LCR components are dependent on the square of η. This indicates a square dependence of the electrode area A_{el} and a g^4 dependence of the gap between the resonator and the electrode. The electrical equivalents of the components are not straightforward to interpret due to complicated relationships between the mass, stiffness and damping of the resonator, as well as complicated relationship due to the electrostatic force.

4.3 The resonance frequency and its implications

4.3.1 The nominal resonance frequency

The natural frequency of the resonator with no voltage applied is given by equation 15 below (Senturia, 2001):

$$f_{0(eff)} = \frac{1}{2\pi}\sqrt{\Lambda_n \frac{k}{m}} \tag{15}$$

where k is the static beam stiffness and m is the static beam mass of the micromechanical system. Λ_n is a constant depending on mode number. A mode is a certain frequency in which the resonator will have a maximum vibration amplitude. A micromechanical resonator may have several modes at distinct frequencies. Λ_n has different values for different modes. For example, Λ_1=1.0302 for mode 1, Λ_2=40.460 for mode 2, Λ_3=317.219 for mode 3 etc. The resonator is operated in the first mode (Λ_1). Both k and m depends on the geometry and structural material of the resonator. The values for Λ_n used here is valid only for the cantilever beam architecture, other types of resonators will have different values of Λ_n.

4.3.2 The effective resonance frequency and Q-factor

The movable parts of the resonator will all vibrate back and forth with the resonance frequency ω_0. The tip of the beam will have a longer distance to move and will thus have a higher velocity $\acute{v}$ compared to the part of the cantilever beam which is closer to the anchor. Because the kinetic energy ($E_k = {}^1\!/_2 m_{eff}\acute{v}^2$) must be the same throughout the beam when it vibrates, the effective mass along the beam in the y-direction in figure 6 will vary. The effective mass is defined as m_{eff} where the largest value appears close to the anchor while the smallest value appears at the tip of the beam. The derivation of m_{eff} is not shown here but can be developed by using the equation for kinetic energy. By using equation 15 and rearranging, the mechanical spring stiffness can be defined as:

$$k_m(y) = \left(2\pi f_{0(eff)}\right)^2 m_{eff}(y) \tag{16}$$

Equation 16 shows the pure mechanical spring stiffness of the beam when it vibrates. $k_m(y)$ varies along the beam in the y-direction with a maximum value close to the anchor and a minimum value close to the tip of the beam. However, when applying a DC voltage V_P to the beam, the total spring stiffness of the beam will be reduced. The resulting effective spring stiffness value k_r is reduced due to an electric spring value k_e. Because of this fact, the resonance frequency of the cantilever beam will be reduced as described in the following equation:

$$f_0 = f_{0(eff)}\sqrt{1 - \frac{k_e}{k_m}} \tag{17}$$

where the relationship ${}^{k_e}\!/_{k_m}$ determines the amount of reduction of the original nominal resonance frequency $f_{0(eff)}$. The effective spring stiffness k_r is defined as:

$$k_r = k_m - k_e \tag{18}$$

where k_r is known as the effective beam stiffness. k_r is the result of subtracting the electrical spring stiffness k_e from the effective mechanical spring stiffness k_m (spring-softening). The effective beam stiffness is more precisely defined as,

$$k_r = \left(2\pi f_{0(eff)}\right)^2 m_{eff}(y) - \int_{W_{e1}}^{W_{e2}} V_P^2 \frac{\varepsilon_0 W_r dy'}{[g(y')]^3} \tag{19}$$

where the second term of equation 19 describes the electrical spring stiffness at a specific location y' centered on an infinitesimal length of the electrode dy'. The k_e part consists of integrating from the start of the electrode (W_{e1}) to the end of the electrode (W_{e2}). The variable part of the k_e equation is the gap which varies along the y-axis throughout the beam length. The

k_e equation is derived from the potential energy equation $U = {}^1\!/_2 C V_P^2$. The gap as a function of y can be described as (Bannon et al., 2000):

$$g(y) = g_0 - \frac{1}{2} V_P^2 \varepsilon_0 W_r \int_{W_{e1}}^{W_{e2}} \frac{1}{k_m(y')[g(y')]^2} \frac{X_{mode}(y)}{X_{mode}(y')} dy' \tag{20}$$

where g_0 is the static electrode-to-resonator gap with $V_P = 0$. X_{mode} is an equation that describes the shape of how the cantilever beam bends. The second term describes the displacement of the resonator towards the electrode at various locations of y. As can be seen in equation 18, if k_e becomes equal to k_m, the resonance frequency should become zero. However, before that would occur, the resonator will enter an unstable state which will pull the beam towards the electrode instead. This effect is known as the "pull-in" effect. Due to the reduction of the original natural frequency of the resonator, the Q-factor will also be reduced in a similar manner. The Q-factor is mainly affected by four factors: Anchor loss, environmental (viscous gas) damping, thermoelastic damping or internal (material) energy loss. The topic of damping mechanisms for MEMS resonators is not trivial, therefore it is typical to do crude estimates for the nominal Q-factor as a starting point for analysis (Bannon et al., 2000).

$$Q_{eff} = Q_{nom} \sqrt{1 - \frac{k_e}{k_m}} \tag{21}$$

From equation 17 and equation 21 we can conclude that when increasing the V_P value, both the resonance frequency and the Q-factor of the resonator are reduced. For oscillators, a high Q-factor is desirable, therefore it is important to also include this reduction of the Q-factor for correct modeling.

4.4 Nonlinear behavior

As described by equation 17, the oscillation frequency is tuned by using V_P. In order to get a good tuneability of the MEMS resonator, it is designed to be soft so that it can operate at low voltages and at the same time have a reasonable tuning range. However, when a beam is too soft, non-linear effects become more dominant. We can classify two different types of resonator non-linearities (Kaajakari et al., 2005; 2004):

- Mechanical non-linearity: Typically non-elasticity due to geometrical and material effects
- Capacitive non-linearity: Introduced due to an inverse relationship between the displacement and the "parallel" plate capacitance

Mechanical non-linearity will be more prominent in other resonator architectures such as the clamped-clamped beam, we will therefore focus on the capacitive non-linearities for this analysis. In order to develop an understanding of the introduction of the capacitive non-linearity, we must take a look at the equation describing the motion of the resonator:

$$m_{eff}\ddot{x} + b\dot{x} + k_r x = F(t) \tag{22}$$

Equation 22 describes the equation of movement of the resonator due to an external force. This equation is basically the same as equation 3 where the external force is the electrostatic force. The equation of movement is related to the effective mass m_{eff}, the damping b (which is inverse proportional to Q), and the effective spring stiffness k_r. In this equation k_r has a mechanical term k_m and an electrical term k_e as described earlier. For a case where k_e is linear, the motion of the amplitude becomes:

$$X_0 = \frac{FQ_{eff}}{k_r} \tag{23}$$

Equation 23 shows the displacement of the tip of the beam at resonance. However, when the resonator has a low mechanical stiffness k_m, and is at the same time operated with large V_P values, the linear k_e model becomes inaccurate. Therefore the following equation is used instead:

$$k_e(x) = k_{e0}\left(1 + k_{e1}x + k_{e2}x^2 + ...k_{en}x^n\right) \tag{24}$$

From equation 24, we can see that the spring stiffness consists of higher order terms that all are related to the displacement x (Kaajakari et al., 2005). The k_{e0} term is the first term and is linear. k_{e1} and k_{e2} are square and cubic electrical spring coefficients respectively:

$$k_{e0} = -\frac{V_P^2 C_0}{g^2}, k_{e1} = \frac{3}{2g}, k_{e2} = \frac{2}{g^2} \tag{25}$$

The $k_e(x)$ terms contribute to reducing or increasing the frequency depending on which term that dominates. When operating the resonator with high vibration amplitudes, the square and cubic spring stiffness terms will become more dominant. Because the amplitude-frequency curve no longer becomes a single valued function, the oscillation may become chaotic once the amplitude is larger than a critical value known as x_c. The maximum usable vibration value is extracted from the largest value that appears before a bifurcation (hysteresis of the curve). The bifurcation amplitude and critical amplitude are respectively (Kaajakari et al., 2005):

$$x_b = \frac{1}{\sqrt{\sqrt{3}Q|\kappa|}}, x_c = \frac{2}{\sqrt{3\sqrt{3}Q|\kappa|}} \tag{26}$$

where

$$\kappa = \frac{3k_{e2}k_{e0}}{8k} - \frac{5k_{e1}^2 k_{e0}^2}{12k^2} \tag{27}$$

Figure 7 is an example of how κ will affect the response out from the resonator. κ_1 is the lowest value and κ_3 is the largest value. In this example, κ is positive and contributes to increase in the resonance frequency as well as tilting the curve to the left. κ_1 is the lowest value and shows less tilting of the curve. When κ is too large (see κ_3), the curve enters a state of hysteresis. At the point when the hysteresis starts, the bifurcation amplitude x_b is reached. For any curve with a hysteresis, the maximum usable amplitude of vibration is x_c as shown in figure 7b. x_c is always larger than x_b and ultimately sets the limit for the maximum vibration amplitude as well as it sets the maximum output current from the resonator. Because κ is a factor which will contribute to a modified resonance frequency due to the spring stiffness non-linearities, the new resonance frequency is therefore expressed as,

$$\omega_{0(effective)} = \omega_0\left(1 + \kappa X_0^2\right) \tag{28}$$

From equations 27 and 28 we can see that κ will either increase (resonator becomes more stiff) the operational resonance frequency or decrease the resonance frequency. The resonator used here will have a positive κ, thus the capacitive non-linearities will contribute to stiffen the

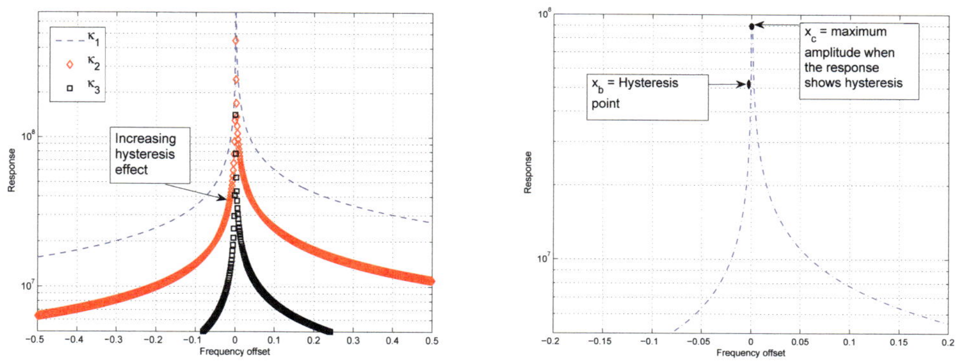

(a) Increasing κ (b) Hysteresis for κ_3

Fig. 7. Bifurcation and critical bifurcation

resonator. Because κ contributes to "stiffen" the output response, more V_P must be applied than first estimated in equation 17. By using equation 26 and 27, an expression for the maximum output current possible from the resonator is developed:

$$i_o^{max} = \eta \omega_0 x_c \tag{29}$$

i_o^{max} sets the limit for how much current that can be registered at the output electrode before bifurcation. The difference between equation 10 and equation 29 is that the maximum current is limited by the critical vibration x_c instead. It is also possible to define the maximum energy stored in the resonator by using x_c in a similar manner.

$$E_{stored}^{max} = \frac{1}{2} k_0 x_c^2 \tag{30}$$

where k_0 is a linear spring constant ($k_0 = k_m - k_{e0}$). The maximum energy stored also determines the energy dissipation out from the resonator which is,

$$P_{dissipated} = R_x i_o^2 = \frac{\omega_0 E_{stored}^{max}}{Q} \tag{31}$$

In order to understand the stability of the resonance frequency, the phase-noise of the system can be evaluated. This is possible by using Leeson's equation to model the phase-noise-to-carrier ratio in an ideal oscillator:

$$\mathcal{L}(\Delta f) = 10log\left[\frac{kT}{\pi E_{stored}^{max}} \frac{Q}{f_0}\left(1 + \left(\frac{f_0}{2Q\Delta f}\right)^2\right)\right] \tag{32}$$

where k is Boltzmann's constant and T is the absolute temperature (Shao et al., 2008). It is common to relate equation 32 to equation 31 and also add a buffer noise source from the amplifier following the resonator as given by (Kaajakari et al., 2004):

$$\mathcal{L}(\Delta\omega) = \frac{2kT}{P_{dissipated}}\left(\frac{\omega_0}{2Q\Delta\omega}\right)^2 + \frac{P_N^{buffer}}{2P_{dissipated}} \tag{33}$$

where P_N^{buffer} is buffer noise from an amplifier source. This value can be set to $-155^{dBm}/_{\sqrt{Hz}}$ (or $v_n = 4n^{V}/_{\sqrt{Hz}}$ for a 50Ω system). The equation for phase noise will be shown in a practical example in section 5.2.

4.5 Integration of MEMS in CMOS

There are three main methods of integrating MEMS in a CMOS process: 1. Insert the MEMS before the CMOS is made. 2. Insert the MEMS in between CMOS process steps. 3. Insert the MEMS after the CMOS has been made. In this demonstration, we will focus on the third step where the MEMS is made after the CMOS has been made which is known as post-CMOS. We will not go into the details of the process here for the sake of simplicity.

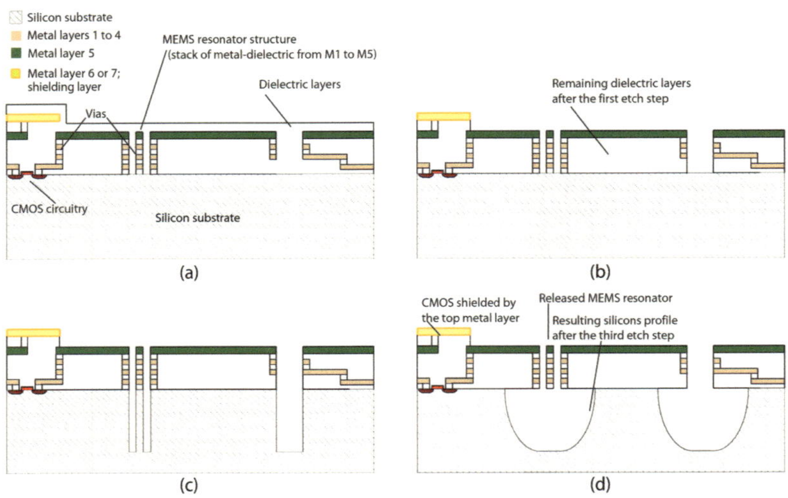

Fig. 8. The CMOS-MEMS process steps

The CMOS-MEMS process demonstrated here is inspired by previous work done at some universities (Ramstad, 2007; Fedder & Mukherjee, 2005; Sun et al., 2009). For low-power applications it is interesting to try to integrate MEMS in a deep sub-micron CMOS process. Figure 8 shows the process steps that have been used for a general deep sub-micron CMOS process. The steps a) to d) consist of the following:

a) The wafer before etching

b) Anisotropic etching of the dielectric

c) Etching of silicon using DRIE

d) Isotropic release-etch of silicon

This list shows the steps performed in order to etch and release MEMS structure(s). From figure 8 it can be seen that the top metal layer will act as a mask and define the MEMS structures. The MEMS resonator and electrodes consist of a stack of metals and dielectrics from metal layer 1 to metal layer 5. Areas that are not to be etched must be protected by a top metal layer (i.e. metal layer 6 or 7). The cross-section reveals that the CMOS must be placed a certain distance away from the open areas where the MEMS structures are etched and defined. The thickness of the resulting MEMS structure depends on the amount of metal layers that are used. The thickness of the metal-dielectric stack influences the smallest possible gap between a resonator and an electrode. There are also rules which define the smallest possible width of a structure and the largest possible width of a structure. There are more CMOS-MEMS rules than discussed here, but these are some of the most important ones when combining CMOS and MEMS on-chip by making MEMS structures from the metal layers offered by a general CMOS process.

4.6 The oscillator circuit

The MEMS resonator described in section 4.1 is made using a conventional 90 nm CMOS process using the same process steps as described in section 4.5. By putting the micromechanical resonator in a feedback loop with an amplifier, we get the basic oscillator circuit as shown in figure 9 below:

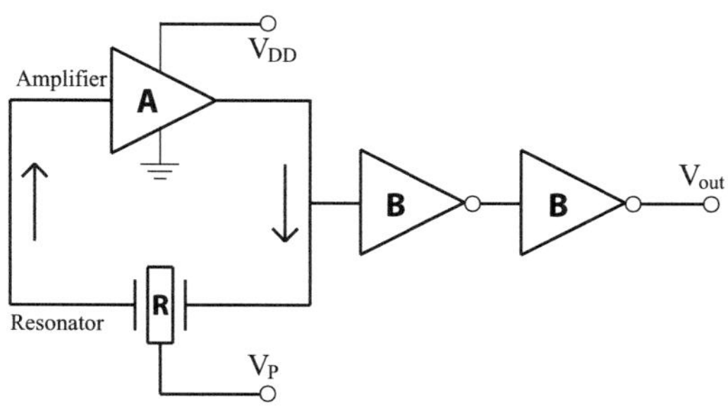

Fig. 9. Basic oscillator circuit

An oscillator is defined as a circuit that produces a periodic output signal at a fixed frequency. The resonator is the element in the circuit which defines the resonance frequency while the amplifier is the active element which sustains oscillation. The bias voltage V_P applied to the resonator is used to tune the frequency of this voltage-controllable oscillator. In this demonstration, the Q-factor of the resulting metal-dielectric MEMS structure is lower compared to state-of-the-art MEMS and will contribute to increase the motional impedance R_x which is seen in series with the amplifier. The low Q-factor will also lead to a large phase-noise. Both these two factors are not critical here as this is a demonstration to show MEMS directly combined with CMOS processing that could lead to future interesting applications. Even though R_x is large, the amplifier will be able to initiate and sustain oscillation. In order for the oscillator to start up the impedance from the amplifier has to be negative and at least three times larger than the total impedance that is in series with the amplifier. The total impedance consists of parasitics in the circuit plus the motional impedance from the resonator. More details of how to start up and sustain oscillation is not described here but can be investigated further in reference (Ramstad, 2007; Vittoz et al., 1998). In figure 9, element A is realized as a Pierce Amplifier, element R is realized as the resonator described in section 4.1, while the two B elements are buffers to amplify the signal for the following FDSM stage.

5. System simulation

In order to investigate the viability of our proposed system, and to discover potential problems, we devised a simulation model of the system. In this section, we first present our simulation of the full FDSM and MEMS system. We then go on to describe our experiment, and finally we discuss the simulation results.

5.1 Method

As the output frequency of the MEMS oscillator in this case is low, a first-order oversampled FDSM as the F/D converter is appropriate. A detailed simulation model would be too computationally demanding to be of practical use. It would also require a mechanical simulation for the MEMS part in co-simulation with the electrical FDSM netlist. We therefore implemented the simulation model using Verilog-A (Accellera Organization, Inc., 2008) building blocks running on a commercial SPICE simulator. An outline of the simulation model is depicted in figure 10. The output from this model is a sampled single-bit bitstream, $y[n]$. The bitstream was then decimated to a stream of output words, which were finally post-processed to compensate for the non-linearity of the MEMS resonator. In the following subsections we describe the components of our simulation model in more detail.

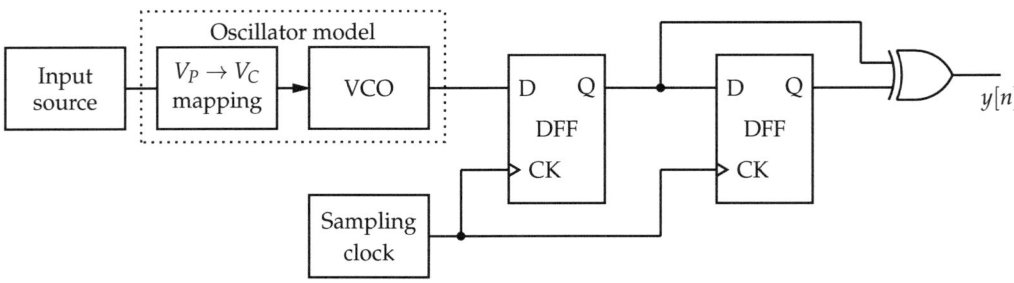

Fig. 10. Simulation model outline

5.1.1 The oscillator circuit

The modeling of the resonator has mostly been done by using analytical scripts from the equations described in section 4. Due to the non-linearity of the MEMS resonator for large values of V_P, the need for a more sophisticated simulation tool became apparent. By using a Finite Element Method (FEM) software tool, an accurate simulation of the resonance frequency and beam displacement as a function of the V_P voltage is performed. The results from the FEM simulations are back annotated into the analytical script in order to develop correct RLC equivalents, resonator output current as well as a correct model of the phase-noise. The total VCO model is then described by using Verilog-A. The VCO model is in itself a linear VCO. The non-linearity (arising from the MEMS resonator) is applied as a pre-distortion of the input signal, mapping the tuning voltage, V_P, to a VCO control voltage, V_C, using a `table_model` construct in Verilog-A code. This gives the designer, flexibility and makes it easy to switch between different VCO characteristics.

Figure 11 shows the implementation of the MEMS resonator where this cantilever beam is 100µm long, 1µm wide and a few microns thick. This is a resonator which is easy to tune in frequency because its mechanical stiffness is rather low. A fixed-fixed beam would allow a higher operational frequency, but is in turn more difficult to tune. A different resonator architecture as a tunable MEMS resonator can be developed, however in this chapter we focus on a simple MEMS architecture in order to point out the non-linearity problem and the resulting phase-noise of this CMOS-MEMS resonator.

The amplifier in the oscillator circuit is a Pierce amplifier which is a single-ended solution. The Pierce amplifier is a simple topology that has low stray reactances and little need for biasing resistors which would lead to more noise. By tuning the bias current in the Pierce amplifier, the gain (or equivalent negative impedance) increases. The MEMS resonator is typically the

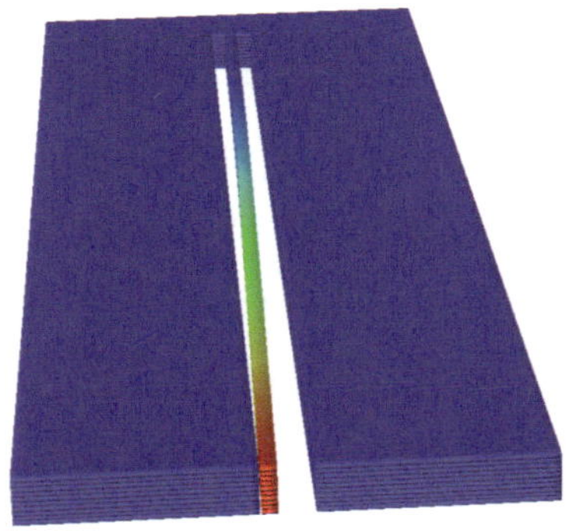

Fig. 11. 3D plot for the 1st vibrational mode of the MEMS resonator

element which limits the phase-noise, not the Pierce amplifier. However, the Pierce amplifier needs to be flexible enough in order to initiate and sustain oscillation of the MEMS resonator. For a variation of Q-factor of the MEMS resonator and possible process variations, the Pierce amplifier has been made to start up oscillation for R_x values up to a few MΩ as the Pierce amplifier can be represented as a negative impedance value of up to around ten MΩ. It would be possible to make a full differential amplifier and resonator configuration for low noise applications, however this has been left out as future work.

5.1.2 FDSM circuit

The FDSM circuit is a first-order single-bit DFF FDSM. The FDSM circuit is made up of two DFFs whose outputs are XOR-ed. The DFFs and XOR gate are implemented as individual Verilog-A components interconnected in a SPICE sub-circuit. The FDSM circuit also contains an ideal sampling clock source.

5.1.3 Decimation and digital post-processing

As we used an FDSM with first order noise shaping, we used a sinc^2 filter with $N = 8$ in the first stage, see figure 12. In the second stage, we used sinc^4 filter with $N = 32$, and finally a FIR filter with a decimation ratio of 2. This is depicted in figure 13. The sinc^4 filter in the second stage was used to give better rejection of excess out-of-band quantization noise. We did not correct for the passband droop incurred by the sinc filters.

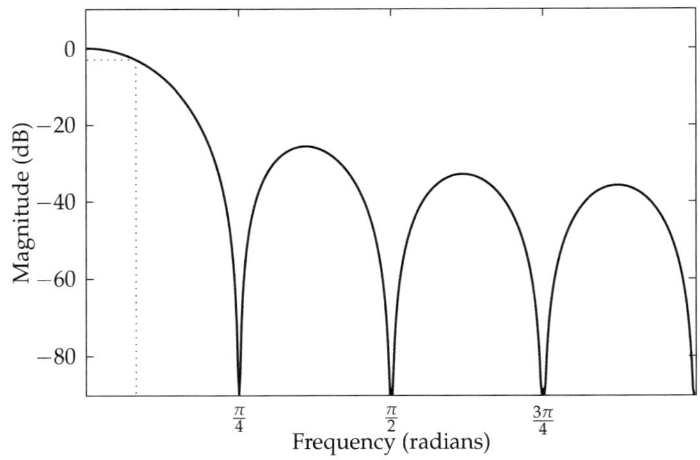

Fig. 12. Magnitude response of the first stage decimation filter

The non-linearity of the oscillator's transfer function gives rise to a significant harmonic distortion, which deteriorates the performance of the ADC. In this case, we used a simple lookup table (LUT) (Kim et al., 2009), to map every possible intermediate output, to a final quantized and corrected value. The non-linearity was characterized by applying a known linear input sequence, which in turn was used to build the inverse mapping LUT.

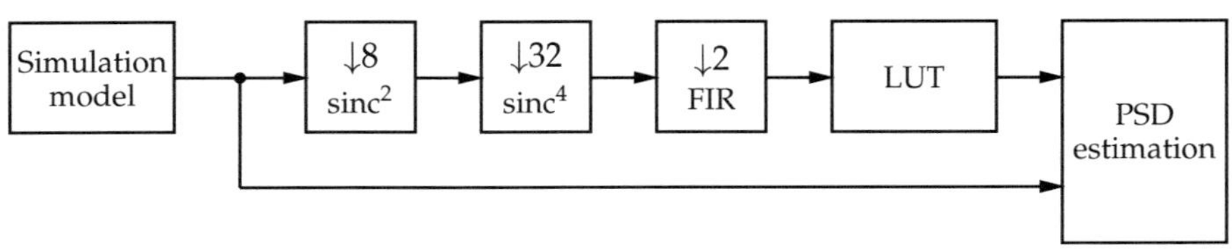

Fig. 13. Bitstream decimation and post-processing

Both decimation and post-processing was implemented outside the simulation model and no quantization was performed until after the post-processing.

5.1.4 Spectral estimation and performance measurement

The output data collected from the simulation model, and from the decimation and post-processing was analyzed using a Fast Fourier Transform (FFT) according to the guidelines in Schreier & Temes (2004).

5.2 Results

In section 4.4, the reason for the critical vibration amplitude x_c was shown and discussed. Varying V_P will eventually make the theoretical amplitude cross the x_c around 6.5V as shown in figure 14a.

If the resonator is initially placed in an environment with some pressure, reducing the pressure to a vacuum state will result in an increase in the Q-factor and x_c can cross the theoretical resonator displacement amplitude x quicker than anticipated. The resonator used here is used in a low-pressure environment, but placing it in vacuum will not increase the Q-factor significantly due to internal material loss. The critical vibration amplitude results in a small

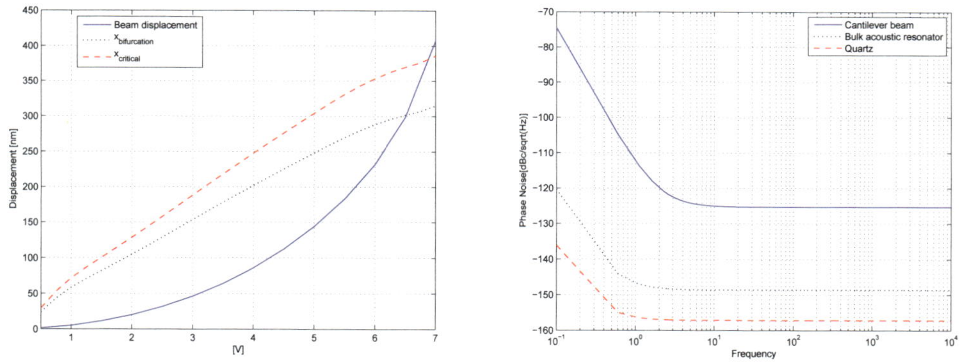

(a) Bifurcation as a function of V_P (b) Phase noise examples

Fig. 14. Bifurcation and phase noise

buffer before the hysteresis amplitude x_b is reached. By using x_c and Leeson's equation for phase noise as shown in section 4.4, we can plot the phase noise as a function of offset from the carrier frequency. Figure 14b shows some examples of other VCO components and how much noise they have compared to the resonator used in this CMOS-MEMS demonstration. The phase-noise example is calculated using equation 33, although this noise model has not been implemented in the total VCO model.

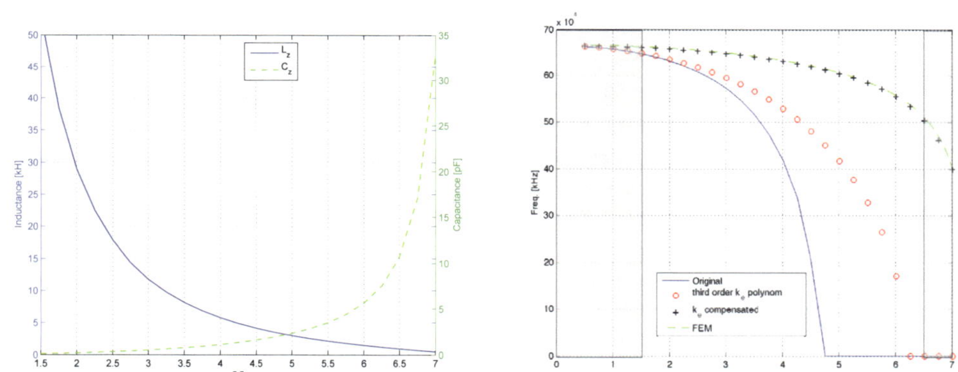

(a) $L_x(V_P)$ and $C_x(V_P)$ (b) $f_0(V_P)$

Fig. 15. Inductance, capacitance and operational frequency as a function of V_P

When varying V_P, the RLC equivalent that represents the MEMS resonator in the oscillator circuit will vary. An example of this is shown in figure 15a where the inductance decreases and the capacitance increases when V_P is increased. The variations of these two components are exactly opposite. From figure 15a, it can be seen that there is an exponential tendency of both values at the ends of the graph. This exponential behavior sets a "starting limit", thus the

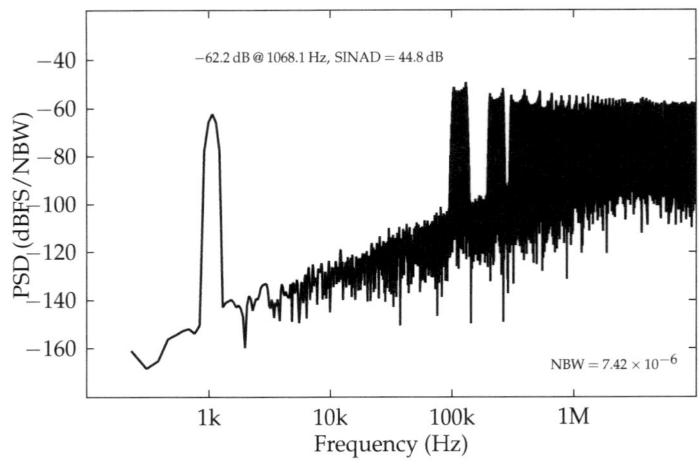

Fig. 16. Reference simulation with linear VCO

critical vibration amplitude x_c ultimately determines the maximum tunable frequency of the VCO as shown in figure 15b.

The k_e compensated term in figure 15b is extracted from the FEM simulation tool in order to develop the correct k_e. A first and third order polynomial k_e is also shown in order to demonstrate that the analytical formulas become too coarse grained for such a soft beam, thus the need for combining FEM results and analytical results becomes more important. The resulting operational area for the VCO gives an input range $V_P = 1.5 \rightarrow 6.5$ V, which gives $f_c = 58546$ Hz, and $f_d = 7743.7$ Hz. We used a sampling frequency, f_s, of 20 MHz for the FDSM circuit, and defined the signal bandwidth, f_b, to be 19 kHz. Equation 2 predicts $\text{SQNR}_{\text{dB}} = 22$ dB. All spectral plots were plotted using 2^{18} samples for the full spectrum, and 2^9 samples for the decimated spectra.

After characterizing the MEMS resonator, we built the LUT by applying 16 equally spaced DC inputs to the system spanning the input range. To estimate the corresponding output codes we averaged each output sequence, which was truncated to 2^9 samples after decimation.

We then simulated the full system for 16.4 ms using a full-scale sine wave input. In the first experiment we used a linear transfer function for the VCO to serve as reference. The result from this experiment is plotted in figure 16. In this case, the signal to quantization noise and distortion (SINAD) ratio is 44.8 dB.

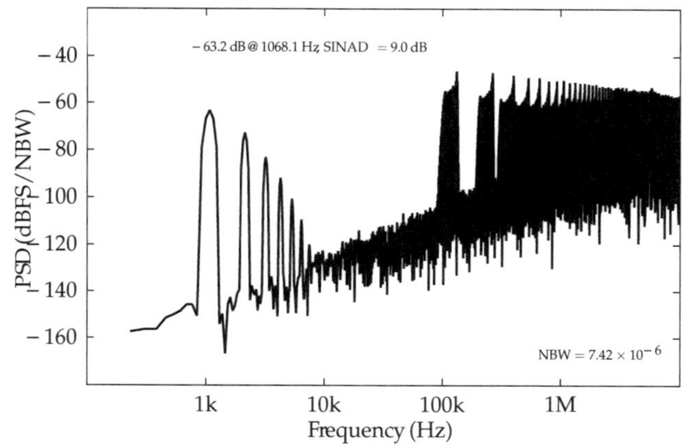

(a) Full spectrum output signal

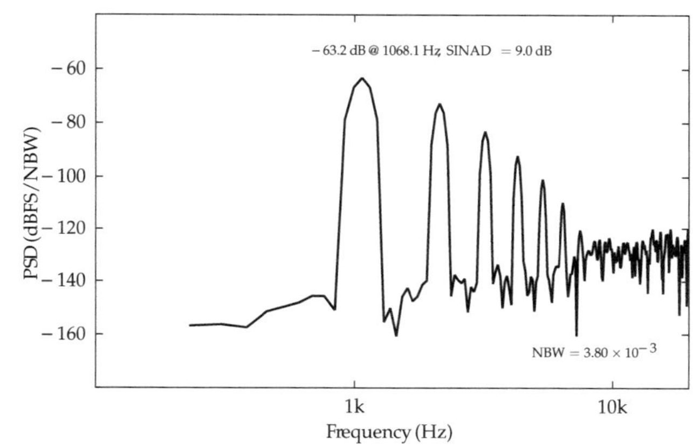

(b) Decimated output signal

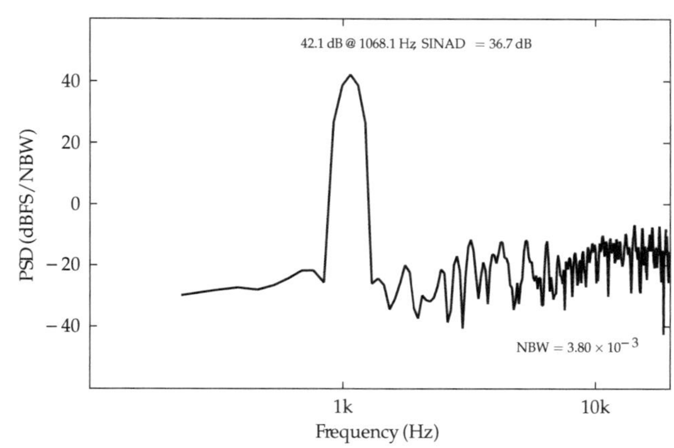

(c) Post-processed and quantized output signal

Fig. 17. Simulations with MEMS resonator non-linearity

In the second experiment we used the transfer function obtained from the MEMS resonator simulation. The results from this experiment are shown in figure 17. The full spectrum is shown in figure 17a, the spectrum after decimation is shown in figure 17b, and the post-processed signal is plotted in figure 17c, quantized to 8 bits. After linearization and quantization, the SINAD is 36.7 dB.

5.3 Discussion

From figure 16, we can see that quantization noise is shaped with a slope of 20 dB/decade as expected and that the spectrum is smooth in the in-band part of the signal. The difference between the simulated SINAD and $SQNR_{dB}$ predicted by equation 2 is 22.8 dB which is significant. However, $f_c/f_s \approx 0.003$, so this discrepancy is supported by the data in figure 4. Given the modest frequency tuning range of the MEMS resonator the overall resolution of the converter is very reasonable, because of the high sampling frequency with respect to the carrier frequency, which compensates for the potential impact on performance. This indicates that the overall system performance can be recovered by shifting the burden to digital circuits—in accordance with the long standing trend in CMOS technology where each new technology generation is geared towards allowing for aggressive performance scaling of digital circuitry, at the expense of analog and mixed signal performance.
As expected, the non-linearity of the MEMS resonator is clearly visible as harmonic distortion in figure 17a and 17b. By comparing figure 17b and 17c, it is evident that the LUT based correction scheme to a large extent recovers overall linearity; approximately one effective bit of resolution is lost. This further supports that relying on digital processing for achieving sufficient resolution is feasible in this system. As explained, the LUT processing scheme was applied before quantization. Thus, in a hardware realization, tradeoffs will have to be made. However, the results presented in this section indicate that given sufficient resources, linearity can to a certain degree be recovered. Another important consideration when using this scheme for linearization is that it gives rise to a non-linear dynamic range—electrical noise will have varying impact on the spectrum due to the non-linear gain.

6. Conclusion

In this chapter, we have presented CMOS MEMS and FDSM as a platform for WSNNs. CMOS MEMS can be used for building a wide range of sensors for use in WSNs, and have application in communication subsystems. FDSM provides a simple and robust means of digitizing the sensor signal. In all, this enables compact low-power WSNNs.
While we have outlined the feasibility of this scheme, more research is needed to further investigate this approach. Currently, we are working on more sophisticated methods for achieving linearity. A higher frequency resonator would enable the application of second order noise shaping, which is beneficial for high resolution, low-power applications. Also, a higher resonator tuning range and better linearity would directly benefit the system's performance. The phase noise needs more attention to investigate the system level impact, and the tuning voltage of the resonator is too high to be compatible with deep sub-micron CMOS transistors. We are currently working towards a prototype implementation of the system.

7. References

Accellera Organization, Inc. (2008). *Verilog-AMS Language Reference Manual.*
Altera Corporation (2007). *Application Note 455: Understanding CIC Compensation Filters.*

Annema, A.-J., Nauta, B., van Langevelde, R. & Tuinhout, H. (2005). Analog circuits in ultra-deep submicron CMOS, *IEEE Journal of Solid-State Circuits* **40**(1): 132–143.

Balestrieri, E., Daponte, P. & Rapuano, S. (2005). A State-of-the-Art on ADC Error Compensation Methods, *IEEE Transactions on Instrumentation and Measurement* **54**(4): 1388–1394.

Bannon, F., Clark, J. & Nguyen, C.-C. (2000). High-Q HF Microelectromechanical Filters, *Solid-State Circuits, IEEE Journal of* **35**(4): 512–526.

Chatterjee, S., Tsividis, Y. & Kinget, P. (2005). 0.5-V analog circuit techniques and their application in OTA and filter design, *IEEE Journal of Solid-State Circuits* **40**(12): 2373–2387.

Chen, F., Brotz, J., Arslan, U., Lo, C.-C., Mukherjee, T. & Fedder, G. (2005). CMOS-MEMS resonant RF mixer-filters, pp. 24–27.

Chen, O.-C., Sheen, R.-B. & Wang, S. (2002). A low-power adder operating on effective dynamic data ranges, *IEEE Transactions on Very Large Scale Integration (VLSI) Systems* **10**(4): 435–453.

Dai, C.-L., Chiou, J.-H. & Lu, M. S.-C. (2005). A maskless post-CMOS bulk micromachining process and its applications, *Journal of Micromechanics and Microengineering* **15**: 2366–2371.

Fedder, G., Howe, R., Liu, T.-J. K. & Quevy, E. (2008). Technologies for Cofabricating MEMS and Electronics, *Proceedings of the IEEE* **96**(2): 306–322.

Fedder, G. K. & Mukherjee, T. (2005). Integrated RF Microsystems with CMOS-MEMS components, *in Proceedings of MEMSWAVE*, pp. 111–115.

Fedder, G. & Mukherjee, T. (2008). CMOS-MEMS Filters, pp. 110–113.

Gerosa, A. & Neviani, A. (2004). A low-power decimation filter for a sigma-delta converter based on a power-optimized sinc filter, Vol. 2, pp. II–245–248.

Hogenauer, E. B. (1981). An Economical Class of Digital Filters for Decimation and Interpolation, *IEEE Transactions on Acoustics, Speech, and Signal Processing* **ASSP-29**(2): 155–162.

Høvin, M., Olsen, A., Lande, T. S. & Toumazou, C. (1995). Novel second-order Δ-Σ modulator frequency-to-digital converter, *Electronics Letters* **31**(2): 81–82.

Høvin, M. E., Wisland, D. T., Marienborg, J. T., Lande, T. S. & Berg, Y. (2001). Pattern Noise in the Frequency $\Delta\Sigma$ Modulator, **26**: 75–82.

Høvin, M., Olsen, A., Lande, T. & Toumazou, C. (1997). Delta-Sigma Modulators Using Frequency-Modulated Intermediate Values, *IEEE J. Solid-State Circuits* **32**(1): 13–22.

Kaajakari, V., Koskinen, J. & Mattila, T. (2005). Phase noise in capacitively coupled micromechanical oscillators, *Ultrasonics, Ferroelectrics and Frequency Control, IEEE Transactions on* **52**(12): 2322–2331.

Kaajakari, V., Mattila, T., Oja, A. & Seppa, H. (2004). Nonlinear limits for single-crystal silicon microresonators, *Microelectromechanical Systems, Journal of* **13**(5): 715–724.

Kim, J. & Cho, S. (2006). A Time-Based Analog-to-Digital Converter Using a Multi-Phase Voltage Controlled Oscillator, *Proc. IEEE International Symposium on Circuits and Systems ISCAS 2006*, pp. 3934–3937.

Kim, J., Jang, T.-K., Yoon, Y.-G. & Cho, S. (2009). Analysis and Design of Voltage-Controlled Oscillator-Based Analog-to-Digital Converter, *IEEE Transactions on Circuits and Systems I: Regular Papers* . Accepted for future publication.

Michaelsen, J. & Wisland, D. (2008). Towards a Second Order FDSM Analog-to-Digital Converter for Wireless Sensor Network Nodes, *NORCHIP, 2008.*, pp. 272–275.

Nguyen, C.-C. (2005). MEMS Technology for Timing and Frequency Control, Vol. 54, p. 11.

Norsworthy, S. R., Schreier, R. & Temes, G. C. (1996). *Delta-Sigma Data Converters*, IEEE Press.

Qu, H. & Xie, H. (2007). Process Development for CMOS-MEMS Sensors With Robust Electrically Isolated Bulk Silicon Microstructures, *Microelectromechanical Systems, Journal of* **16**(5): 1152–1161.

Raghunathan, V., Ganeriwal, S. & Srivastava, M. (2006). Emerging techniques for long lived wireless sensor networks, *IEEE Communications Magazine* **44**(4): 108–114.

Ramstad, J. E. (2007). *System-on-Chip micromechanical vibrating resonator using post-CMOS processing*, Master's thesis, University of Oslo, Department of Informatics.

Ramstad, J. E., Kjelgaard, K. G., Nordboe, B. E. & Soeraasen, O. (2009). RF MEMS front-end resonator, filters, varactors and a switch using a CMOS-MEMS process, pp. 170–175.

Sauerbrey, J., Tille, T., Schmitt-Landsiedel, D. & Thewes (2002). A 0.7-V MOSFET-only switched-opamp Sigma Delta modulator in standard digital CMOS technology, *IEEE Journal of Solid-State Circuits* **37**(12): 1662–1669.

Schreier, R. & Temes, G. C. (2004). *Understanding Delta-Sigma Data Converters*, Wiley-IEEE Press.

Senturia, S. (2001). *Microsystem Design*, Springer Science and Business Media, Inc., chapter 7-10.

Shao, L., Palaniapan, M., Khine, L. & Tan, W. (2008). Nonlinear behavior of Lamé-mode SOI bulk resonator, pp. 646–650.

Soeraasen, O. & Ramstad, J. E. (2008). From MEMS Devices to Smart Integrated Systems, *Microsystem Technologies, Journal of* **14**(7): 895–901.

Sun, C.-M., Wang, C., Tsai, M.-H., Hsieh, H.-S. & Fang, W. (2009). Monolithic integration of capacitive sensors using a double-side CMOS MEMS post process, *Journal of Micromechanics and Microengineering* **19**(1): 15–23.

Taillefer, C. & Roberts, G. (2007). Delta-Sigma Analog-to-Digital Conversion via Time-Mode Signal Processing, *Proc. IEEE International Symposium on Circuits and Systems ISCAS 2007*, pp. 13–16.

Vittoz, E., Degrauwe, M. & Bitz, S. (1998). High-Performance Crystal Oscillator Circuits: Theory and Application, *Solid-State Circuits, IEEE Journal of* **23**(3): 774–783.

Wang, A. & Chadrakasan, A. (2005). A 180-mV Subtreshold FFT Processor Using a Minimum Energy Design Methology, *IEEE Journal of Solid-State Circuits* **40**(1): 310–319.

Wismar, U., Wisland, D. & Andreani, P. (2007). A 0.2V, 7.5µW, 20kHz $\Sigma\Delta$ modulator with 69 dB SNR in 90 nm CMOS, *Proc. ESSCIRC 33rd European Solid State Circuits Conference*, pp. 206–209.

Wismar, U., Wisland, D. T. & Andreani, P. (2006). A 0.2V 0.44µW Audio Analog to Digital $\Sigma\Delta$ Modulator with 57 fJ/conversion FoM, *Proceedings of the 32nd European Solid-State Circuit Conference*, Switzerland, pp. 187–190.

Yang, H. & Sarpeshkar, R. (2005). A Time-Based Energy-Efficient Analog-to-Digital Converter, *IEEE J. Solid-State Circuits* **40**(8): 1590–1601.

Øysted, K. & Wisland, D. (2005). Piezoresistive CMOS-MEMS Pressure Sensor with Ring Oscillator Readout Including Δ-Σ Analog-to-Digital Converter On-chip, *Proc. Custom Integrated Circuits Conference the IEEE 2005*, pp. 511–514.

3

Addressing Non-linear Hardware Limitations and Extending Network Coverage Area for Power Aware Wireless Sensor Networks

Michael Walsh[1] and Martin Hayes[2]
[1]*CLARITY: Centre for Sensor Web Technologies*
Tyndall National Institute,
University College Cork,
Cork, Ireland,
[2]*University of Limerick*
Limerick, Ireland

1. Introduction

Heterogeneous Wireless Sensor Network (WSN) technology will soon emerge from the research laboratories around the world and become embedded in everyday life. Here it will actuate, sample and organize at a scale previously thought impossible. WSNs offer an alternative to the wired communications network or can be deployed rapidly in a previously un-serviced area where they provide the ability to observe physical phenomena at a fine resolution over large spatio-temporal scales.

A wireless sensor is in essence a miniature computer which can be placed anywhere or attached to anything. Typically it is powered by a battery that should be small and ideally need replacement as infrequently as possible. These ubiquitous or pervasive devices are typically in-expensive, miniature, and capable of independent computation, communication and sensing. Continuing improvements in affordable and efficient integrated electronics is having a considerable impact on the technology, that can underpin the sensor network itself and to that end, a number of state of the art sensor node platforms are now readily available.

The WSN can be viewed in two ways, firstly as a decentralised group of wireless sensor nodes each limited in terms of memory, computation and functionality. Alternatively and as is more commonly the case, a WSN can be viewed as the sum of its parts. The addition of nodes to a network therefore increases the overall capabilities of the network, while the distributed manner in which these nodes are added allows the network to retain its ability to self-heal and organise.

The application space for WSNs is quite large and continues to expand vigorously encompassing habitat, ecosystem, seismic and industrial process monitoring, security and surveillance as well as rapid emergency response and wellness maintenance. This unsurprisingly has generated significant attention within the research community where the question of performance robustness and optimisation appears to be a recurring theme. The engineer is therefore presented with many challenges when designing an effective deployment.

2. Wireless Sensor Network Challenges

There are numerous challenges that must be addressed when designing a WSN. There follows a brief look at a number of problems, general in the wireless context, to which systems science can provide a useful solution.

2.1 Reliable Quality of Service

In a survey carried out amongst possible users of industrial wireless technology (IMS Research, 2006), 43% of the surveyed suggested that communications reliability was a major barrier to the uptake of wireless solutions in industry. The provision for Quality of Service (QoS) is therefore a key requirement if any form of WSN market penetration is to be generated. QoS has a number of different associated meanings (Goldsmith, 2006; Rappaport, 2002). In this work, QoS is taken, where specified, to imply one or both of the following

1. QoS implies that the transmitted signal will exhibit certain minimum signal strength at the receiver. This in turn will guarantee pre-specified levels of Bit Error Rate (BER) and improve demodulation at the point of access.
2. System connectivity must be ensured under the assumption that the communication link will be severed if some reliable measurable link quality metric falls below a minimum threshold value. Below this threshold the QoS is deemed unacceptable in terms of BER and the associated probability of outage in service.

2.2 Energy Efficiency

Although some guaranteed level of QoS is a clear necessity, for service provision issues such as energy consumption, battery life and size are proving to be important factors when it comes to increasing the uptake of new WSN systems. Placing an upper bound on power consumption in order to maximise operational longevity is therefore also a requirement. This poses a difficult challenge as many factors can contribute to energy consumption for any given WSN deployment. However one suggestion was made in (Otto et al., 2006) where empirical evidence attributed 95% of the overall energy consumed by a wireless sensor node to communication. To narrow the focus further it was highlighted in (Zurita Ares et al., 2007) that 70% of the energy consumed by widely available WSN platforms is as a result of data transmission alone. It therefore stands to reason that minimising the time spent transmitting or optimising transceiver output power can aid greatly in energy efficiency.

2.3 Network Coverage Area

In (Mobihealthnews, 2009) it was suggested that wireless networks in healthcare applications need to perform to "mission critical perfection", where the end user must have no concerns over network coverage. It was highlighted that real service should not be "homebound" in nature but rather some level of ambulatory motion must be provided, without any technical concerns about information loss being a factor. As WSN technology is for the most part a low range solution, some design consideration must be given to provision for the need to extend network coverage area. A multi-hop hierarchy is a clear solution to this problem, however when mobility is considered the need for handoff is introduced as a by-product. Whether it is between access points within a network or between networks, handoff must appear seamless to the user and the service must where possible remain uninterrupted.

2.4 Hardware Constraints

Practical limitations are a feature of any WSN. Without exception each wireless technology is bandwidth limited and is therefore prone to congestion under heavy workloads. However empirical evidence would suggest that hardware limitations will inevitably become a factor prior to the impingement of bandwidth constraints. For instance, the IEEE 802.15.4 standard specified at 2.4 GHz supports a bandwidth of 250 kbps (IEEE 802.15.4 Standard, 2006). However, the state-of-the-art 802.15.4 compliant Tmote Sky platform can achieve only 125 kbps maximum upload and 150 kbps download over the air, as a result of microcontroller process saturation (Polastre, 2005).
Other practical hardware constraints must also be considered. Transceiver output power limitations are an omnipresent feature of the WSN device. This nonlinearity can severely degrade network performance when encountered and can potentially destabilize the system entirely. Quantisation is also invariably present in a wireless communications system. Generally, a radio transceiver has a discrete number of output power levels and switching between these levels introduces unwanted quantisation noise into the system. This undesirable additional noise signal can impact negatively on communications quality. While each of these constraints is unavoidable, in practice, it is vital that their negative impact on the communication quality should be limited in an efficient manner.

3. A Solution in Systems Science

This work proposes a number of novel systems science based solutions tackling the challenges outlined above. The wireless architecture illustrated in Fig. 1 is envisaged. The IEEE 802.15.4 standard is referred to throughout as a benchmark technology, although each of the proposed methodologies presented is extendable to the general case.

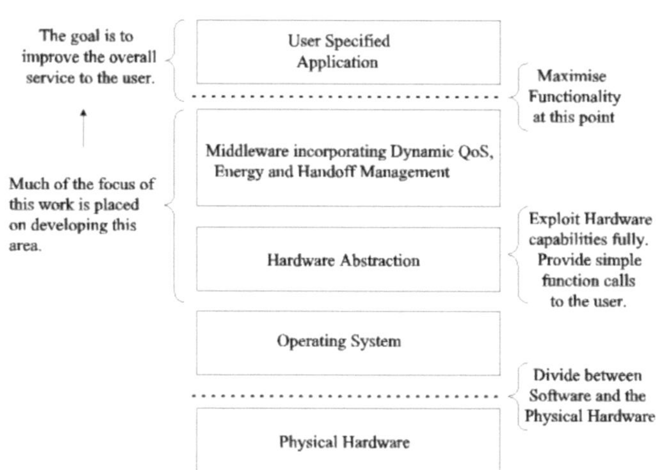

Fig. 1. Envisaged Wireless Sensor Network Architecture

A layered approach is adopted where the goal is to exploit fully the hardware and software capabilities of the employed technology, to improve the overall service to the user. This is achieved by firstly providing suitable hardware abstractions completely exposing the functionality of the WSN hardware devices. This functionality is presented to the upper layers in the form of simple function calls. Systems science based middleware solutions are then proposed utilizing the hardware abstraction. In this regard, robust dynamic power and handoff schemes are designed and implemented on a fully compliant 802.15.4 benchmark testbed. Quantifiable improvements are reported in terms of QoS, energy efficiency and network coverage. The emphasis is placed on modularity where code reuse is encouraged sparing valuable network resources.

3.1 Closed Loop Feedback Control over Wireless Networks

The goal of any closed loop feedback system is to firstly measure a feedback metric employing a sensor of some type to do so. This measurement is compared with a predefined reference value. A subsequent control command update is generated using the difference between these two signals as an input to the controller and the plant actuators are adjusted accordingly. In traditional feedback control systems, the feedback loop and the connection between the controller and the plant are fixed or wired in nature as in Fig. 2.
Closed loop control over wireless networks differs in that, the feedback loop and/or the control command update link are/is wireless in nature. This places an additional constraint on the system as the wireless radio channel is typically affected by exogenous, uncertain factors that must necessarily have an adverse impact on system performance. This inevitably makes the controller design and implementation more difficult. However, with a more detailed understanding of wireless channel behaviour, robust control design techniques can be extended to the WSN case and can in turn improve overall operating efficiency.

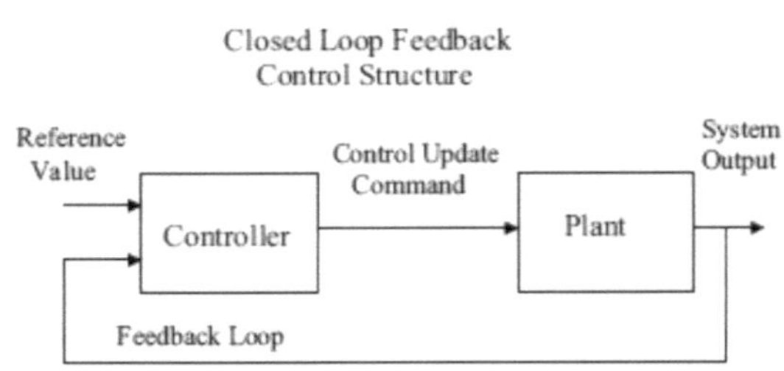

Fig. 2. The Closed Loop Feedback Structure

4. A Canonical Closed-loop Distributed Power Control Structure for WSNs

The goal of this scheme is to dynamically adjust device transmitter power, from a finite list of available levels, in a distributed manner so that the power consumption is minimized while also maintaining sufficient transmission quality. The received signal strength indicator (RSSI) is selected as the dynamic variable to manage this objective. In the past, it has been suggested that RSSI was a less than ideal metric for control. This claim however was based on experimentation with early platforms that used radios, e.g. the Texas Instruments CC1000, where hardware miscalibration or drift was often a problem. However, in recent times the use of RSSI has undergone something of a renaissance, with newer radios such as the 802.15.4 compliant TI CC2420 exhibiting highly stable performance. For example, in (Srinivasan and Levis, 2006), RSSI was proven to exhibit quite insignificant time variability as long as it stayed above an a priori defined threshold level. Recent empirical evidence would also suggest this to be the case (Alavi et al., 2008; Walsh et al., 2008; Walsh et al., 2009).

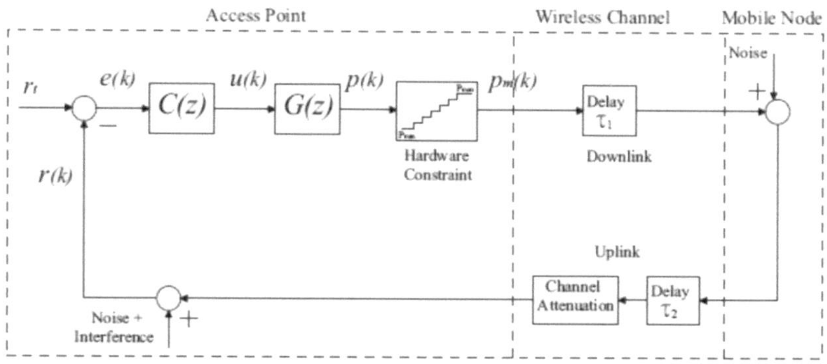

Fig. 3. Block diagram of the WSN Closed Loop Distributed Power Control structure based on RSSI measurement.

The proposed canonical closed loop WSN power control structure is illustrated in Fig. 3. A decentralized scheme is envisaged where the RSSI *r(k)* is measured at the access point or coordinator and compared with a target value *rt*. The difference or error *e(k)* is then fed into the controller *C(z)*, a number of realisations for which are presented in subsequent sections. The controller outputs a command update which in turn is passed to the plant *G(z)*. The plant outputs a power update which is limited by the inherent quantisation and saturation constraints. The resultant command *pm(k)* is transmitted to the mobile node where the new output power value is applied. In this scheme τ1 and τ2 represent downlink and uplink transmission delays respectively.
The objective therefore is to design *C(z)* such that *rt* is efficiently tracked, thusly guaranteeing QoS while minimising power consumption. *C(z)* must be robust to time varying stochastic channel uncertainties and interference which are modelled in this paradigm as an output disturbance. This simplifies controller design to some extent, as when the worst case interference and uncertainty scenarios are considered in the synthesis routine, exact information in relation to these difficult to quantify metrics is not required in realtime (Alavi et al., 2008). The hardware constraints must also be addressed in a manner so as to limit their impact on system performance. It is also worthwhile noting that almost all computational work is carried out at the access point. This allows for star topological deployments where the mobile nodes may be Reduced Functional Devices (RFDs).

4.1 Relating Received Signal Strength to Signal-to-Interference plus Noise Ratio

Working under the assumption that noise is correctly filtered at the receiver, (Zurita Ares et al., 2007) introduced a method to directly estimate the signal to noise plus interference ratio (SINR) using RSSI measurements. This approach denotes RSSI as,

$$\bar{r}(k)=\bar{p}(k)+\bar{g}(k)-\bar{I}(k)+\kappa+30 \tag{1}$$

where $\bar{r}(k)$ is the RSSI value, $\bar{p}(k)$ and $\bar{g}(k)$ are output power and attenuation respectively and $\bar{I}(k)$ contains path-loss, shadowing, fading, interference and noise. The addition of the scalar term 30 accounts for the conversion from dBm to dB and κ is the measurement offset determined empirically to be 45 dB. From (Zurita Ares et al., 2007) the SINR $\bar{\gamma}(k)$, in terms of RSSI can be described as,

$$\bar{\gamma}(k) = \bar{r}(k) - \kappa - 30 \tag{2}$$

This relationship is useful for a number of reasons. Firstly expressing RSSI in terms of SINR which in turn can be related to PER, is a suitable means of guaranteeing pre-specified levels of QoS in the closed loop system. To expand a target or reference RSSI value can be selected and related directly to PER, as outlined in the 802.15.4 standard (IEEE 802.15.4 Standard, 2006). The bit error rate (BER) for the 802.15.4 standard operating at a frequency of 2.4GHz is given by,

$$BER = \frac{8}{15} \times \frac{1}{16} \times \sum_{k=2}^{16} -1^{k} \binom{16}{k} e^{20 \times SINR \times (\frac{1}{k} - 1)} \tag{3}$$

and given the average packet length for this standard is 22 bytes, the PER can be obtained from,

$$PER = 1 - (1 - BER)^{PL} \tag{4}$$

where PL is packet length including the header and payload. PER is more useful here given the transceiver used to practically implement the proposed methodology, is a wideband transceiver, transmitting and receiving data in packet rather then bit format. Establishing a relationship between RSSI, SINR, BER and subsequently PER can therefore help to pre-specify levels of system performance. The relationship can also be used for comparative purposes, given control algorithms employing SINR, as a feedback metric can be directly applied to the WSN closed loop power control structure in Fig. 3. This is a useful tool in evaluating the performance of the proposed power control solution that follows.

4.2 Practical Hardware Limitations

Practical hardware limitations are a feature of any hardware platform and can result in severe performance degradation if not handled correctly. Addressing these constraints in parallel with improving reliability and power awareness is therefore a worthwhile endeavour.

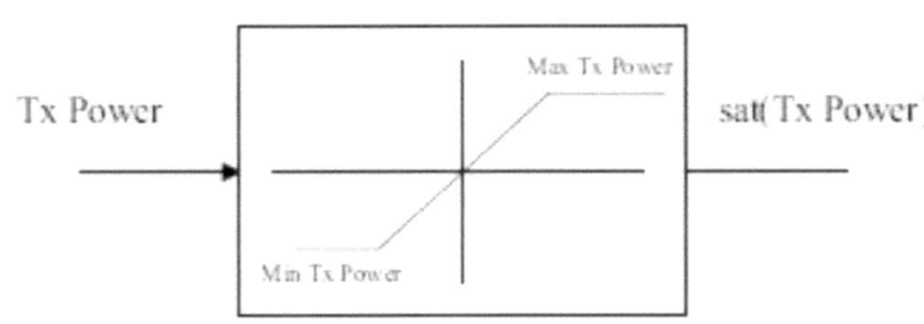

Fig. 4. Transceiver Output Power Saturation Nonlinearity

There is a maximum and minimum power at which any transceiver can transmit. These limits introduce a nonlinear saturation element to the system. The saturation nonlinearity *sat(.)* is illustrated in Fig. 4 and can be represented by equation (5).

$$sat(Tx\ Power) = \begin{cases} Max\ Tx\ Power, & \text{for } Tx\ Power > Max\ Tx\ Power \\ Tx\ Power, & \text{for } Min\ Tx\ Power \leq Tx\ Power \leq Max\ Tx\ Power \\ Min\ Tx\ Power, & \text{for } Tx\ Power < Min\ Tx\ Power \end{cases} \tag{5}$$

Without exception, there are also constraints placed on the system by the discrete nature of a transceiver's power levels. The impact switching between each discrete power level can adversely affect system performance as quantisation error is introduced. This additional input is normally modelled as noise. Generally, this signal is small in magnitude when compared with the channel variation associated with propagation effects; however it should be considered in any effective control design solution. The quantization and saturation nonlinearities are illustrated in Fig. 5.

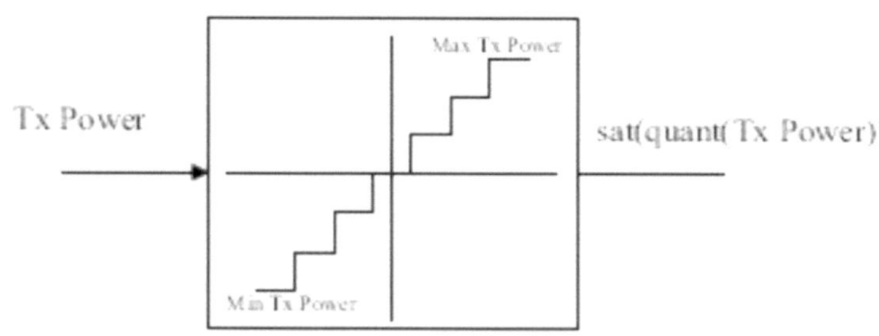

Fig. 5. Transceiver Output Quantisation Nonlinearity

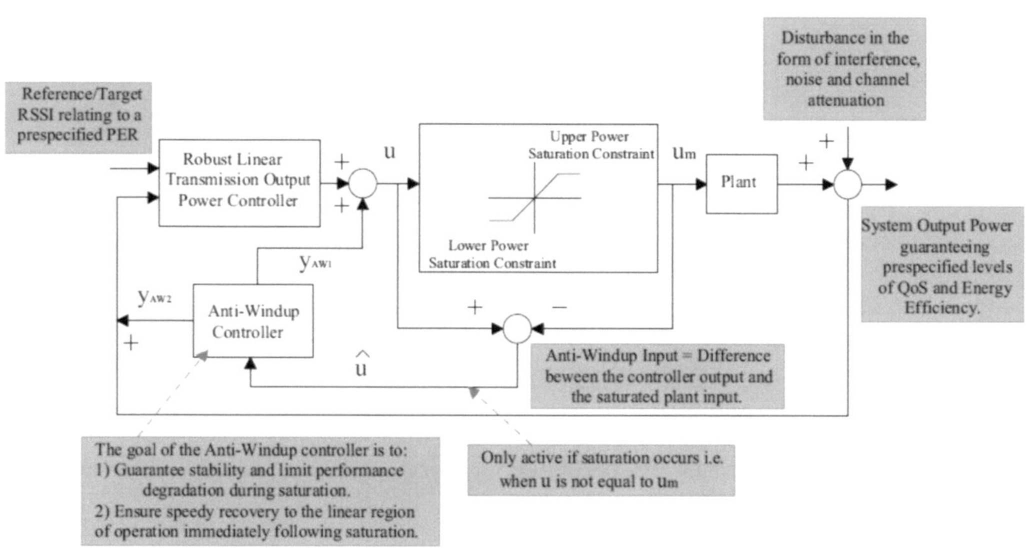

Fig. 6. The Anti-Windup approach as it applies to the Wireless Sensor Network Power Control Problem

5. An Anti-Windup solution to Robust Power Control

Consider a WSN implementing power control in a distribute manner and subject to practical hardware limitations as per any deployment of this nature. The focus here is placed on assessing the effect that the limited power transmission capabilities of a typical mobile node, within a practical sensor network, will have on performance. These natural hardware constraints will impose saturation type limits that will obviously severely degrade network performance. In this chapter, a two step Anti-Windup (AW) design procedure is introduced to tackle this problem. The first step is to design a linear controller, ignoring the inherent nonlinear constraints that are placed on the system that uses a Quantitative Feedback Theory (QFT) approach to provide both robust stability and nominal performance in the linear region of operation. A feature of this first step is that it naturally bounds the time domain response of the system for a particular power level and provides a basis for assessing how a change in the quantisation noise caused by power level selection will affect performance. The second step, shown in Fig. 6, incorporates recent advances in AW theory to minimize performance degradation in the face of actuator constraints.

5.1 The Simplified System Model

A systems science representation of a single access point communicating to a single mobile node is illustrated in Fig. 7. The system has reference input *r(k)* (reference RSSI), the value for which is determined using (2), (3) and (4) above, guaranteeing a predefined PER. *q(k)* is quantization noise introduced as a result of switching between discrete power levels. The controller *K(z)* has controller output *u(k)* and takes the form $K(z) = [K_1(z)\ K_2(z)]$, a standard two degree of freedom structure.

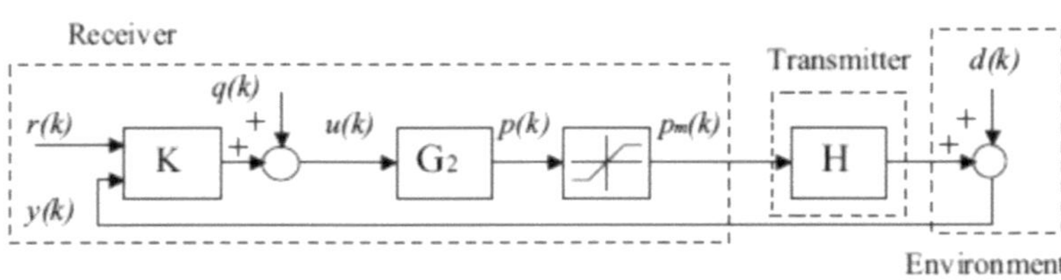

Fig. 7. Wireless System Model with saturation block at the output.

The plant *G(z)* is represented by $G(z) = [G_1(z)\ G_2(z)]$, where $G_1(z)$ and $G_2(z)$ are the disturbance feedforward and feedback parts of *G(z)* respectively. Given no structured disturbance model is available in the form of a transfer function, $G_1(z)$ is taken to be $G_1 = I$, where I is the identity matrix. The approach adopted regard to modelling $G_2(z)$ is similar to that suggested by (Gunnarsson et al., 1999) where the plant model for the WSN device is no longer represented by an integrator. However, rather than replace the plant model with a direct feedthrough term, (i.e., for a device *G* and power command update p_i, the plant output is $G(p_i) = p_i$), the plant is herein modelled as a low pass filter possessed of sufficient available bandwidth to be robust to a particular level of quantization noise. $G_2(z)$ is therefore selected as,

$$G_2(z) = \frac{1}{1.1z - 0.9} \tag{6}$$

$G_2(z)$ outputs a power level update $p(k)$, which in turn is transmitted to the mobile node. The mobile node transmitter has inherent upper and lower bounds on hardware transmission power output, represented in Fig. 7 by the saturation block, the output for which is saturated output power or $p_m(k)$. H represents the hardware switch in the mobile node's transceiver and is taken here to be the identity matrix or $H = I$. $d(k)$ is a disturbance to the system and comprises of channel attenuation, interference and noise.

5.2 Mapping the Saturation Function

For this scenario, a problem presents itself in that the saturation constraint is located at the output of the system and while there have been some advances in control design theory to deal with this type of output constraint for instance (Grandhi et al., 1995; Andersin et al., 1998), there is a vast literature covering the treatment of linear systems subject to input saturation constraints, see (Bernstein and Michel, 1995) and references therein. A solution therefore lies in the mapping of the output saturation constraint to the input of the plant or the output of the controller. The saturation function is defined as,

$$p_m(k) = sat(p(k)) \tag{7}$$

where $sat(p(k)) = sign(p(k)) \times \min\{|p(k)|, \bar{p}_m(k)\}$ and $\bar{p}_m(k)$ is the output power saturation limit. Note the *sat(.)* function in (6), belongs to sector [0, 1] and is assumed locally Lipschitz. The following set is defined,

$$\mathrm{P} = [-\bar{p}_m(k), \bar{p}_m(k)] \tag{8}$$

where $sat(p(k)) = p(k), \forall p(k) \in \mathrm{P}$. This is the set in which the saturation behaves linearly i.e. if there is no saturation present $p(k) = p_m(k)]$ and the nominal closed loop system conditions are exhibited. Fig. 8 portrays the system with the saturation block mapped from the output of the system to the input where $u_m(k)$ is the input to the plant. To represent the mapped saturation function we define the new set,

$$\Psi = [\frac{-\bar{p}_m(k)}{h_{G_2}}, \frac{\bar{p}_m(k)}{h_{G_2}}] \tag{9}$$

where h_{G_2} is the gain of the transfer function G_2. Recent advances in the antiwindup literature can now be applied to the problem at hand, ensuring minimal performance degradation during saturation and speedy recovery following saturation.

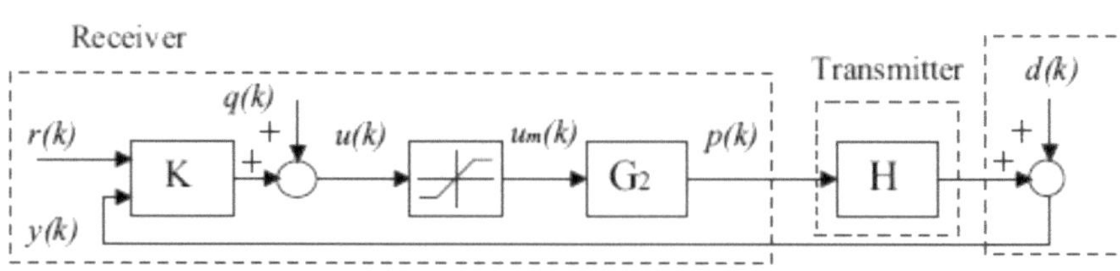

Fig. 8. Wireless System Model with saturation block mapped from the output to the input of the system.

5.3 Robust Linear Power Tracking Controller Design

Quantitative feedback theory (QFT) provides an intuitively appealing means of guaranteeing both robust stability and performance and is essentially a Two-Degree-of-Freedom (2DOF) frequency domain technique, as illustrated in Fig. 8. The scheme achieves client-specified levels of desired performance over a region of parametric plant uncertainty, determined a priori by the engineer. The methodology requires that the desired time-domain responses are translated into frequency domain tolerances, which in turn lead to design bounds in the loop function on the Nichols chart. In a QFT design, the responsibility of the feedback compensator, $K_2(z)$, is to focus primarily on attenuating the undesirable effects of uncertainty, disturbance and noise. Having arrived at an appropriate $K_2(z)$, a pre-filter $K_1(z)$, is then designed so as to shift the closed-loop response to the desired tracking region, again specified a priori by the engineer. The approach requires that the designer select a set of desired specifications in relation to the magnitude of the frequency response of the closed-loop system, thusly achieving robust stability and performance. The design procedure in its entirety is omitted here due to space constraints, however the interested reader is directed to (Horowitz, 2001) and references therein. Using this technique, $K_2(z)$ was found to be,

$$K_2(z) = \frac{z - 0.6622}{0.7103z - 0.7103} \tag{10}$$

guaranteeing a phase and gain margin equal to 50° and 1.44, respectively. The closed-loop transfer function is shaped using $K_1(z)$ ensuring the system achieves steady state around the target value of $5 \le t_{ss} \ge 25(s)$ and a damping factor of ξ = 0.5 is selected to reduce outage probability at the outset of communication. The resultant $K_1(z)$ is,

$$K_1(z) = \frac{1.4127z}{z - 0.4127} \tag{11}$$

5.4 Weston Postlethwaite Anti-Windup Synthesis

Consider the generic AW configuration shown in Fig. 9. As illustrated above the plant takes the form $G(z) = [G_1(z)\ G_2(z)]$, the linear controller is represented by $K(z) = [K_1(z)\ K_2(z)]$, and $\Theta = [\theta_1(z)\ \theta_2(z)]$ is the AW controller becoming active only when saturation occurs. Given the difficulty in analyzing the stability and performance of this system we now adopt a framework first introduced in (Weston and Postlewaite, 2000) for the problem at hand. This approach reduces to a linear time invariant Anti-Windup scheme that is optimized in terms of one transfer function M(z) shown in Fig.10. It was shown by (Weston and Postlewaite, 2000) that the performance degradation experienced by the system during saturation is directly related to the mapping $T : u_{lin} \to y_d$. This may not be clear at first glance, however if one looks at the equivalent representation of the system illustrated in Fig.11 and derived in (Weston and Postlewaite, 2000), it can be seen that the decoupled system is divided into three sections: the nominal linear system, the disturbance filter and the nonlinear loop. Note that from Fig. 11, $M - I$ is considered for the stability of T and G_2M determines the system recovery after saturation. This decoupled representation clearly illustrates how this mapping can be utilized as a performance measure for the AW controller. To quantify this an AW controller is selected such that the l_2-gain, $\|T\|_{i,2}$, of the operator T,

$$\|T\|_{i,2} = \sup_{0 \neq u_{lin} \in l_2} \frac{\|y_d\|_2}{\|u_{lin}\|_2} \tag{12}$$

where the l_2 norm $\|x\|_2$ of a discrete signal $x(h),(h=0,1,2,3,\ldots)$ is,

$$\|x\|_2 = \sqrt{\sum_{h=0}^{\infty} \|x(h)\|^2} \tag{13}$$

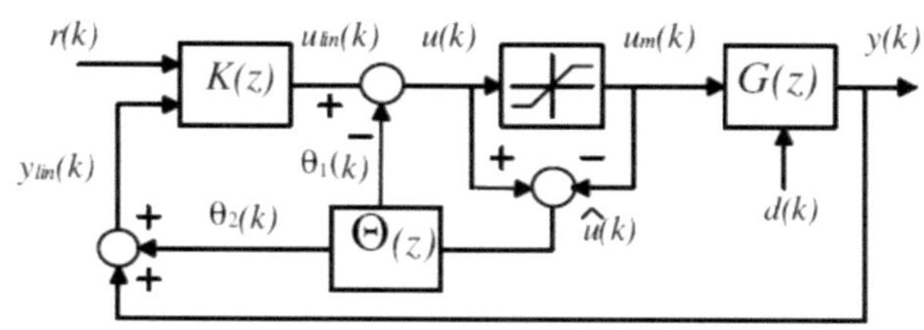

Fig. 9. A generic anti-windup scenario.

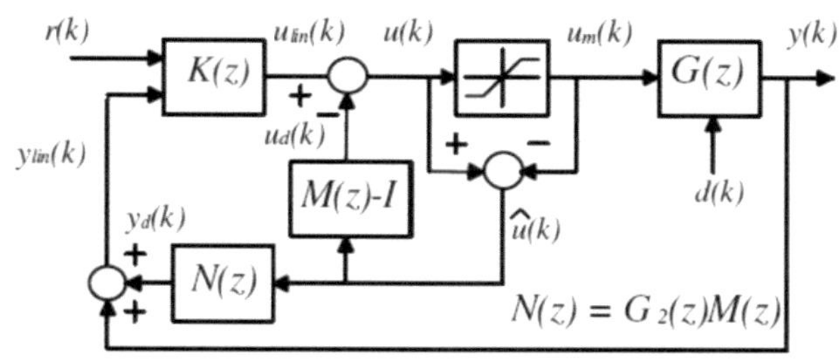

Fig. 10. Weston Postlethwaite Anti-Windup conditioning technique.

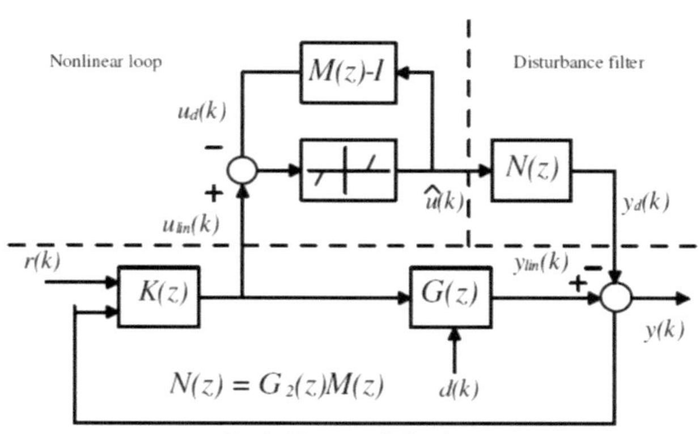

Fig. 11. Equivalent representation WPAW conditioning technique.

5.5 Static anti-windup synthesis

Static AW has an advantage in that it can be implemented at a much lower computational cost and adds no additional states to the closed loop system. Full order AW synthesis or AW with order equal to the plant will often lead to less response deterioration during saturation, however significant computation is required. This is often unacceptable, especially in systems that are of higher order and where additional states are undesirable. For this reason, it is common practice that most windup problems are suppressed using static compensators, see for example (Hanus et al., 1987). Using the aforementioned conditioning technique via *M(z)*, outlined in (Turner and Postlethwaite 2004), from Fig. 9 is given by,

$$\begin{bmatrix} \theta_1 \\ \theta_2 \end{bmatrix} = \Theta \hat{u} = \begin{bmatrix} \Theta_1 \\ \Theta_2 \end{bmatrix} \hat{u} \tag{14}$$

where u is derived from Figs. 9 and 10, respectively as,

$$\begin{aligned} u &= K_1 r + K_2 y - [(I - K_2 G_2)M - I]\hat{u} \\ u &= K_1 r + K_2 y - (K_2 \Theta_2 - \Theta_1)\hat{u} \end{aligned} \tag{15}$$

Thus, *M(z)* can be written as,

$$M = (I - K_2 G_2)^{-1}(-K_2 \Theta_2 + \Theta_1 + I) \tag{16}$$

The goal of the static AW approach is therefore to ensure that extra modes do not appear in the system. Since this will inevitably be the case, it must be ensured that minimal realizations of the controller and plant are used (Turner and Postlethwaite 2004). A state space realization can then be formed,

$$\begin{bmatrix} M(z) - I \\ N(z) \end{bmatrix} \approx \begin{bmatrix} \dot{\overline{x}} \\ u_d \\ y_d \end{bmatrix} = \begin{bmatrix} \overline{A} & B_0 + \overline{B}\Theta \\ \overline{C}_1 & D_{01} + \overline{D}_1 \Theta \\ \overline{C}_2 & D_{02} + \overline{D}_2 \Theta \end{bmatrix} \begin{bmatrix} \overline{x} \\ \hat{u} \end{bmatrix} \tag{17}$$

where $\Theta = [\Theta_1(z)\ \Theta_2(z)]$ is a static matrix and $\overline{x}$, $\overline{A}$, B_0, $\overline{B}$, $\overline{C}_1$, D_{01}, $\overline{D}_1$, $\overline{C}_2$, D_{02} and $\overline{D}_2$ are minimal realizations given in Appendix A. A solution is obtained for the Linear Matrix Inequality (LMI) in (18) with $Q>0, U = \mathrm{diag}(v_1, \ldots, v_c)>0$, $L \in \Re^{(c+n)\times n}$ (where $c=n$), and the minimized l_2 gain $\|T\|_{i,2} < \gamma$ (where γ is the l_2 gain bound on T). In this instance, Θ is given by $\Theta = L\Theta^{-1}$ using which, the

$$\begin{bmatrix} -Q & -Q\overline{C}'_1 & Q\overline{A}'_1 & 0 & Q\overline{C}'_2 \\ - & -X & UB_0 + L'\overline{B}' & I & UD'_{01} + L'\overline{D}'_2 \\ - & - & -Q & 0 & 0 \\ - & - & - & -\gamma I & 0 \\ - & - & - & - & -\gamma I \end{bmatrix} < 0 \tag{18}$$

where $X = 2U + D_{01}U + \overline{D}_1 L + UD'_{01} + L'\overline{D}'_1$. Such an l_2 design ensures that during saturation closed-loop performance is achieved by staying close to the nominal design while the time spent in saturation is also jointly minimized. Applying this synthesis routine to our plant given by (6) and linear controller (18), the resultant controller is $\Theta = [-0.2049 \ \ 0.6377]'$ obtained using the LMI toolbox in Matlab.

6. An Anti-Windup approach to Power Aware Seamless Handoff

A major WSN challenge lies in maximizing network coverage area. Given that many of the "off-the-shelf" sensor node platforms operate using low power wireless sensor technologies, transmission range is extremely limited, especially in the indoor environment. A multihop or mesh network topology is often proposed in order to extend coverage area necessitating the introduction of a handoff protocol that is power aware. Fig. 12 illustrates the type of scenario that is envisaged whereby subject X is being monitored and is wearing (perhaps a number of) wireless biometric devices. Initially X is in communication with base station BS_1. When X moves to an adjoining area in an ambulatory fashion, data must at some point be transmitted via BS_2 rather than BS_1. It is crucial that the QoS and energy efficient properties of the network be retained in such a scenario. This chapter proposes a Bumpless Transfer (BT) scheme to optimize this naturally nonlinear switching process. In any BT scheme, the global controller oversees multiple local loop devices that are designed to ensure the network is both power and QoS aware. Depending on certain performance requirements, a sequence of switches is necessary between each controller. In essence, one controller will be operational or "on-line" while the other candidate controller(s) must be deemed "off-line" at any instant. Clearly, it is necessary to be able to switch between these controllers (located at adjacent base stations or access points) in a stable fashion. Sufficient conditions must therefore be established to ensure that the induced transient signals are bounded, thereby satisfying network stability requirements. To achieve this smoothly, the gap between the off and on-line control signals must be bounded so that the control signal driving the plant cannot induce instability.

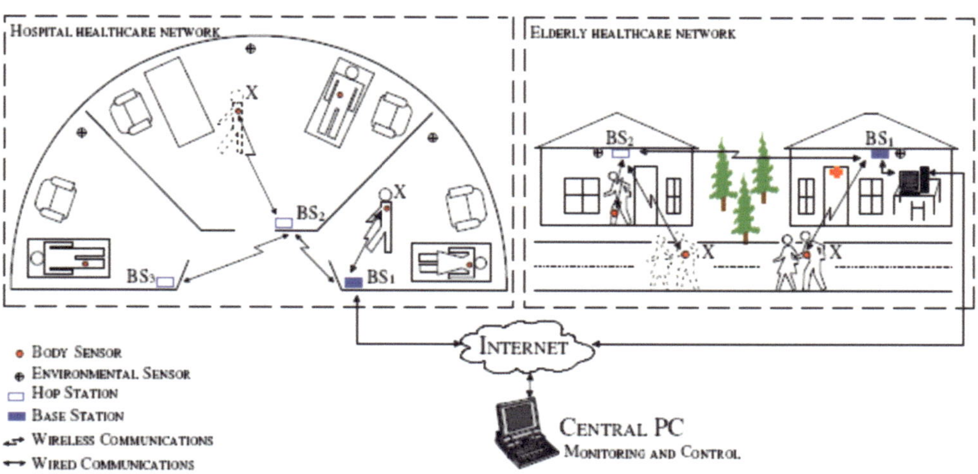

Fig. 12. The ambient healthcare environment. Power control for X is initially handled by BS_1. Subject X then moves in an ambulatory fashion and handoff occurs between BS_1 and BS_2. Data is now multi-hopped via BS_2 to BS_1 and BS_2 handles power control for X. Hence, power controller handoff has occurred between BS_1 and BS_2.

The overall solution therefore requires both AW and BT to operate in tandem for the first time in a practical WSN, thereby providing effective control of the signal entering the 'plant' (in this case the node transceiver) at any instant. For the remainder of the work, the term Anti-Windup-Bumpless-Transfer or AWBT will denote the new technique. Traditional AWBT schemes require that the gap between the feedback measurement observed at the off-line controller(s), is (are) sufficiently close in magnitude to the signal observed at the on-line controller. This is unlikely to be the case in the closed loop canonical WSN power control structure considered here as the RSSI observed at each access point will differ dramatically. To this end a specific modification is now proposed that delivers an AWBT scheme capable of compensating for the differing feedback signals that naturally arise and are unique to the wireless communications problem at hand. In the first instance, the problem is treated for a 2 base station scenario and is subsequently extended to the general case.

6.1 Formal Statement of the Handoff Problem: Two Base Station Scenario

To determine when handoff should occur, the filtered downlink RSSI signal is considered at the mobile node. It is assumed that each base station or access point will transmit at a pre-defined maximum power level within some pre-defined quantization structure at any instant. Initially, a two node mobile ad-hoc WSN structure depicted in Fig. 13 is considered. When the network initializes, it is assumed that the Mobile Node (MN) is unaware of its position and is transmitting data at the maximum power level to all "listening" base stations Fig. 13(i).

The network connects and implements a handoff protocol illustrated in Fig. 14. The MN will subsequently receive data packets from each base station within range (in this scenario limited to BS_1 and BS_2). A downlink RSSI is now calculated for each received packet and this signal is subsequently filtered to remove any multipath or high frequency component, using a digital filter, *F(z)*. In the experiment presented in this work, the following filter was found to be satisfactory.

$$F(z) = \frac{0.25z}{z - 0.75} \tag{19}$$

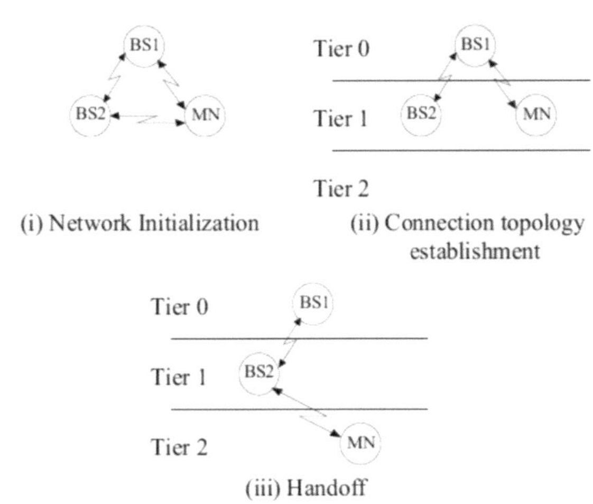

Fig. 13. Simple WSN multihop handoff scenario.

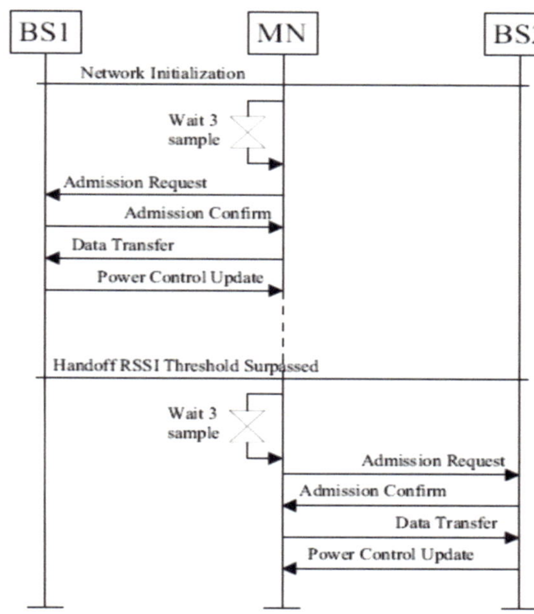

Fig. 14. The handoff procedure based on filtered downlink RSSI.

Fig. 15 illustrates how, subsequent to filtering the downlink RSSI signal, the pathloss component remains. This element is shown here, (and earlier by other authors e.g. (Goldsmith, 2006)) to be sufficiently distance dependant to be a useful metric for real time control. The MN now executes the algorithm presented in Fig. 16 comparing the resultant filtered signals, $RSSI_{DownlinkBS1}$ and $RSSI_{DownlinkBS2}$ over three sample periods. The signals are also compared with a predefined threshold value, selected here to be -40 dBm. This threshold ensures that the base station is located in the highest possible tier of the WBAN hierarchy and is also within range of the mobile node that is currently enjoying routing precedence, thereby satisfying a minimal latency requirement within the network.

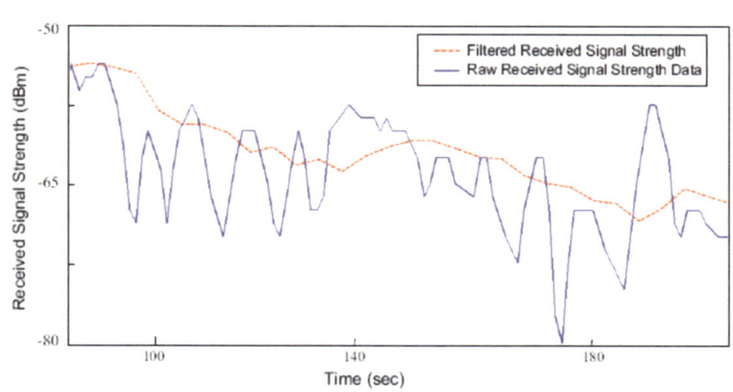

Fig. 15. Received signal strength filtered to remove the high frequency component.

Base Station 1 (BS1 located in tier 0)	Base Station 2 (BS2 located in tier 1)
Network Initialization	Network Initialization
Downlink RSSI for BS1 recorded at MN	Downlink RSSI for BS2 recorded at MN
For number of sample periods = 1 to 3	For number of sample periods = 1 to 3
If $RSSI_{DownlinkBS1} > RSSI_{DownlinkBS2}$	**If** $RSSI_{DownlinkBS2} > RSSI_{DownlinkBS1}$
Or If $RSSI_{DownlinkBS1} > -40dBm$	**And If** $RSSI_{DownlinkBS1} < -40dBm$
Use power level updates from Base Station 1	Use power level updates from Base Station 2

Fig. 16. Pseudo code for handoff algorithm: 2 base station example.

An admission request is then sent to the base station whose downlink RSSI satisfies the handoff criteria (BS_1 following network initialization). Following receipt of a confirmation message, the mobile node implements any power level updates received from this base station. Filtering the RSSI provides the added advantage of preventing any handoff chatter, i.e., that might occur due to deep fades in the RSSI that can be a characteristic of the MN position at any instant. Furthermore, the three sample period delay prior to the transmission of an admission request ensures that jitter is not present in the system. From Fig. 12(ii) and following network initialization, MN is now located in Tier 1 of the network hierarchy and BS_1, located in Tier 0, dynamically manages the MN's power based on the uplink RSSI observed at BS_1. At some future sampling instant, due to MN mobility, handoff is required based on the handoff algorithm of Fig. 16, again by a consideration of the filtered downlink RSSI values, $RSSI_{DownlinkBS1}$ and $RSSI_{DownlinkBS2}$ and the threshold value -40 dBm. Subsequently MN joins Tier 2 in the hierarchy; see Fig. 13(iii) and a floor performance level of power control for MN should now be immediately achieved employing the uplink RSSI at BS_2 as a feedback metric.

6.2 The Handoff Problem

Fig. 17 illustrates a simplified handoff problem for a two base station, one mobile node scenario. K_{BS1} and K_{BS2} are two degree of freedom controllers. Initially and without loss of generality, assume base station 1 is on-line and is therefore controlling the mobile node's transmission power at the sample instant k. The problem at hand when switching is necessary between base station 1 and 2, is to avoid the jump discontinuity that may arise between $p_1(k)$ and $p_2(k)$ at the time of switching. This jump can occur due to e.g., incompatible initial conditions and can induce an unwanted transient and even instability in the system. This can lead to insufficient floor levels in the flow of information in the network.

Conditions for stable Handoff:

Assumption 1: Given $G_2 = (Ap,Bp,Cp,Dp)$ in state space format and that $H(z)$ is the identity matrix, if $|\lambda_{\max}(A_p)| < 1$, where λ_{max} is the maximum eigenvalue, then asymptotic stability will be attained.

Assumption 2: It is assumed that the poles of $(1-K_{BS1}G_2H)(z)$ and $(1-K_{BS2}G_2H)(z)$ are in the open unit disc, ensuring that both nominal closed loops are stable.

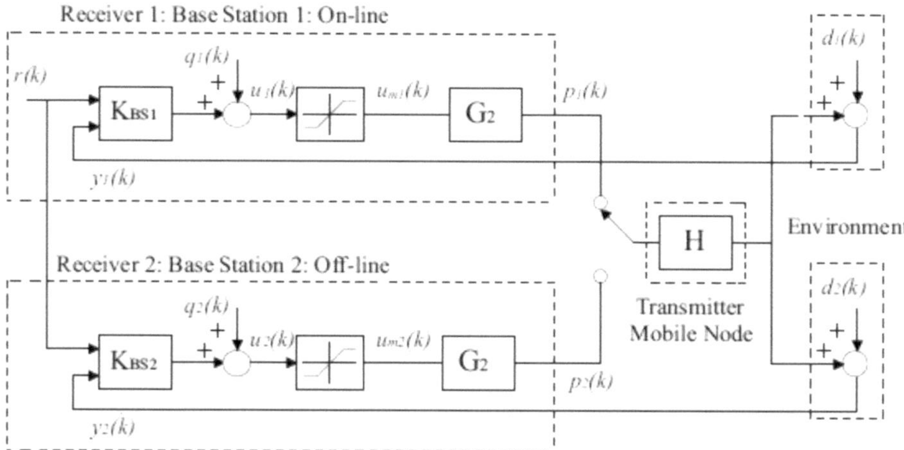

Fig. 17. Wireless System Model with power controller handoff.

When the above two necessary conditions are met, then the stability of the switched system will be guaranteed if the control signals, $u_{m1}(k)$ and $u_{m2}(k)$ are sufficiently close to each other. An AWBT approach that satisfies this performance criterion therefore provides a stable solution to the handoff problem. $p_1(k)$ will be close enough to $p_2(k)$ and should handoff occur, a large potentially destabilising transient will not be induced in the system. One particular difficulty arises in the wireless case. In order that AWBT be effective, the feedback measurement observed at the off-line controller must be sufficiently close in magnitude to the feedback measurement observed by the on-line controller. Clearly from Fig. 17, $d_1(k) \neq d_2(k)$ due to differing propagation environments. This disparity can mean AWBT will be unable to compensate for the difference between $u_{m1}(k)$ and $u_{m2}(k)$.

6.3 Modified Anti-Windup-Bumpless-Transfer Design

The following modification compensates for the inherent discrepancy in feedback RSSI signals between the off-line and the on-line controllers. Figure 18 illustrates the modification.. Consider the off-line controller base station 2, where an additional signal $y_{diff2}(k)$ is added the feedback signal. This signal is now,

$$y_{diff2}(k) = -y_{online}(k)W(z) + y_{lin2}(k)W(z) \tag{20}$$

where $W(z)$ is a low pass filter that removes the high frequency component present in each of the feedback RSSI signals. Note that $y_{online}(k)$ is determined by which base station is on-line. Therefore $y_{online}(k) = y_{lin}1$ when BS_1 is on-line. The signal driving the off-line controller then becomes,

$$\begin{aligned} y_{\mathrm{mod}\,2}(k) &= y_{lin2}(k) - y_{diff2}(k) = y_{lin2}(k) + y_{lin1}(k)W(z) - y_{lin2}(k)W(z) \\ y_{\mathrm{mod}\,2}(k) &= y_{lin1}(k)W(z) + y_{lin2}(k)(1 - W(z)) \end{aligned} \tag{21}$$

which comprises the DC or low frequency component of the on-line feedback signal or $y_{lin1}(k)W(z)$ plus the high frequency component of the off-line control signal $y_{lin2}(k)(1-W(z))$.

Each of these signals is incorporated in the design for different reasons. Firstly, driving the off-line controller with the DC component of the on-line control signal will ensure both controller outputs will be approximately equal or $u_1(k) \approx u_2(k)$. Retaining the high frequency component of the off-line feedback signal enables the off-line controller with the ability to compensate for deep fades in the associated feedback signal. Should handoff then occur, a large transient is avoided as the feedback conditions are sufficiently close to each other.

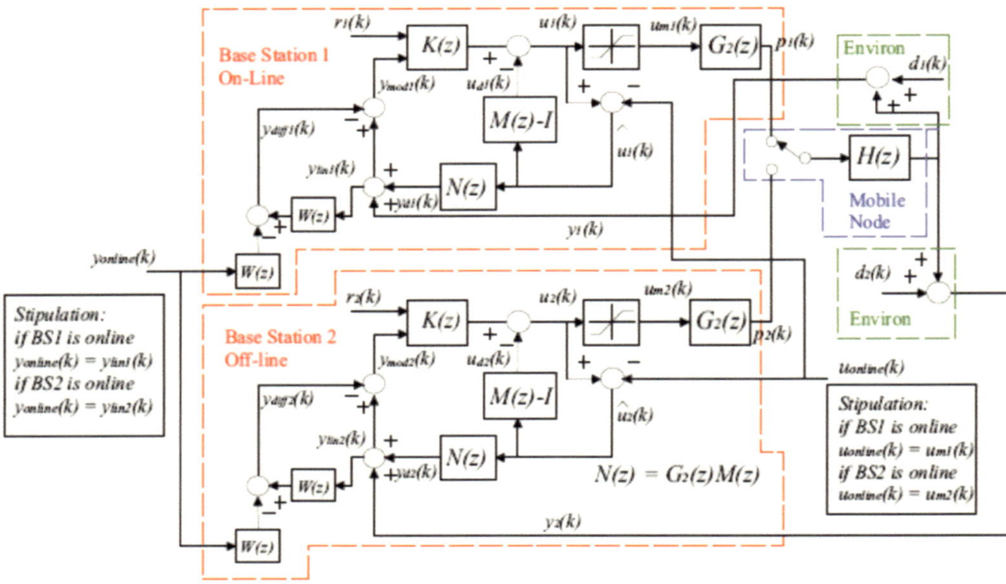

Fig. 18. The proposed modified WP-AW scheme, 2 Base Station Scenario.

Should base station 2 become on-line equation (21) becomes,

$$y_{\text{mod}\,2}(k) = y_{lin2}(k) - y_{diff\,2}(k) = y_{lin2}(k) + y_{lin2}(k)W(z) - y_{lin2}(k)W(z) = y_{lin2}(k) \qquad (22)$$

hence the modification will have no effect on the system and the AWBT scheme operates as normal. This approach adds a filtered additional disturbance to the system that is intuitively appealing given that a perturbation of the disturbance feedforward portion of the plant G_1 will have no bearing on the stability properties of the system (Turner et al., 2007).

7. An 802.15.4 Compliant Testbed for Practical Validation

Employing the IEEE 802.15.4 compliant Tmote Sky platform (Polastre et al., 2007) operating using TinyOS, the goal is to construct a testbed for realistic highly repeatable and rigorous experiments. A fully scalable realistic scenario is envisaged where Line-Of-Sight (LOS) and non-LOS occurrences are frequently observed inducing a Ricean and Rayleigh fading channel respectively. The testbed must therefore include randomly located obstructions. Stationary or embedded deployments are used to analyze the Additive White Gaussian Noise channel and mobility must be introduced to examine multipath fading characteristics.

The physical makeup of the testbed is illustrated in Fig. 19 where the idea is to emulate a scaled model of a building. The structure measures 2 meters squared and has re-configurable partitioning to introduce obstructions for non-LOS experiments. This simple scenario consists of three stationary nodes, a coordinator connected to a PC and two nodes mounted on autonomous robots thereby introducing mobility into the system. Up to five of mobiles can be introduced at any one time. A versatile robot, the MIABOT Pro, fully autonomous miniature mobile robot is employed for this purpose. Dataflow withing the network is illustrated in Fig. 20.

Fig. 19. Testbed Architecture

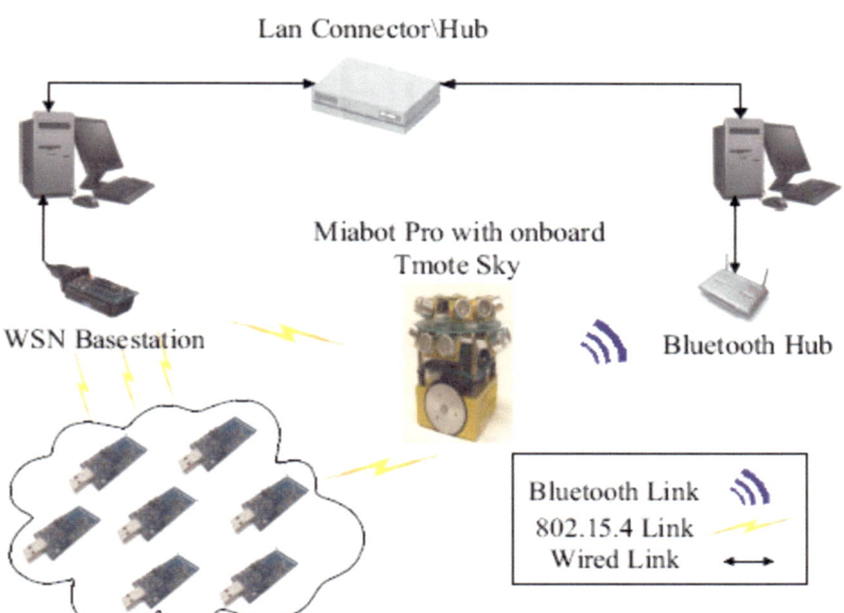

Fig. 20. Dataflow within the nework.

7.1 Topological Support

As outlined in the IEEE 802.15.4 standard, the testbed must be capable of both star and peer-to-peer type topological deployments.

Star Topology

To enable realtime control and data management over a star topological deployment, an interface between Matlab and TinyOS has been established using TinyOS-Matlab tools written in Java. The dataflow within the WBAN is illustrated in Fig. 21. The WSN nodes gather sensor data from their surrounding environment. This information is then forwarded to the PAN coordinator in packet format. The PAN coordinator upon receiving a packet, takes a channel quality measurement e.g., RSSI or data-rate and attaches the result to the packet. The packet is then bridged over a USB/Serial connection to a personal computer. The realtime Matlab application identifies this connection by its phoenixSource name, e.g., 'network@localhost:9000' or by its serial port name, e.g., 'serial@COM3:tmote' and imports the packet directly into the Matlab environment for further processing. The channel quality measurement taken by the coordinator is then used to implement a control strategy, the result of which is packaged in a suitable message and forwarded via the PAN coordinator to the WSN node. The node can subsequently update its control variable e.g. transceiver output power or transmission frequency. An advantage of using this approach lies in the fact that most of the processing occurs within the Matlab environment and at the PAN coordinator. Reduced Functional Devices (RFDs) nodes can therefore be employed if required by the application.

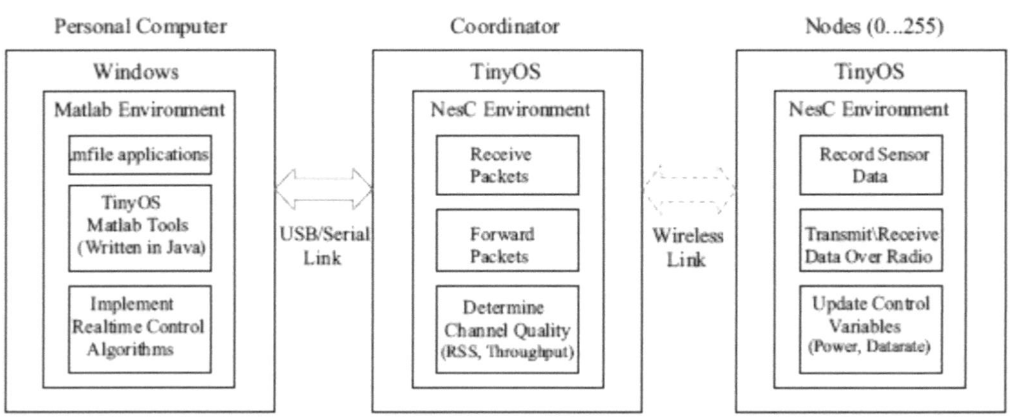

Fig. 21. IEEE 802.15.4 Testbed Dataflow with Matlab/TinyOS interface for Star Topology.

Peer-to-Peer Topology

The peer to peer configuration is also supported by the testbed. Fig. 22 illustrates a simple peer-to-peer network scenario where C is the PAN coordinator again assumed to be connected to a PC. N_1 and N_2 are Full Functional Devices (FFD) capable of communicating with any device in the network. Initially in Fig. 22, both N_1 and N_2 are communicating with C therefore the PAN coordinator is responsible for processing forwarded information and implementing control strategies for both devices. N_2 then becomes mobile and moves out of range of C. Subsequently, N1 multihops N_2's sensor readings to the PAN coordinator.

Handoff has therefore occurred between C and N_1, who now also has the responsibility for implementing control decisions based on channel quality measurements taken when a packet is received from N_2. Each FFD in the network is therefore programmed with similar capabilities to that of the PAN coordinator.

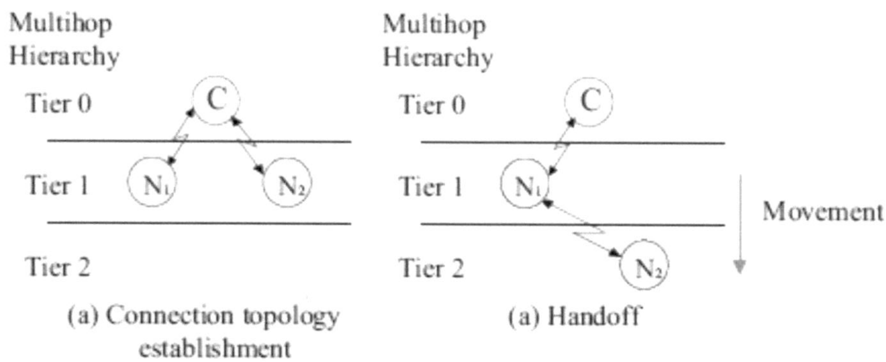

Fig. 22. Simple Peer to Peer Topology Handoff Scenario.

8. Practical Evaluation of the Proposed Methodologies

This section is organized as follows: Firstly, a number of system parameters and performance criteria specific to this scenario are outlined. Experimental results are then presented to highlight the improvements afforded by AWBT. Simulation is employed to emphasize how the modified AWBT scheme can improve performance at handoff, when the inherent saturation constraints are ignored. Further, practical validation of the modified AWBT scheme is then carried out on the testbed introduced previously. Where applicable, the system response is analysed firstly without AWBT, then with AWBT in place and finally with the modified AWBT design in place. Note: The QFT pre-filter and feedback controllers in equations (10) and (11) and the AW controller (17) are tested in these experiments.

8.1 System Parameters and Performance Criteria

A sampling frequency of $T_s = 1(sec)$ is used throughout and a target RSSI value of −55dBm is selected as a tracking floor level, guaranteeing a PER of $< 1\%$, verified using equations (2), (3) and (4). The standard deviation of the RSSI tracking error is chosen as the performance criterion in this work.

$$\sigma_e = \frac{1}{S}\left\{\sum_{k=1}^{S}[r(k) - RSSI(k)]^2\right\}^{\frac{1}{2}} \quad (23)$$

where S is the total number of samples and k is the index number of the sample. Outage probability is defined as,

$$P_o(\%) = \frac{numberoftimesRSSI < RSSI_{th}}{k} \times 100 \quad (24)$$

where $RSSI_{th}$ is selected to be −57dBm, a value below which performance is deemed unacceptable in terms of PER. This can be easily verified again using equations (2), (3) and (4). To fully assess each paradigm, some measure of power efficiency is also necessary and here the average power consumption in milliwatts is defined as,

$$Pav = 10^{\left[\left\{\frac{1}{S}\sum_{k=1}^{S} p_{dBm}(k)\right\}/10\right]}(mW) \tag{25}$$

where $p_{dBm}(k)$ is the output transmission power in dBm, S is the total number of samples and k is the index of these samples.

8.2 Justification and Improvements afforded by Anti-Windup

To validate the use of AWBT, a number of experiments were conducted using the repeatable scenario outlined above. Firstly, in order to justify the use of the standard deviation performance criterion (23), the results for a single experiment are shown in Fig. 23. This experiment consists of one mobile node and uses the QFT controller design without AW but with pre-filter. It can be observed that, without AWBT, the controller output when saturated begins to increase or `wind-up' and as a result the system upon re-entry to the linear region of operation, a substantial period of time is necessary for the actuator signal to 'unwind' back down to normal levels. This results in performance degradation in terms of standard deviation away from the setpoint. This feature wherein the operation of the system is in linear mode but the actuator variable is still higher than is necessary, translates into real energy loss that can be treated using AW methods.

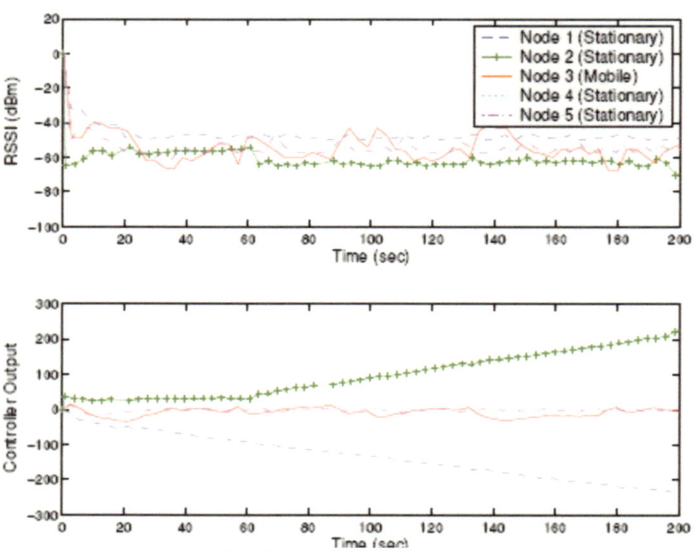

Fig. 23. System response without AWBT.

Fig. 24 displays the results of the same experiment with AW in place. It is clear that while saturation cannot be avoided, the 'wind-up' exhibited previously without AW is no longer present. Note: there is no handoff induced in this experiment therefore the modified AWBT scheme is not required for validation purposes.

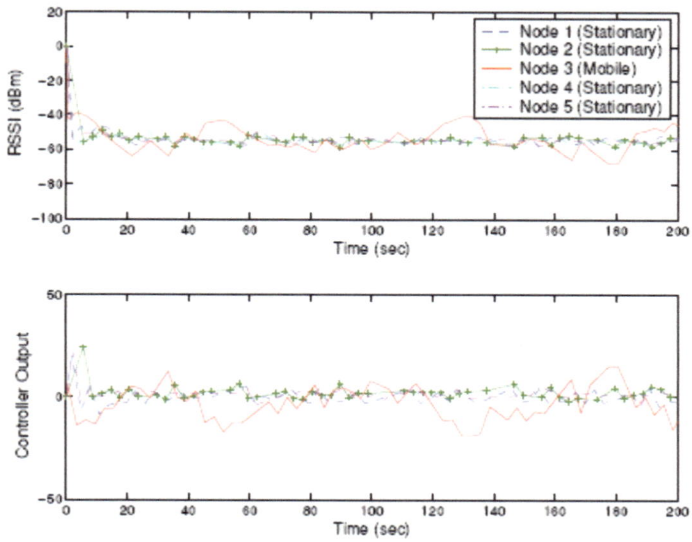

Fig. 24. System response with AWBT.

8.3 Benchmark Comparative Study

In this section the performance of the AWBT methodology is compared with fixed step, H∞/LMI and adaptive step active power control methods. A brief description of these alternative methods is now presented.

Fixed Step (Conventional) Size Power Control

This method is widely used in CDMA IS-95 systems due to its rapid convergence (Goldsmith, 2006). This strategy also assumes that the plant is modelled as an integrator. The approach is implemented using the following power control law

$$y(k) = y(k-1) + \delta(r(k) - RSSI(k)) \tag{26}$$

where *y(k)* is the transmission power and δ is the fixed step size (1 for the purposes of this experiment).

H∞/LMI Power Control

The LMI based approach outlined by (Ho, 2005) is also included in the study. Given the relative low order of the proposed distributed system, this approach will yield the controller *K = 1*, this is equivalent to the conventional approach with step size equal to one. These two methods are therefore analyzed as one.

Adaptive Step Size Power Control

This method uses the same power control law as the fixed step approach (Goldsmith, 2006), however the parameter δ needs to be updated depending on local system requirements according to the following,

$$\delta(k) = [\alpha\delta^2(k-1) + (1-\alpha)\sigma_e^2]^{\frac{1}{2}} \tag{27}$$

where as before σ_e, is the sampled standard deviation of the power control tracking error and a is the forgetting factor, (assumed to be 0.95 here), introduced to smooth the measured RSSI signal which may be corrupted by noise.

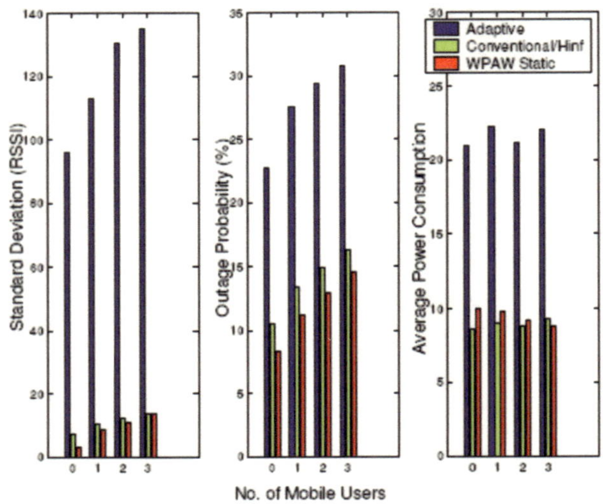

Fig. 25. Comparison between adaptive, conventional/H∞ and AWBT Hybrid schemes.

Benchmark Comparative Study Results

Fig. 25 illustrates how the proposed AWBT system performs when compared with the approaches outlined above. Clearly the hybrid design outperforms the adaptive approach for all of the stated criteria and exhibits substantial improvement over a conventional/H∞ approach in terms of standard deviation and outage probability when low levels of mobility exist in the system. However, with fewer mobile nodes in the system, the conventional/H∞ approach consumes less power. This is due to the aggressive action of the pre-filter that results in improved tracking performance. As the number of mobile users is increased the standard deviations of the AWBT design and the conventional/H∞ converge, however the hybrid design continues to exhibit improved outage probability.

The average power consumption for the three approaches also converges, highlighting the improved power efficiency characteristics that are achieved for the hybrid design with increased levels of mobility. This is to be expected given that AW inherently seeks to dynamically decrease the magnitude of the controller output. It should be noted that the vast majority of the complexity of the proposed hybrid solution lies in the synthesis routine,and that very little additional computational overhead was a feature of the practical implementation. Empirical evidence suggests little or no difference between the AWBT approach and a more conventional adaptive step size power control approach in terms of microcontroller activity during realtime experiments.

8.4 Stand-Alone Bumpless Transfer performance

Due to the naturally occurring output power saturation constraints that arise in the system, which cannot be removed, it is difficult to ascertain the performance improvements afforded by the BT method as a stand alone handoff scheme. Simulation can be a useful tool in this regard. Fig. 25 illustrates some results where at time index 35 sec, handoff occurs between two base stations. In this instance there is a difference of 20 dBm in the RSSI, between the signal received at the on-line base station and the RSSI signal observed at the off-line base station. As mentioned earlier, this dissimilarity in observed RSSI is due to the propagation environment and is a realistic value based on the experimental observations in the indoor environment that was used in this study.

From Fig. 25, it is clear that the system without AWBT exhibits an extremely large transient response and following handover never achieves steady state prior to the completion of the simulation. The system with AWBT in place exhibits some improvement, however there is significant time spent below RSSIth and as a result outage probability is still at an unacceptable level. When the modified AWBT solution is added, the outage probability is dramatically reduced highlighting the improved performance afforded by the new approach. The modified solution also improves the transient response by considering the off-line high frequency component and compensating accordingly. The performance is summarized in Table 1.

	Without AWBT (QFT Only)	**With AWBT**	**Modified AWBT**
Standard Deviation σ_e	30.59	4.445	1.603
Outage Probability P_o	63.77	31.88	8.696
Average Power Consumption P_{av}	1	0.199	0.158

Table 1. Simulation Results.

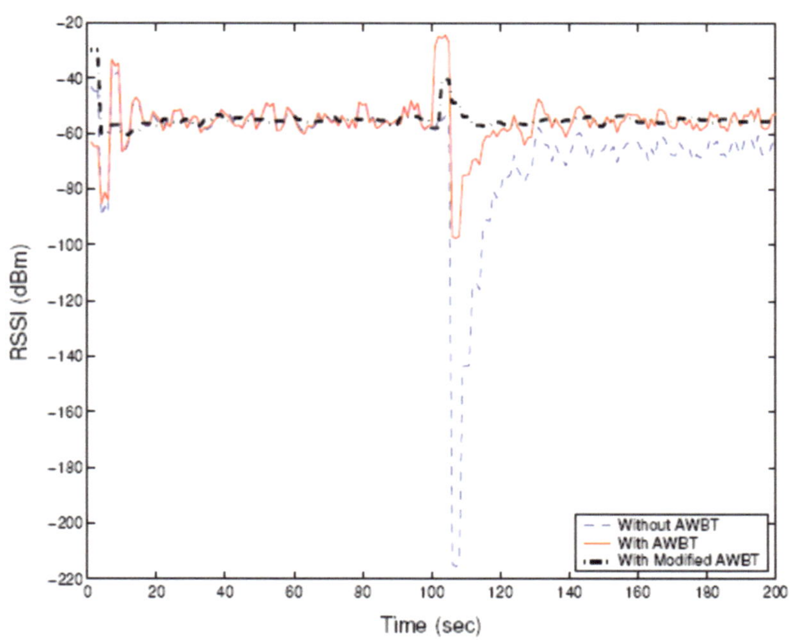

Fig. 26. Modified AWBT performance ignoring saturation constraints and where handoff occurs at 100 (sec)

8.5 Modified Anti-Windup-Bumpless-Transfer performance

Fig. 26 illustrates the experimental system response without AWBT or with QFT only. Clearly, without AWBT there is significant integral windup in the system, keeping both the controller at BS_1 and at BS_2 saturated for the entire duration of the experiment and making it impossible for the system to track its reference RSSI accurately. In Fig. 27, AWBT is added to the system and some improvement is observed in tracking performance, however upon closer inspection it is apparent that when handoff occurs an undesirable transient is imposed on the system. The off-line controller output also exhibits an undesirable increase in magnitude, for instance the controller at BS_2 between 0 and 50 (sec). This is due to the discrepancy in the feedback signals or as $d_1(k) \neq d_2(k)$ and results in excess power consumption in the network.

Fig. 28 highlights significant improvement when the modified AWBT solution is employed. Windup is almost entirely eliminated and the transient overshoot that occurs at handover is decreased. This can be attributed to the ability of the modified compensator, when off-line, to keep its control signal sufficiently close in magnitude to the signal entering the plant despite the presence of uncertainty in the feedback signal. The results are summarized in Fig. 29.

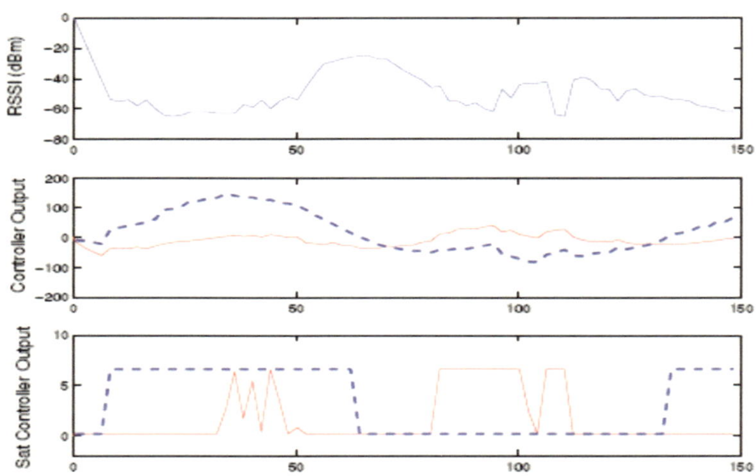

Fig. 27. Experimental results without AWBT where RSSI is the overall tracking signal, the dashed (bold) line is the saturated/actual controller output for BS_1 and the solid line is the saturated/actual controller output for BS_2.

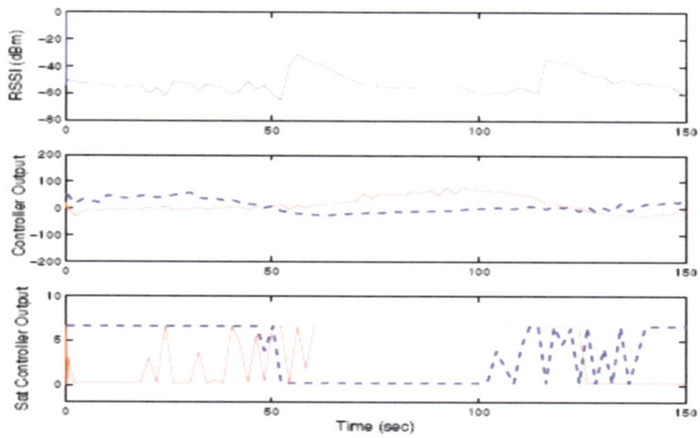

Fig. 28. Experimental results where RSSI is the overall tracking signal, the dashed (bold) line is the saturated/actual controller output for BS_1 and the solid line is the saturated/actual controller output for BS_2. System response with AWBT compensation

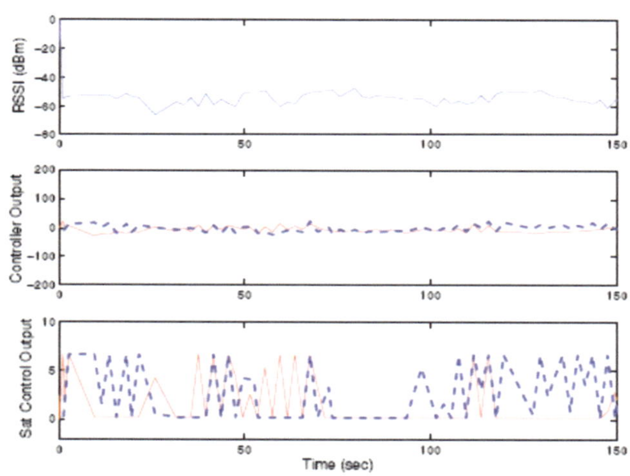

Fig. 29. Experimental results where RSSI is the overall tracking signal, the dashed (bold) line is the saturated/actual controller output for BS_1 and the solid line is the saturated/actual controller output for BS_2. System response with modified AWBT compensation

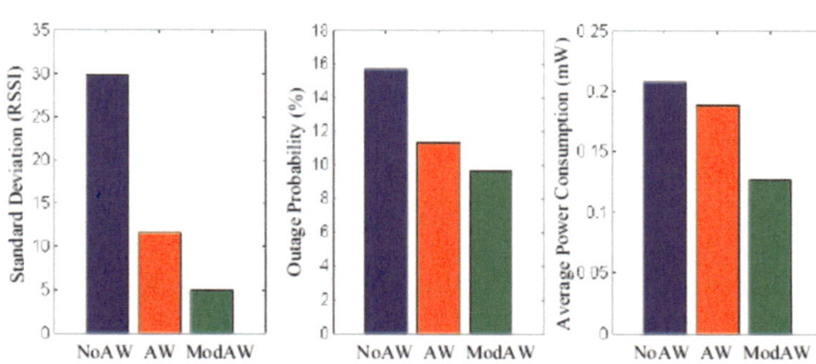

Fig. 30. Results in terms of the performance criteria. Standard deviation has units dBm. Average power consumption is given in milliwatts.

9. Conclusion

This chapter has presented a new strategy for power control in WSNs where operational longevity is an issue. An a priori level of performance is achieved in terms of packet error rate using minimum power where significant quantisation noise exists in the selection of the appropriate transmission power. Robustness to a variety of communication constraints have been illustrated using an AWBT scheme. The new approach provides a methodology for the rigorous assessment of the effect that a general class of static memory-less nonlinearity can have on overall system performance in a wireless power control problem setting.
Also presented in this chapter was a novel modified AWBT scheme that enables smooth, power aware handoff. The new technique facilitates floor levels on the flow of information to be maintained in a wireless network that arises quite naturally in an ambulatory setting. Feedback discrepancies, hardware limitations and propagation phenomena that are posed by the use of commercially available wireless communication devices were addressed using new signal processing and robust AW design tools. The technique was validated using a fully scalable 802.15.4 compliant wireless testbed that has been a feature of this work. The new AWBT schemes have exhibited significant performance improvements, particularly in terms of transient behaviour at handoff, when compared with analogous systems operating with simple dynamic control only or when AW methods alone were applied within the testbed.

10. Acknowledgements

This work is supported by Science Foundation Ireland under grant 07/CE/I1147 and by the IRCSET Embark Initiative.

11. References

Alavi S.M.M., Walsh M. J. and Hayes M. J. (2008). Distributed power control technique for 802.15.4 wireless sensor networks, based on quantitative feedback theory. Proc. IET Irish Signals and Systems Conference, Pages 260-267, Galway, Ireland.

Andersin M., Rosberg Z., and Zander J. (1998). Distributed discrete power control in cellular pcs, Wireless Personal Communications, Vol. 3, No. 6.

Bernstein D.S. and Michel A.N. (1995). A chronological bibliography on saturating actuators, International Journal of Robust and Nonlinear Control, Vol. 5, Pages 375-380.

Goldsmith A. (2006). Wireless Communications. Cambridge University Press, 2006.

Grandhi S. A., Zander J., and Yates R. (1995). Constrained power control, Wireless Personal Communications, Vol. 2, No. 3.

Gunnarsson F., Gustafsson F. and Blom J. (1999). Pole placement design of power control algorithms, In Proc. IEEE Vehicular Technology Conference, Houston, TX, USA.

Hanus R, Kinnaert M, Henrotte J. (1987) Conditioning technique a general anti-windup and bumpless transfer method. *Automatica, Vol. 23, Pages* 729–739.

Ho Y., lee C. and Chen B. (2006). Robust Hind Power Control for CDMA Cellular Communication Systems, IEEE Transactions on Signal Processing, Vol. 54, No. 10, Pages 3947-3956.

Horowitz I. (2001). Survey of quantitative feedback theory (QFT), *Int. J. Robust Nonlinear Control*, Vol. 11, Pages 887-921.

IEEE 802.15.4 Standard (2006). Wireless lan Medium Access Control (MAC) and Physical layer (PHY) specifications for Low-Rate Wireless Personal Area Networks (LR-WPANs), IEEE Std 802.15.4.

IMS Research (2009). Wireless in industrial systems: Cautious enthusiasm. Industrial Embedded Systems, Winter, 2006, Available: http://www.industrial-embedded.com/columns/Market_Pulse/2006/FallWinter/.

Mobihealthnews. Analyst: Wireless health can't be homebound. March, 2009, Available: http://mobihealthnews.com/1008/analyst-wireless-health-cant-be-homebound/ [Accessed March 2009].

Otto C., Milenkovi A., Sanders C., and Jovanov E. (2006). System architecture of a wireless body area sensor network for ubiquitous health monitoring. Journal of Mobile Multimedia, Vol. 1, No. 4, Pages 307-326.

Polastre J., Szewczyk R., and Culler D. (2005). Telos: enabling ultra-low power wireless research. Proceedings of the 4th international symposium on Information processing in sensor networks, Los Angeles, California, USA.

Rappaport T.S. (2002). Wireless Communications principles and practice. Prentice Hall, second edition.

Srinivasan K. and Levis P. (2006). RSSI is Under Appreciated, Third Workshop on Embedded Networked Sensors (EmNets)

Turner M., Herrmann G. and Postlethwaite I.. (2007). Incorporating robustness requirements into anti-windup design, IEEE Transactions on Automatic Control, Vol. 52, No. 10, Pages 1842-1855.

Turner M, Postlethwaite I. (2004). A new perspective on static and low-order anti-windup synthesis. *International Journal of Control, Vol. 77, Pages* 27–44.

Walsh M., Alavi S. M. M. and Hayes M. (2008). On the effect of communication constraints on robust performance for a practical 802.15.4 Wireless Sensor Network Benchmark problem. Proc. 47th IEEE Conference on Decision and Control (CDC08), Pages 447-452, Cancun, Mexico.

Walsh M. J., Alavi S.M.M. and Hayes M. J. Practical assessment of hardware limitations on power aware 802.15.4 wireless sensor networks- an anti- wind up approach. International Journal of Robust and Nonlinear Control (in press 2009).

Weston P. F. and Postlewaite I. (2000). Analysis and design of linear conditioning schemes for systems containing saturating actuators, Automatica, Vol. 36, No. 9.

Zurita Ares B., Fischione C., Speranzon A., and Johansson K. H. (2007). On power control for wireless sensor networks: system model, middleware component and experimental evaluation. European Control Conference, Kos, Greece.

4

Cooperative Beamforming and Modern Spatial Diversity Techniques for Power Efficient Wireless Sensor Networks

Tommy Hult
Lund University
Sweden

Abbas Mohammed and Zhe Yang
Blekinge Institute of Technology
Sweden

1. Introduction

Wireless Sensor Networks (WSN) have been attracting great attention recently. They are relatively low cost to be deployed and to be used in many promising applications, such as biomedical sensor monitoring (e.g., cardiac patient monitoring), habitat monitoring (e.g., animal tracking), weather monitoring (temperature, humidity, etc.), low-performance seismic sensing, environment preservation and natural disaster detection and monitoring (e.g., flooding and fire) Lewis (2004); Tubaishat & Madria (2003); Stankovic et al. (2003); Akyildiz (2002); Rashid-Farrokhi et al. (1998).

The WSN applications analyzed in this chapter have a topology where a large number of wireless sensor nodes are spread out over a large or small geographic area (e.g., disaster regions, indoor factory, large sports event areas, etc.). In this topology, an inefficient use of bandwidth and transmitter power resources is resulted if each wireless sensor is transmitting its measurement data to the base station (processing central). In this case, each sensor node would have to be assigned its own frequency channel and, if the base station is located a long distance from the sensor nodes, it would also demand a higher than average sensor node transmitter power. By using a coordinating cluster head, for each cluster of wireless sensor nodes, we can instead use the combined transmitter power of the node cluster through the use of beamforming to increase the transmitter-receiver separation and/or to improve the signal-to-noise ratio (SNR) of the communication link. Another advantage of using this cooperative transmission is that we can exert power control to minimize the power consumption of each individual sensor node, and thus maximizing network lifetime. In addition, in a cooperative network the measurement data could be sent by using Time Division Multiplexing (TDM) instead of Frequency Division Multiplexing (FDM) which improves the overall bandwidth efficiency of the system.

The spatial properties of wireless communication channels are extremely important in determining the performance of the systems. Thus, there has been great interest in the application of beamforming and modern spatial diversity techniques (or multiantenna systems) since they can offer a broad range of ways to improve wireless systems performance. For instance, diversity techniques such as multiple-input single-output (MISO), single-input multiple-output (SIMO) and multiple-input multiple-output (MIMO) can enhance the capacity, coverage, quality and energy efficiency of of wireless systems.

Energy efficiency is one of the key requirements in many WSN applications. This is particularly crucial for WSN deployed in inaccessible or disaster environments in which battery recharging and replacement is not a viable option. Thus, in this chapter we first propose to use a cooperative beamforming approach in wireless sensor networks to increase the transmission range, minimize power consumption and maximize network lifetime. This will be of particular interest for outdoor applications, especially when monitoring remote areas using aerial vehicle, such as a High Altitude Platform (HAP) or Unmanned Aerial Vehicle (UAV), as a platform for the data collecting base station. We will investigate how the required transmitter power of each sensor node is affected by the number of cooperating transmission nodes in the network. In addition, we present a comparison in the use of beamforming with the different forms of modern spatial diversity techniques for the same purpose of achieving a longer transmission distance (or range) while maintaining a low energy consumption. Beamforming can of course be interpreted as a form of MISO system although it differs from the normal view of how a diversity system operates.

This chapter is organized as follows: Section 2 presents an overview and analysis of cooperative beamforming using a large aperture random array. In section 3, the MISO, SIMO and MIMO diversity schemes are introduced and analysed using the Rician fading channel employed in the simulations. Section 4 present numerical results and comparisons of the simulated beamformer and modern diversity systems. Finally, section 5 concludes the chapter.

2. Traditional Cooperative Beamforming

In this chapter we use the delay-and-sum beamforming technique which is the oldest and simplest algorithm for Space-Time processing. This beamforming is done through coherent excitation/reception of amplitude and phase of the signal transmitted/received from each individual antenna element in a collection or cluster of similar antenna elements also known as an antenna array Johnson (1993). Antenna arrays can have different configurations (e.g., linear, planar, circular, triangular, rectangular or spherical). Extensive research has been done on uniform array beamforming using one (linear) or two (planar) dimensional equi-distant element arrays Johnson (1993); Hansen & Woodyard (1938); Drane (1968). In addition, there is also work done on beamforming using circular, triangular and rectangular arrays Johnson (1993); Balanis (1997).

The antenna array formed by individual sensor node antennas is assumed to be a planar array, of randomly positioned sensor node antennas, which is parallel with the plane containing all sensor nodes so that the sensor nodes are only extended in x and y direction and not in z direction. This is a valid assumption in most cases since the elongation of the networks in z direction in most cases is very small compared to the distance between the network cluster and the base station we want to communicate with Jenkins (1973). The design of this type of cooperative array is similar to the design of large aperture arrays where we have an inter-element spacing that is random and larger than half the wavelength. There are no known simplifying techniques for synthesis of randomly spaced arrays, like Schelkunoffs polynomial method Johnson (1993); Balanis (1997) or the Fourier Transform method Johnson (1993); Balanis (1997). In the random array all properties, e.g., array pattern, beamwidth, sidelobe level and gain are stochastic variables.

In figure 1, we show a scenario with $N = 50$ sensor nodes deployed inside a circular boundary in the x-y plane with a radius R. The sensor nodes are independent and uniformly distributed within the cluster area. The nth sensor then has the polar coordinates (r_n, ϕ_n) .

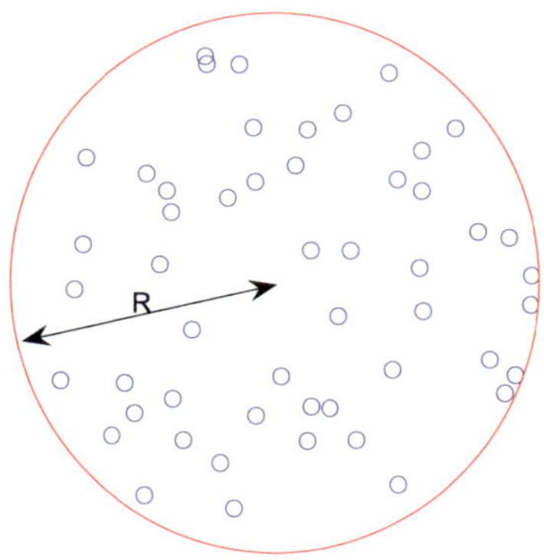

Fig. 1. The positioning of the employed sensor nodes within a cluster area of radius R according to an independent uniform distribution.

The signal $y_n(t)$ at the array sensor node n can then be expressed as,

$$y_n(t) = s(t - \boldsymbol{\alpha_0} \cdot \mathbf{x_0}), \tag{1}$$

where $s(t)$ is the signal to be transmitted/received and the nth sensor at location x_n transmits/receives the electromagnetic signal $y_n(t)$. The slowness vector $\boldsymbol{\alpha_0}$ is the required delay for each sensor to steer the array in a specific direction toward the signal source or target, and is defined as,

$$\boldsymbol{\alpha_0} = \frac{\mathbf{d_0}}{c} \tag{2}$$

where $\mathbf{d_0}$ is the direction of the wave propagation and c is the speed of light. The total output of the delay-and-sum algorithm can be expressed by,

$$z(t) = \sum_{n=0}^{N-1} w_n s(t + (\boldsymbol{\alpha} - \boldsymbol{\alpha_0}) \cdot \mathbf{x_0}), \tag{3}$$

where w_n is the amplitude weights of the array tapering and $\boldsymbol{\alpha}$ is the slowness vector for the direction of observation. If we assume that all the sensor nodes are approximately located in the same plane (i.e., the x-y plane) and the source/target is located at the spherical coordinates $\mathbf{d_0} = (d_0, \phi_0, \theta_0)$ in the far-field, and we are transmitting a narrow band signal then we can approximate equation (3) as, (see appendix)

$$G(\phi, \theta) = \frac{1}{N} \sum_{n=0}^{N-1} w_n e^{j\omega(t - \frac{r_n}{c}(\cos(\phi_n)u + \sin(\phi_n)v)}, \tag{4}$$

where $u = \sin(\theta)\cos(\phi) - \sin(\theta_0)\cos(\phi_0)$ and $v = \sin(\theta)\sin(\phi) - \sin(\theta_0)\sin(\phi_0)$ for the direction of the incoming/outgoing wave (ϕ_0, θ_0) and the direction of observation (ϕ, θ). The function $G(\phi, \theta)$ is then one ensemble of the array amplitude gain function for one set of stochastic sensor locations. To find the ensemble mean of the array amplitude gain functions, we assume an independent uniform distribution of the sensor locations within the radius R,

$$\mathrm{E}\{G(\phi,\theta)\} = \iint G(\phi,\theta) p_{R,\phi}(r_n,\phi_n), \tag{5}$$

where $p_{R,\phi}(r_n,\phi_n)$ is the probability density function (PDF) of the sensor locations.
In figure 2 we show the absolute squared average array gain function $|\mathrm{E}\{G(\phi,\theta)\}|^2$ of 250 realizations of the array amplitude gain function $G(\phi,\theta)$, and in figure 3 we show the standard deviation for the distribution of the amplitude sidelobe levels. From figure 2 we can also estimate a mean sidelobe level that will converge toward ≈ -17 dB which is consistent with the theoretical value, N^{-1}. The average signal-to-noise ratio of the array is defined as $SNR_{array} = SNR_{node} \cdot G(\phi,\theta)$ which means that the array average SNR is $SNR_{array} = N \cdot SNR_{node}$ when we are aiming the array toward the incoming assumed plane wave. The SNR_{array} is a Gaussian distributed parameter with a mean of 17 dB, and a 95% confidence that the SNR of the array will be higher than 7 dB.

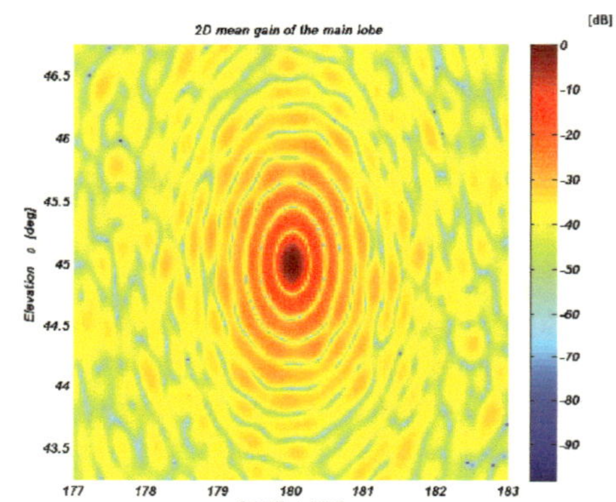

Fig. 2. The absolute squared average array pattern of 250 realizations of the random sensor locations. Only a small part around the main lobe is shown in the figure.

3. Modern Spatial Diversity Techniques

Another recently popular technique to improve the signal to noise ratio of the long range transmission is to use some form of spatial multiantenna diversity system. In this chapter, we employ modern diversity techniques which have gained great interest in the past decade or so. These are: multiple-input single-output (MISO), single-input multiple-output (SIMO) and multiple-input multiple-output (MIMO) antenna systems. Multiple transmit and receive antenna systems allow increased data rates and enhanced link reliability of wireless communication systems while reducing the transmission power requirements. In the following analysis of these diversity techniques, we will assume a perfect knowledge of the propagation channel.

3.1 Cooperative Multiple-Input Single-Output

Consider a frequency flat fading propagation model with N_{tx} antenna elements at the transmitter and one antenna element at the receiver. To take full advantage of the antenna transmit

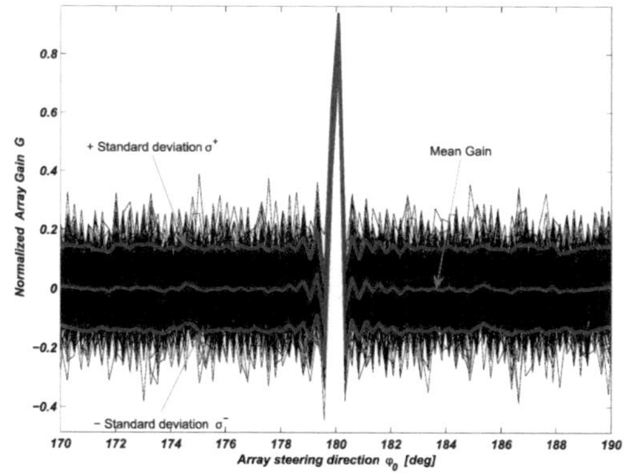

Fig. 3. A plot showing a cross-section of the main lobe of all 250 realizations of the array amplitude gain pattern.

diversity we send multiple weighed copies of the signal sample through all the transmitting antenna elements. The received baseband signal sample can then be expressed as,

$$r[m] = \sqrt{\frac{E_s}{N_{tx}}} \sum_{l=0}^{L-1} h_l w_l s[m] + n[m], \tag{6}$$

where $r[m] \in \mathbb{C}$ is the received sample, $s[m] \in \mathbb{C}$ is the transmitted sample and $n[m]$ is a noise sample with $n[m] \sim \mathcal{CN}(0, \sigma_n^2)$. The coefficient w_l is the channel weight for channel l and E_s is the transmitted average symbol energy. This can be expressed in vector notation as,

$$r = \sqrt{\frac{E_s}{N_{tx}}} \mathbf{hw}s + n, \tag{7}$$

where $\mathbf{h} \in \mathbb{C}^{N_{tx} \times 1}$ is the frequency of flat fading channel vector with a Rice distribution. The normalized Rician channel vector $\mathbf{h}$ can then be defined as, (McKay et al., 2006)

$$\mathbf{h} \triangleq \sqrt{c_1}\mathbf{l} + \sqrt{c_2}\mathbf{R}_{tx}\mathbf{h}_n, \tag{8}$$

where $\mathbf{l}$ is the line-of-sight (LOS) component represented as a mean value that satisfies the condition $|\mathbf{l}|^2 = N_{tx}$, and $\mathbf{R}_{tx}$ is the transmit correlation vector. $\mathbf{R}_{tx}$ is assumed to be positive definite full rank matrix. $\mathbf{h}_n \sim \mathcal{CN}_{N_{tx}}(\mathbf{0}_{N_{tx}}, \mathbf{1}_{N_{tx}})$ is a complex valued Gaussian vector representing the non line-of-sight (NLOS) component. The coefficients $c_1 = K/(K+1)$ and $c_2 = 1/(K+1)$ are normalizing factors, where K is the Rice factor which represents the power ratio between the LOS and NLOS components. The weight vector $\mathbf{w}$ that maximizes the received SNR is given by,

$$\mathbf{w} = \sqrt{N_{tx}} \frac{\mathbf{h}^H}{||\mathbf{h}||}, \tag{9}$$

which is the transmit maximum ratio combining (MRC) method and is also known as matched beamforming. The SNR of the received signal can then be expressed as,

$$\gamma_{rx} = \frac{E_s \cdot |\mathbf{h}|^2}{N_0}. \tag{10}$$

3.2 Cooperative Single-Input Multiple-Output

The second type of spatial diversity is receive diversity in which we are utilizing a single-input multiple-output (SIMO) frequency flat fading propagation channel model with N_{rx} receiving antenna elements and a single transmitting antenna element. To fully exploit the receive diversity we will receive multiple copies of the transmitted signal through all the N_{rx} receiving antenna elements. The received baseband signal sample can then be expressed as,

$$r[m] = \sqrt{\frac{E_s}{N_{rx}}} \sum_{l=1}^{L} (w_l h_l) s[m] + \sum_{l=1}^{L} w_l n_l[m], \tag{11}$$

where $r_l[m] \in \mathbb{C}$ is the received sample from receiving antenna element l, $s[m] \in \mathbb{C}$ is the transmitted sample and $n_l[m]$ is a noise sample at receiving antenna element l with $n_l[m] \sim \mathcal{CN}(0, \sigma_n^2)$. the coefficient w_l is the channel weight at receiving antenna element l and E_s is the transmitted average symbol energy. This can be expressed in vector notation as,

$$r = \sqrt{E_s} \mathbf{w}^H \mathbf{h} s + \mathbf{w}^H \mathbf{n}, \tag{12}$$

where $\mathbf{h} \in \mathbb{C}^{N_{tx} \times 1}$ is the frequency flat fading channel vector with a Rice distribution. The normalized channel vector $\mathbf{h}$ can then be defined as, (McKay et al., 2006)

$$\mathbf{h} \triangleq \sqrt{c_1} \mathbf{l} + \sqrt{c_2} \mathbf{R}_{rx} \mathbf{h}_n, \tag{13}$$

where $\mathbf{l}$ is the line of sight (LOS) component represented as a mean value that satisfies the condition $|\mathbf{l}|^2 = N_{rx}$, and $\mathbf{R}_{rx}$ is the receive correlation vector. $\mathbf{R}_{rx}$ is assumed to be a positive definite full rank matrix. $\mathbf{h}_n \sim \mathcal{CN}_{N_{rx}}(\mathbf{0}_{N_{rx}}, \mathbf{1}_{N_{rx}})$ is a complex valued Gaussian vector representing the nnon-line-of-sight (NLOS) component. The weight vector $\mathbf{w}$ that maximize the received SNR at each antenna element is given by,

$$\mathbf{w} = \sqrt{N_{rx}} \frac{\mathbf{h}^H}{||\mathbf{h}||}. \tag{14}$$

The SNR of the received signal after we have performed a maximum ratio combining (MRC) can then be expressed as

$$\gamma_{rx} = \frac{E_s \cdot |\mathbf{h}|^2}{N_0}. \tag{15}$$

3.3 Cooperative Multiple-Input Multiple-Output

By combining the MISO and SIMO diversity techniques we create a system of (N_{tx} and N_{rx}) transmitting and receiving antenna elements, respectively, which is known as a multiple-input multiple-output (MIMO) system. If we consider a frequency flat fading ($N_{tx} \times N_{rx}$) MIMO propagation model, the received signal can be written in vector notation as,

$$r = \sqrt{\frac{E_s}{N_{tx}}} \mathbf{w}_{rx}^H \mathbf{H} \mathbf{w}_{tx} s + \mathbf{w}_{rx} \mathbf{n}. \tag{16}$$

In the MIMO case, the Rice distributed channel matrix $\mathbf{H}$ can be derived as,

$$\mathbf{H} \triangleq \sqrt{c_1} \mathbf{L} + \sqrt{c_2} \mathbf{R}_{rx}^{\frac{1}{2}} \mathbf{H}_n \mathbf{R}_{tx}^{\frac{1}{2}}, \tag{17}$$

where $\mathbf{L}$ represents the LOS component and is the arbitrary rank mean value matrix with the condition that $\mathrm{Tr}(\mathbf{L}\mathbf{L}^H) = N_{rx} \cdot N_{tx}$, $\mathbf{R}_{rx}$ and $\mathbf{R}_{tx}$ are the correlation matrices on the transmitter and receiver side respectively. $\mathbf{H}_n \sim \mathcal{CN}_{N_{rx},N_{tx}}(\mathbf{0}_{N_{rx}\times N_{tx}}, \mathbf{I}_{N_{rx}} \otimes \mathbf{I}_{N_{rx}})$.
To maximize the combined SNR at the receiver antenna elements we maximize,

$$\gamma_{rx} = \frac{E_s}{N_0} \cdot \frac{\left\|\mathbf{w}_{rx}^H \mathbf{H} \mathbf{w}_{tx}\right\|^2}{N_{tx} \left\|\mathbf{w}_{rx}\right\|^2}. \tag{18}$$

γ_{rx} is then maximized when $\mathbf{w}_{rx}$ and $\mathbf{w}_{tx}/N_{tx}$ are equal to the singular input and output vectors of the channel matrix $\mathbf{H}$ corresponding to the maximum singular value of the channel matrix $\mathbf{H}$. Equation 16 can then be written as,

$$r[m] = \sqrt{E_s}\sigma_{max}s[m] + n[m]. \tag{19}$$

where σ_{max} is the maximum singular value of the channel matrix $\mathbf{H}$ and since σ_{max}^2 is the same as the maximum eigenvalue λ_{max} of $\mathbf{H}\mathbf{H}^H$. We can now express the received SNR of the MIMO diversity technique as,

$$\gamma_{rx} = \frac{E_s}{N_0} \cdot \lambda_{max}. \tag{20}$$

4. Simulation Results

In this section we assess the performance of beamforming technique and modern spatial diversity techniques and compare the results with the nondiversity single antenna (or SISO) system. If we consider a base station mounted on an aerial platform such as a HAP or a UAV to collect data from remote sensor networks, then the amount of obstructions in the transmission path would depend on the type of environment at the sensor locations, although it can still generally be assumed that the number of obstructions will increase with a decreasing antenna elevation angle. Therefore, the propagation effect of the change in elevation can be translated into a change of the Rice distribution K-factor.

In the presented simulations, the Rician K-factor was varied over an interval of $K \in [1 \cdot 10^{-8}, 1 \cdot 10^{+8}]$, where the low value represents a channel with no LOS component and very little correlation between the different signal paths and therefore resembles a Rayleigh fading channel. When the Rician K-factor is gradually increased the correlation between the signal paths will increase and the Direction of Departure (DoD)/Direction of Arrival (DoA) of the signals will narrow into a smaller and smaller angular sector, until the K-factor asymptotically goes toward infinity and all signal paths will be correlated and pointing in the same direction.

In figure 4 we see the comparison between the ordinary random array beamformer performance and the MISO/SIMO diversity systems performance. Inspecting figure 4, we can see that the MISO/SIMO diversity system seems to maintain a constant low node transmitter power P_{tx} even in a NLOS scenario by spreading the energy over multiple paths instead of transmitting it all in one direction. Furthermore, we can see from figure 4 that if the distance between the transmitting nodes and the basestation is increased from 1 km to 10 km, the nodes need a 100 fold increase of the total transmitted power to maintain the same capacity. This is independent of whether we are using the nodes as a beamforming array or a diversity system, which is consistent with the inverse square law of the free space loss.

Finally, we assess the performance of the full multiantenna diversity system (or MIMO) where we have multiple antenna nodes on both the transmitting and receiving end of the link. In addition, we compare the results with the conventional array beamformer, with its subsets (SIMO/MISO) and the nondiversity single antenna (or SISO) system. In the results shown in figure 5 we increase the number of receiving antenna nodes to be equal to the number of transmitting antenna nodes to get a (50 × 50) MIMO system which will increase the array and diversity gains even further. This effect can clearly be seen in figure 5 where the performance of the MIMO system outperforms the other systems in *both* LOS and NLOS scenarios. It is also clear from this figure that the nondiversity SISO system and the conventional beamformer will not function properly in this setting and in particular in NLOS conditions. These initial results suggest that the application of modern spatial diversity systems is expected to improve the energy efficiency, lifetime and the overall performance of the wireless sensor network.

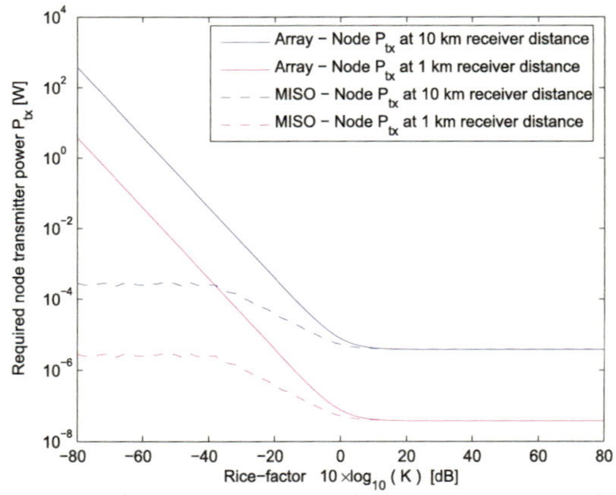

Fig. 4. Comparison between of the Array Beamformer and MISO/SIMO system for different K-factor values for a distance from the base station of 1 km and 10 km, respectively.

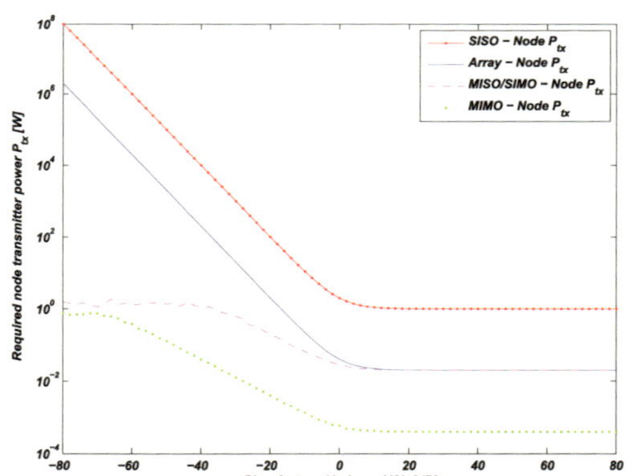

Fig. 5. Performance of the Array Beamformer, MISO/SIMO and MIMO systems for different K-factor values and compared with a single antenna SISO system. The performance results are *normalised* against SISO in this figure.

5. Conclusions

In this chapter we have investigated how the required transmitter power of each sensor node is affected by the number of cooperating transmission nodes in a traditional random beamformer array. Due to the randomness of the sensor node positions, there is no simple algorithm for mitigation of interference from a fixed direction. This is because the sidelobe levels and the sidelobe positions are random. A comparison in the use of beamforming with modern diversity systems such as MISO/SIMO and MIMO for the same purpose of achieving a longer transmission distance or maintaining a low energy consumption is also presented. It is clear from these investigations that the MISO/SIMO and MIMO diversity systems are superior in performance to both the SISO link and the traditional form of array beamforming, especially when the LOS component is small or non-existent. Even one extra antenna at the receiving base station will increase the performance of the system two-fold in a LOS scenario and give an improved performance in NLOS as well. The best performance though, is given by the MIMO system where we have multiple antenna nodes on both the transmitting and receiving end of the link. Initial results suggest that the application of modern spatial diversity systems is expected to improve the energy efficiency and lifetime of wireless sensor network.

Appendix: Derivation of Equation (3)

The slowness vector $\boldsymbol{\alpha}$ in (2) is defined as,

$$\boldsymbol{\alpha} = \frac{\mathbf{d}}{c}. \tag{21}$$

The $\mathbf{d}$ vector represents the direction of observation and can be expressed in cartesian coordinates as,

$$\mathbf{d} = d \cdot \{-\sin(\theta)\cos(\varphi), -\sin(\theta)\sin(\varphi), \cos(\theta)\}. \tag{22}$$

Assuming that the sensor nodes are only distributed in the $x-y$ plane. In addition, if we assume a far-field plane wave solution, then the individual propagation induced time delay Δt_n is calculated from the slowness vector $\boldsymbol{\alpha}$ and the position vector $\mathbf{x}_n$ of each node n as,

$$\Delta t_n = \boldsymbol{\alpha} \cdot \mathbf{x}_n = \frac{r_n}{c}\left(-\sin(\theta)\cos(\varphi)\cos(\varphi_n) - \sin(\theta)\sin(\varphi)\sin(\varphi_n) - 0\right) \tag{23}$$

$$\Delta t_n = -\frac{r_n}{c}\left(\sin(\theta)\cos(\varphi - \varphi_n)\right) \tag{24}$$

The actual direction of propagation $\mathbf{d}_0$ is used to calculate the slowness vector $\boldsymbol{\alpha}_0$ of the centre point of the array,

$$\Delta t_0 = \frac{r_n}{c}\left(\sin(\theta)\cos(\varphi_0 - \varphi_n)\right) \tag{25}$$

Substituting (24) and (25) into (3) results in,

$$z(t) = \sum_{n=0}^{N-1} w_n s(t - \frac{r_n}{c}((\sin(\theta)\cos(\varphi - \varphi_n)) - (\sin(\theta)\cos(\varphi_0 - \varphi_n)))). \tag{26}$$

Denoting $u = \sin(\theta)\cos(\phi) - \sin(\theta_0)\cos(\phi_0)$, $v = \sin(\theta)\sin(\phi) - \sin(\theta_0)\sin(\phi_0)$ and assuming a sinusoidal signal $s(t)$, (26) can be expressed as a time harmonic solution,

$$G(\phi,\theta) = \frac{1}{N}\sum_{n=0}^{N-1} w_n e^{j\omega(t - \frac{r_n}{c}(\cos(\phi_n)u + \sin(\phi_n)v)}. \tag{27}$$

6. References

Akyildiz, I.; Su, W.; Sankarasubramaniam, Y. & Cayirci, E. (2002). A survey on wireless sensor networks. *IEEE Communications Magazine*, Vol. 40, No. 8, August 2002, 102-114.

Balanis, C. (1997). *Antenna Theory: Analysis and Design*, Chapter 6, John Wiley, 1997.

Cui, S.; Goldsmith, A. & Bahai, A. (2004). Energy-efficiency of MIMO and cooperative MIMO techniques in sensor networks. *IEEE Journal on Selected Areas in Communications*, Vol. 22, No. 6, August 2004, 1089-1098.

Drane, C. Jr. (1968). Useful approximations for the directivity and beamwidth of large scanning Dolph-Chebyshev arrays. *IEEE Proceedings*, Vol. 56, No. 11, November 1968, 1779-1787.

Hansen, W. & Woodyard, R. (1938). A New Principle in Directional Antenna Design, *Proceedings IRE*, Vol. 26, No. 3, March 1938, 333-345.

Jenkins, J. (1973). Some properties and examples of random listening arrays. *IEEE Oceans*, Vol. 5, September 1973, 466-469.

Johnson, D. & Dudgeon, D. (1993). *Array Signal Processing: concepts and techniques*, Prentice-Hall Inc., 1993, ISBN 0-13-048513-6.

Lewis, F. (2004). Wireless sensor networks, *In: Smart Environments: Technologies, Protocols, and Applications*, Cook, D. & Das, S. (Editors), John Wiley, New York, 2004.

McKay, M. & Collings, I. (2006). Improved general lower bound for spatially-correlated Rician MIMO capacity. *IEEE Communications Letters*, Vol. 10, No. 3, March 2006, 162-164.

Pillutla, L. & Krishnamurhty, V. (2005). Joint rate and cluster optimization in cooperative MIMO sensor networks. *IEEE 6th Workshop on Signal Processing Advances in Wireless Communications*, June 2005, 265-269.

Rashid-Farrokhi, F.; Tassiulas L. & and Liu, K. (1998). Joint power control and beamforming in wireless networks using antenna arrays. *IEEE Transactions on Communication*, Vol. 46, No. 10, October 1998, 1313-1324.

Stankovic, J.; Abdelzaher, T.; Lu, C.; Sha, L. & Hou, J. (2003). Realtime communication and coordination in embedded sensor networks. *Proceedings of The IEEE*, Vol. 91, No. 7, July 2003, 1002-1022.

Tubaishat, M. & Madria, S. (2003). Sensor networks: an overview. *IEEE Potentials*, Vol. 22, No. 2, April 2003, 20-23.

5

Energy Efficient Cooperative MAC Protocols in Wireless Sensor Networks

Mohd Riduan Ahmad[1], Eryk Dutkiewicz[2] and Xiaojing Huang[3]
[1]Universiti Teknikal Malaysia Melaka, [2]Macquarie University, [3]CSIRO ICT Centre
[1]Malaysia, [2,3]Australia

1. Introduction

Multiple sensor nodes can be used to transmit and receive cooperatively and such a configuration is known as a cooperative Multiple-Input Multiple-Output (MIMO) system. Cooperative MIMO systems have been proven to reduce both transmission energy and latency in Wireless Sensor Networks (WSNs). However, most current work in WSNs considers only the energy cost for the data transmission component and neglects the energy component responsible for establishing a cooperative mechanism. In this chapter, both transmission and circuit energies for both components are included in the performance models.

Furthermore, in previous work, all sensor nodes are assumed to be always on which could lead to a shorter lifetime due to energy wastage caused by idle listening and overhearing. Low duty cycle Medium Access Control (MAC) protocols have been proposed to tackle this challenge for non-cooperative systems. In this chapter, we propose a new Cooperative low duty cycle MAC protocol (CMAC) for two cooperative MIMO schemes: Beamforming ($CMAC_{BF}$) and Spatial Multiplexing ($CMAC_{SM}$). Performance of the proposed CMAC protocol is evaluated in terms of total energy consumption and packet latency for both synchronous and asynchronous scenarios. All the required energy components are taken into consideration in the system performance modelling and a periodic monitoring application model is used. The impact of the clock jitter, the check interval and the number of cooperative nodes on the total energy consumption and latency is investigated. The $CMAC_{BF}$ protocol with two transmit nodes is suggested as the optimal scheme when operating at the 250 ms check interval with the clock jitter difference below $0.6Tb$ where Tb is the bit period corresponding to the system bit rate.

The rest of the chapter is organized as follows. In Section 2, we briefly describe the related work. Section 3 describes the system model considered in this chapter and explains the low duty cycle protocols that we propose for cooperative transmission. Sections 4 and 5 model the system performances and the analytical results for the two cooperative MIMO schemes (BF and SM) in terms of total energy consumption and latency are presented in Section 6. Finally, in Section 7 we conclude the chapter.

2. Related Work

A practical MAC that can suit cooperative transmission is required. Also, a combination of a practical MAC protocol and an efficient MIMO scheme for cooperative transmission leads to a more energy efficient and lower latency cooperative MIMO system. A combination of a MAC protocol and a virtual SM scheme for cooperative MIMO transmission has been proposed in (Yang et al., 2007) where the combined scheme achieves significant energy efficiency and lower latency. Further study has been done in (Ahmad et al., 2008a) evaluating the MAC protocol in (Yang et al., 2007) using the other two cooperative schemes: BF and Space-Time Block Coding (STBC). The authors in (Ahmad et al., 2008a) proposed that the optimal scheme for the Cooperative always on MAC ($CMAC_{ON}$) is the BF scheme with $M = 2$. However, the MAC protocols for all the schemes considered the transceivers as always being on and the networks are perfectly synchronized. Although the transmission energy is reduced and the deep fading threat is reduced, the idle listening problem is not tackled in previous research work. Also, the imperfect synchronization due to clock jitter is not considered.

Most of the duty cycle MAC protocols are designed for non-cooperative Single-Out Single-In (SISO) schemes. Polastre in 2004 introduces B-MAC or Berkeley MAC (Polastre et al., 2004). The protocol is a variant of Carrier Sense Multiple Access (CSMA) with a preamble sampling mechanism. The preamble sampling is improved with a selective sampling method where only energy above the noise floor is considered as useful. However B-MAC experiences a long preamble problem which leads to higher transmission and reception powers. In order to reduce the long preamble problem, X-MAC (Buettner et al., 2006) proposed the use of a series of short preamble packets with the destination address embedded in the packet. The X-MAC protocol provides more energy efficient and lower latency operation by reducing the transmission energy and period burdens, idle listening at the intended receiver and overhearing by the neighbouring nodes. One concern is that the gaps between transmissions of a series of preamble packets can be mistakenly understood by the other contending nodes as an idle channel and they would start to transmit their own preamble packets which can lead to collision. One solution is to ensure that the length of the gaps must be upper bounded by the length of the listen interval.

In the same year, SpeckMAC (Wong & Arvind, 2006) was introduced as a variation of B-MAC with the idea of redundant transmission of short packets and an embedded destination address. There are two variants: SpeckMAC-Back-off (SpeckMAC-B) and SpeckMAC-Data (SpeckMAC-D). SpeckMAC-B sends short wake-up frames with an embedded target destination address many times. The problem with this scheme is that the sender wastes its transmission power by still sending the short frames although the receiver has already received it. Meanwhile, SpeckMAC-D sends the data packet which is preceded with a short preamble many times until the packet hits the receiver.

In this chapter, we propose redundant transmission of Ready-to-Send (RTS) and Clear-to-Send (CTS) packets to hit the intended receiver. The cyclic RTS-CTS transmission scheme is used also for other purposes such as collision avoidance, cooperative nodes selection and channel state information (CSI) sharing between nodes. A combination of low duty cycle MAC with cyclic RTS-CTS transmission scheme is believed to reduce further the energy consumption in cooperative MIMO transmission. In addition, an imperfect synchronisation scenario due to clock jitter differences is investigated. The major contribution of this chapter is the proposal of CMAC with embedded low duty cycle mechanism which implements cyclic RTS-CTS transmission scheme and acknowledgement (ACK) reply to ensure higher reliability. The CMAC is suggested to be used with two cooperative schemes: optimal BF and Spatial Multiplexing. We compare the performance of both these schemes in terms of energy consumption and latency. We also include a comparison with $CMAC_{ON}$, B-MAC and always on SISO MAC. The impact of the jitter difference, the check interval and the number of cooperative nodes on the total energy consumption and latency are investigated.

3. System Model

3.1 System Description

The baseline system for cooperative MIMO communication with the transceivers being always on is equipped with $CMAC_{ON}$ protocol as proposed and evaluated in (Jagannathan et al., 2004). Meanwhile, the baseline system for cooperative MIMO with a periodic wake-up cycle for the transceiver is equipped with the CMAC protocol as proposed and explained in sub-section 3.2. The baseline MAC for the SISO scheme with the transceiver being always on is CSMA-CA with RTS-CTS and ACK packets transmissions. For simplicity of notation, we denote the SISO scheme with this MAC protocol as the SISO always on protocol or $SISO_{ON}$ protocol. Also in this chapter we consider the impact of imperfect synchronization which is caused by clock jitter alone. The detailed modelling of the impact of clock jitter is given in sub-section 3.3.

The network configurations for all the schemes considered in this work are as shown in Figures 1 and 2. The network is assumed to be distributed without any infrastructure. A new node can join or leave the network at any time because the knowledge of neighbours is not important due to the fact that the selection of cooperative nodes is done during the control packets communication. We assume that there are M cooperative transmitting nodes and one receiving node. A special case for the spatial multiplexing scheme is used where the number of the cooperative receivers is assumed to be N. Both the source and destination nodes have n neighbours in their vicinity. The distance between the cooperating nodes either at the transmitting or receiving side is assumed to be very small compared to the distance between the source node and the destination node, d. In the case of the cooperative BF scheme, the channel information is estimated and optimized from the CTS packet by all the M nodes. As for the cooperative SM scheme, the recovered data from N-1 nodes is forwarded to the destination node. Both schemes utilize a Maximum Likelihood (ML) detector and use a coherent receiver.

3.2 Protocol Description

The proposed CMAC protocol combines the advantages of the cooperative MAC with always on radios and a low duty cycle mechanism. The basic structure of the protocol is given in Algorithm 1. A node may respond to three events for the case of the BF scheme ($CMAC_{BF}$) and to four events for the case of the SM scheme ($CMAC_{SM}$). In case a node has a data packet to send where the node is acting as the source node, the basic operations for both schemes are shown in Algorithm 2.

A node starts by sending RTS packets followed by an inter-frame spacing (IFS) for a period of the length of the check interval, T_i after sensing the channel idle. When a CTS packet is received, the source sets a timer to wake up later (the sleep duration is T_i -T_{cts} -$T_{transient}$) in order to transmit a broadcast packet at source (BS) immediately followed by the data packet

(DATA), to its M-1 neighbours. Transmission of BS and DATA packets occurs at low transmission power due to the very short distance, d_m between the source and its M-1 neighbours. The BS packet is broadcasted by the source node to recruit its neighbours for cooperative transmitting operation and the DATA packet is the original data packet provided by the sensor device. When the sending timer expires (included in the BS packet), M nodes cooperatively transmit the data packet to the destination. After cooperatively transmitting the data, the source waits for an ACK packet. If an ACK is not received, the whole process is repeated. The number of RTS and CTS packets to be transmitted is given by:

$$R = \frac{T_i + T_{ifs\_rts}}{T_{rts} + T_{ifs\_rts}} \tag{1}$$

and

$$C = \frac{T_i + T_{ifs\_cts}}{T_{cts} + T_{ifs\_cts}} \tag{2}$$

where T_{rts}, T_{cts}, T_{ifs_rts}, and T_{ifs_cts} are the duration of one RTS and CTS packet and the IFS intervals for RTS and CTS, respectively. The latter are given as:

$$T_{ifs\_rts} = T_{ifs\_cts} \le T_{listen} \tag{3}$$

where the value T_{listen} is given in (Polastre et al., 2004). The operation of the destination node is shown in Algorithm 3 for both schemes. On receiving the RTS packet, the destination estimates the time to wake up in order to transmit CTS packets followed by IFS for a period of the length of the check interval, T_i. The sleep duration is T_i - ($S_{eq}N_{um}$ x T_{rts} + ($S_{eq}N_{um}$-1) x T_{ifs_rts}) - $T_{transient}$. After all the CTS packets are transmitted, the destination sets the timer to wake up at T_{Bs} + T_{data} - $T_{transient}$ to receive the data packet. In the case of the SM scheme, the destination broadcasts the broadcast packet BR at the receiver (BR packet is broadcasted by the destination to recruit its neighbours for cooperative receiving operation.)

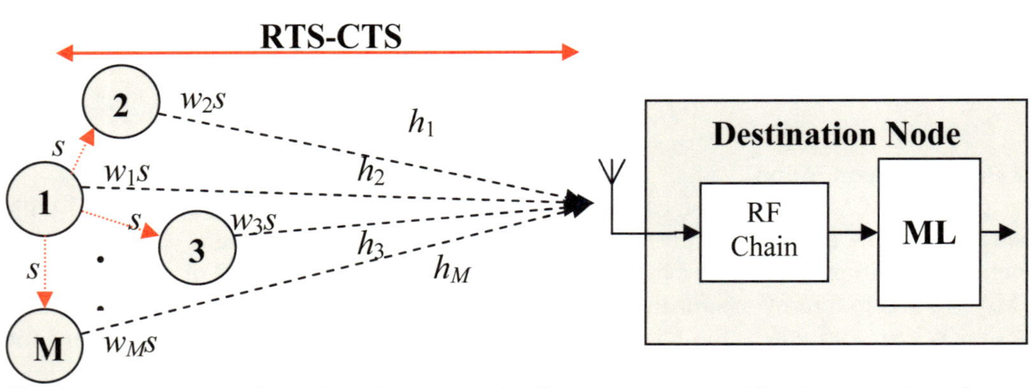

Fig. 1. A cooperative beamforming transmit diversity system with M transmit nodes and destination

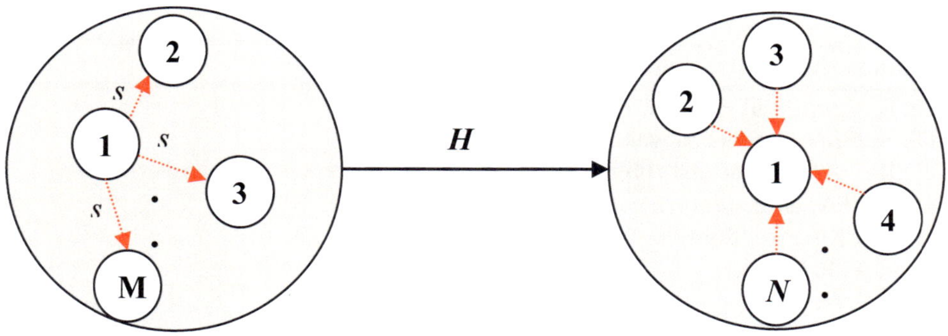

Fig. 2. A cooperative spatial multiplexing system with M transmit nodes and N receive nodes

first and then goes to sleep for the duration of $T_{Bs} + T_{data} - T_{Br} - T_{transient}$. After receiving the data packet, the destination sends an ACK packet immediately. In the case of the SM scheme, the destination waits for its neighbours to forward the data packets and does the final decoding of the packet based on all the received copies of the data packet from its neighbours.

The operations of cooperative sending and receiving nodes are shown in Algorithm 4 and 5. The selection of cooperative nodes is done during the control packets transmission where a node which receives RTS is informed to wake up at $T_i - (S_{eq}N_{um} \times T_{rts} + (S_{eq}N_{um}-1) \times T_{ifs\_rts}) - T_{transient}$ to receive CTS. The time waiting for CTS packet is denoted as T_{wfcts}. If a node receives CTS, it is informed to wake up at $T_i - T_{cts} - T_{transient}$ to receive BS for both schemes and BR for the SM scheme. The time waiting for the BS packet is denoted as $T_{wfbsdata}$. The time waiting for the BR packet is the same as the time waiting for the BS packet. A node is chosen to be one of the cooperative nodes when it receives the broadcast packet. By using this mechanism, we can ensure that the network is scalable and no prior knowledge about neighbours is required for cooperative transmitting and receiving. Also, any node which does not receive CTS after receiving RTS or does not receive a broadcast packet after receiving CTS needs to go to sleep. This mechanism avoids the problems of hidden nodes. The timers' settings are described in more detail in the timing diagrams in Figures 3 and 4 for the BF and SM schemes, respectively.

Algorithm 1: Cooperative MIMO MAC Protocol

STATE: LISTEN node listens to the channel after it wakes up
if Packet ready to be sent **then**
  go to **Algorithm 2**
end if
if receive RTS **then**
  go to **Algorithm 3**
end if
if receive BSDATA **then**
  go to **Algorithm 4**
end if
if receive BR **then**
  go to **Algorithm 5**

end if

Algorithm 2: Node is the source

STATE: RTS sends all RTS packets and receives CTS packet
STATE: SLEEP sets timer to wake up and goes to sleep
STATE: BSDATA broadcasts BS followed by DATA packet with low power
STATE: DATA sends data when the sending timer expires
if receive ACK packet **then**
  go to **STATE: LISTEN**
else
  go to **STATE: RTS**
end if

Algorithm 3: Node is the destination for BF scheme

STATE: LISTEN receives RTS and sets timer to wake up
go to **STATE: SLEEP**
STATE: CTS sends CTS packet for a period of check interval
STATE: SLEEP the node sets timer to wake up and goes to sleep
if data packet is received **then**
  go to **STATE: ACK**
else if
  go to **STATE: LISTEN**
STATE: ACK node sends ACK packet
go to **STATE: LISTEN**

Algorithm 3: Node is the destination for SM scheme

STATE: LISTEN receives RTS packet and sets timer to wake up
go to **STATE: SLEEP**
STATE: CTS sends CTS packet for a period of check interval
STATE: BR sends broadcast packet to neighbours
STATE: SLEEP sets timer to wake up and goes to sleep
if data packet is received **then**
  go to **STATE: COLLECTION**
else if
  go to **STATE: LISTEN**
STATE: COLLECTION set timer to wait for data packets
if packet is not received correctly **then**
  go to **STATE: LISTEN**
end if
STATE: ACK node sends ACK packet
go to **STATE: LISTEN**

Algorithm 4: Cooperative sending node

STATE: COOPERATIVE_SENDING nodes transmit data packet when sending timer expires
go to **STATE: LISTEN** listens for channel activity

Algorithm 5: Cooperative receiving node

STATE: COOPERATIVE_RECEIVING set expiration timer
if data packet received **then**
  go to **STATE: COLLECTION**
else if
  go to **STATE: SLEEP** after timeout
end if
STATE: COLLECTION sends data to destination node
go to **STATE: SLEEP**

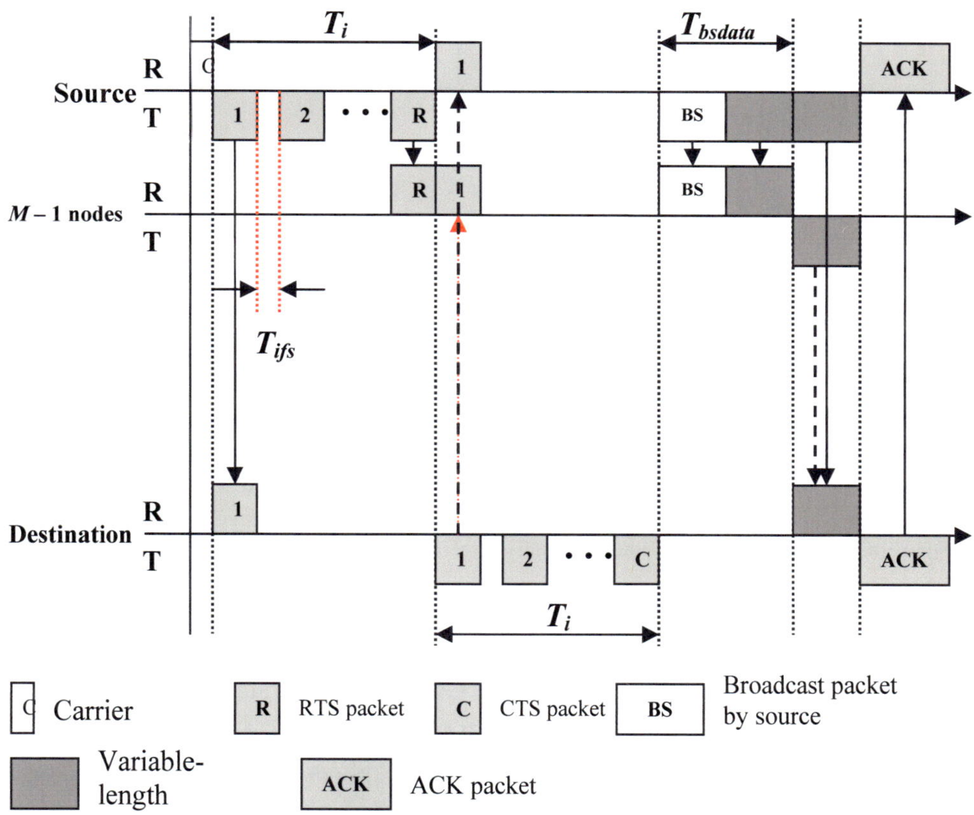

Fig. 3. Timing diagram of $CMAC_{BF}$ cooperative transmission

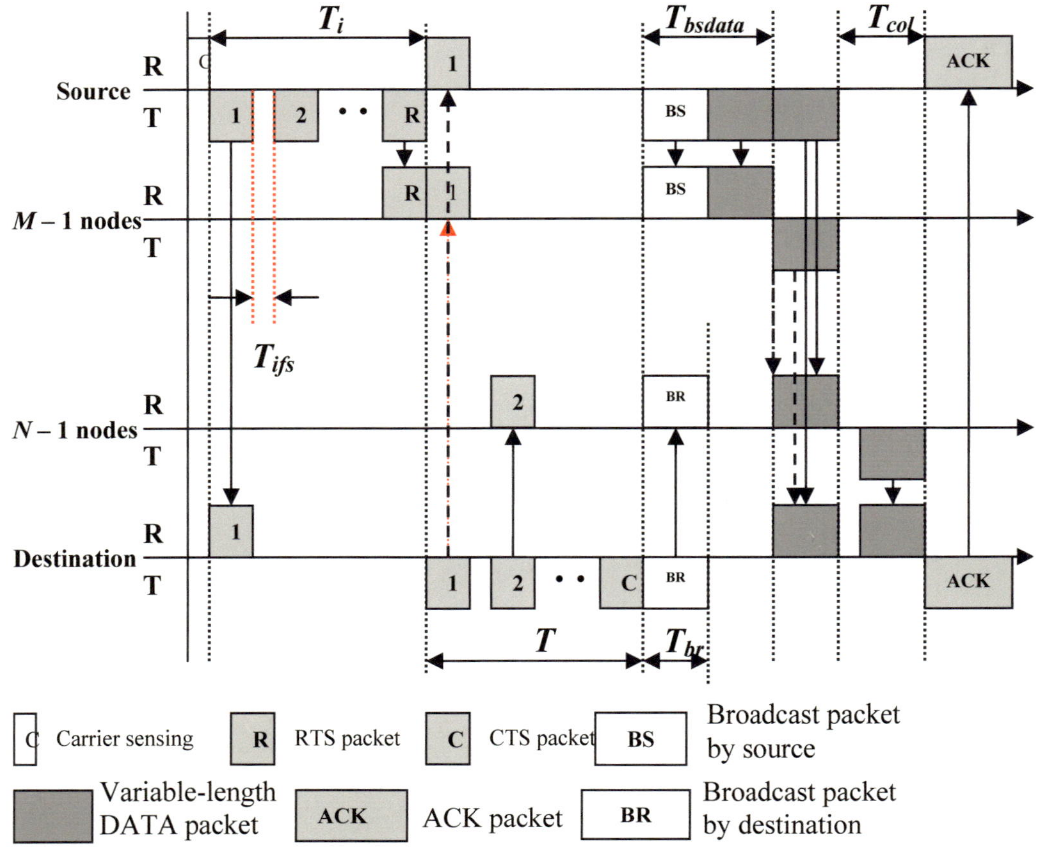

Fig. 4. Timing diagram of $CMAC_{SM}$ cooperative transmission

3.3 Timing Error Model

We consider the impact of imperfect synchronization which is caused by clock jitter alone. Each cooperative sending nodes experiences clock jitter with the jitter around a reference clock, T_o denoted as T_j^m where $1 \leq m \leq M$. The worst case scenario is considered here with only 2 cooperative transmitting nodes where the clock jitters are fixed at the extreme ends, $T_j^1 = -\frac{\Delta T_b}{2}, T_j^2 = +\frac{\Delta T_b}{2}$ where $0 \leq \Delta T_b \leq T_b$ and T_b is the bit duration. Thus the clock jitters difference is $\Delta T_j = T_j^1 - T_j^2 = \Delta T_b$. The effect of imperfect synchronization can be modelled as a degrading function of the bit period which consequently degrades the received bit energy. Therefore the timing error as a function of the bit period and clock jitters difference is given as:

$$T_e = T_b - \Delta T_j \tag{4}$$

4. Energy Consumption Performance Model

In this section, three analytical models are developed and analyzed: $SISO_{ON}$, $CMAC_{ON}$ with the optimal BF scheme and CMAC with 2 variants, $CMAC_{BF}$ and $CMAC_{SM}$. The total energy consumption of each model is analysed and compared. The retransmission rate is modelled as a function of PER where the detailed models and analysis can be found in (Ahmad et al., 2008a).

We consider a periodic sampling application with a uniform sampling period, T_s which has been discussed in detail (Polastre et al., 2004). In general, the energy consumed by a sensor node can be categorized into five major parts (Cui et al., 2004): energy expended during data sampling by sensor, E_{sensor}, energy expended during running the transceiver circuits, E_c, energy expended during packet transmission, E_t, energy expended during packet reception, E_r and energy expended while idle listening, E_{idle}.

For the case of the system with the CMAC protocol, additional energy must be considered: energy expended during sleeping, E_{sleep}, listen energy after waking up, E_{listen} and transient energy, $E_{transient}$. The cooperative mechanism establishment energy cost is included in the transmission and reception energy models. Therefore, all the energy components must be considered when comparing the total energy consumption of the cooperative MIMO and SISO transmission schemes.

4.1 SISO System

The total energy consumption in the SISO system, in general, is given as:

$$E_{siso} = (E_{rx} + E_{cr}) + (E_{tx} + E_{ct}) + E_{sensor} + E_{idle} \quad (5)$$

where E_{rx} and E_{tx} are the energy spent during reception and transmission, and E_{cr} and E_{ct} are the energy spent by the receiver and transmitter circuits. The transmission energy model for the SISO system which includes both the radiated power and circuit power is the same as discussed in (Ahmad et al., 2008a). Consequently, the reception energy model can be obtained directly from the transmission energy model in (Ahmad et al., 2008a). The total time a node spends during successful transmission is given as:

$$T_{tx\_s} = r_s \times (N_{rts} + N_{cts} + N_{data} + N_{ack}) \times T_{tx\_b} \quad (6)$$

and the total time a node spends during unsuccessful transmission is given as:

$$T_{tx\_u} = r_s \times (N_{rts} + N_{cts} + N_{data}) \times T_{tx\_b} \quad (7)$$

where r_s is the sampling frequency and can be obtained by the inverse of the sampling period, T_{tx_b} is the transmit period per bit, and $N_{rts}, N_{cts}, N_{data}$ and N_{ack} are the lengths of the RTS, CTS, DATA and ACK packets. The total time a node spends during successful reception is given as:

$$T_{rx\_s} = r_s \times \left(n \cdot N_{rts} + n \cdot N_{cts} + N_{data} + N_{ack}\right) \times T_{rx\_b} \tag{8}$$

and the total time a node spends during unsuccessful reception is given as:

$$T_{rx\_u} = r_s \times \left(n \cdot N_{rts} + n \cdot N_{cts} + N_{data}\right) \times T_{rx\_b} \tag{9}$$

where T_{rx_b} is the receive period per bit. The total time a node spends idle for successful communication is given as:

$$T_{idle\_s} = 1 - T_{tx\_s} - T_{rx\_s} - T_{sensor} \tag{10}$$

and the idle time for unsuccessful communication is given as:

$$T_{idle\_u} = 1 - T_{tx\_u} - T_{rx\_u} \tag{11}$$

where T_{sensor} is the period of a sensor to start, initialise, and collect data as discussed in (Mainwaring et al., 2002; Polastre et al., 2004). Thus, the total energy consumption for successful SISO system communication can be obtained as:

$$E_{siso\_s} = \left(P_{pa} + P_{ct}\right) \cdot T_{tx\_s} + \left(P_r + P_{cr}\right) \cdot T_{rx\_s} + P_{idle} \cdot T_{idle\_s} \tag{12}$$

and the total energy consumption for unsuccessful SISO system communication can be obtained as:

$$E_{siso\_u} = \left(P_{pa} + P_{ct}\right) \cdot T_{tx\_u} + \left(P_r + P_{cr}\right) \cdot T_{rx\_u} + P_{idle} \cdot T_{idle\_u} \tag{13}$$

Therefore, the total energy consumption for the SISO system can be modelled as a function of the retransmission rate:

$$E_{siso} = \left(\frac{P_{pSISO}}{1 - P_{pSISO}}\right) E_{siso\_u} + E_{siso\_s} + E_{sensor} \tag{14}$$

where P_{pSISO} is the packet error probability of the SISO system which can be obtained from (Ahmad et al., 2008a).

4.2 Cooperative Always On MIMO System

In this sub-section, we analyze total energy consumption for the optimal cooperative BF scheme with the CMAC$_{ON}$ protocol. The transmission energy model for the cooperative always on MIMO system which includes the radiated power, circuit power and cooperative mechanism power is the same as discussed in (Ahmad et al., 2008a). Consequently, the reception energy model can be obtained directly from the transmission energy model in (Ahmad et al., 2008a).

In order to provide better understanding about the energy models for cooperative MIMO systems in this chapter, we categorize both the transmission and reception total time into three categories which are based on packet types, namely: control, cooperative mechanism and data categories. The total time a node spends during successful control packet transmission is given as:

$$T_{tx\_s\_control} = r_s \times \left(N_{rts} + N_{cts} + N_{ack}\right) \times T_{tx\_b} \tag{15}$$

and the total time a node spends during cooperative mechanism transmission for optimal BF scheme is given as:

$$T_{tx\_Bsdata} = r_s \times \left(N_{Bs} + N_{data}\right) \times T_{tx\_b} \tag{16}$$

and the total time a node spends during data packet transmission is given as:

$$T_{tx\_data} = r_s \times M \times \left(N_{data}\right) \times T_{tx\_b} \tag{17}$$

Thus, the total time a node spends during successful transmission in cooperative always on MIMO system with optimal BF scheme can be given as:

$$T_{tx\_s\_BF} = T_{tx\_s\_control} + T_{tx\_Bsdata} + T_{tx\_data} \tag{18}$$

and the total time a node spends during unsuccessful transmission is given as:

$$T_{tx\_u\_BF} = T_{tx\_s\_BF} - \left(r_s \times N_{ack} \times T_{tx\_b}\right) \tag{19}$$

where N_{Bs} is the length of the broadcast packet at the source node. The total time a node spends during successful control packet reception is given as:

$$T_{rx\_s\_control} = r_s \times \left(n \cdot N_{rts} + n \cdot N_{cts} + N_{ack}\right) \times T_{rx\_b} \tag{20}$$

and the total time a node spends during cooperative mechanism reception is given as:

$$T_{rx\_Bsdata} = r_s \times \left(M - 1\right) \times \left(N_{Bs} + N_{data}\right) \times T_{rx\_b} \tag{21}$$

and the total time a node spends during data packet reception is given as:

$$T_{rx\_data} = r_s \times \left(N_{data}\right) \times T_{rx\_b} \tag{22}$$

Thus, the total time a node spends during successful reception in cooperative always on MIMO system with optimal BF scheme can be given as:

$$T_{rx\_s\_BF} = T_{rx\_s\_control} + T_{rx\_Bsdata} + T_{rx\_data} \tag{23}$$

and the total time a node spends during unsuccessful reception is given as:

$$T_{rx\_u\_BF} = T_{rx\_s\_BF} - r_s \times (N_{ack}) \times T_{rx\_b} \tag{24}$$

The total time a node spends idle for successful communication is given as:

$$T_{idle\_s\_BF} = 1 - T_{tx\_s\_BF} - T_{rx\_s\_BF} - T_{sensor} \tag{25}$$

and the idle time for unsuccessful communication is given as:

$$T_{idle\_u\_BF} = 1 - T_{tx\_u\_BF} - T_{rx\_u\_BF} \tag{26}$$

Thus, the total energy consumption for successful cooperative always on MIMO system communication can be obtained as:

$$\begin{aligned} E_{BF\_s} &= (P_{pa} + P_{ct}) \cdot T_{tx\_s\_control} + (P_{paBs} + P_{ct}) \cdot T_{tx\_Bsdata} + \\ &(P_{paBF} + P_{ct}) \cdot T_{tx\_data} + (P_r + P_{cr}) \cdot T_{rx\_s\_control} + (P_{rBs} + P_{cr}) \cdot T_{rx\_Bsdata} + \\ &(P_{rBF} + P_{cr}) \cdot T_{rx\_data} + P_{idle} \cdot T_{idle\_s\_BF} \end{aligned} \tag{27}$$

and the total energy consumption for unsuccessful cooperative always on MIMO system communication can be obtained as:

$$\begin{aligned} E_{BF\_u} &= (P_{pa} + P_{ct}) \cdot T_{tx\_u\_control} + (P_{paBs} + P_{ct}) \cdot T_{tx\_Bsdata} + \\ &(P_{paBF} + P_{ct}) \cdot T_{tx\_data} + (P_r + P_{cr}) \cdot T_{rx\_u\_control} + (P_{rBs} + P_{cr}) \cdot T_{rx\_Bsdata} + \\ &(P_{rBF} + P_{cr}) \cdot T_{rx\_data} + P_{idle} \cdot T_{idle\_u\_BF} \end{aligned} \tag{28}$$

Therefore, the total energy consumption for the cooperative always on MIMO system can be modelled as a function of the retransmission rate:

$$E_{BF\_on} = \left(\frac{P_{pBF}}{1 - P_{pBF}} \right) E_{BF\_u} + E_{BF\_s} + E_{sensor} \tag{29}$$

where P_{pBF} is the packet error probability of the cooperative BF system which can be obtained from (Ahmad et al., 2008a).

4.3 Cooperative Low Duty Cycle MIMO System

In this sub-section, we analyze the total energy consumption for the cooperative BF and SM schemes equipped with the proposed cooperative low duty cycle MAC protocol. The only modifications on the total energy consumption model are the definition of the control packets intervals which should be depended on the length of the check interval where the *R* and *C* terms are included and the addition of sleep energy. Also, the idle listening cost still exists when a node is in listening and waiting states. The transient energy is included in the total listening energy cost as explained in details in (Polastre et al., 2004). The total time a node spends during successful control packet transmission in cooperative low duty cycle MIMO system is given as:

$$T_{tx\_s\_control} = r_s \times \left(R \cdot N_{rts} + C \cdot N_{cts} + N_{ack}\right) \times T_{tx\_b} \quad (30)$$

The total time a node spends during cooperative mechanism transmission at the transmitting side for both BF and SM schemes in a cooperative low duty cycle MIMO system is the same as given by Equation (16). The total time a node spends during cooperative mechanism transmission at the receiving side by the SM scheme in a cooperative low duty cycle MIMO system can be given as:

$$T_{tx\_Br} = r_s \times \left(N_{Br}\right) \times T_{tx\_b}$$
$$T_{tx\_col} = r_s \times \left(N-1\right) \times \left(N_{data} \cdot T_{tx\_b} + \left(\frac{\sum_{BE=1}^{\max BE} T_{BO} + T_{CCA}}{5} \right) \right) \quad (31)$$

where N_{Br} is the length of broadcast packets at the destination node. T_{BO}, T_{CCA} and BE are the average back-off duration, the clear channel assessment (CCA) analysis duration and the back-off exponent value with all the values derived in detail in (Kohvakka et al., 2006; Kuorilehto et al., 2007). The total time a node spends during data packet transmission for both BF and SM schemes in a cooperative low duty cycle MIMO system is the same as given by Equation (17).

Thus, the total time a node spends during successful transmission for the BF scheme is the same as given in Equation (18) and the total time a node spends during successful transmission for the SM scheme in a cooperative low duty cycle MIMO system can be obtained as:

$$T_{tx\_s\_SM} = T_{tx\_s\_BF} + T_{tx\_Br} + T_{tx\_col} \quad (32)$$

and the total time a node spends during unsuccessful transmission is the same as in Equation (19) for cooperative BF scheme and is given as:

$$T_{tx\_u\_SM} = T_{tx\_s\_SM} - \left(r_s \times N_{ack} \times T_{tx\_b}\right) \quad (33)$$

for the cooperative SM scheme. The total time a node spends during successful and unsuccessful receptions for both cooperative schemes are the same as in Equations (20) to (24) with an addition for the total time of cooperative mechanism reception at the receiving side by the cooperative SM scheme which is given as:

$$
\begin{aligned}
T_{rx\_Br} &= r_s \times (N-1) \times N_{Br} \times T_{tx\_b} \\
T_{rx\_col} &= r_s \times (N-1) \times N_{data} \times T_{tx\_b}
\end{aligned} \tag{34}
$$

The total time a node spends idle for successful communication for both cooperative schemes is given as:

$$
\begin{aligned}
T_{idle\_s\_BF} &= T_{ifs\_rts} + T_{ifs\_cts} + T_{wfcts} + T_{wfbsdata} \\
T_{idle\_s\_SM} &= T_{idle\_s\_BF}
\end{aligned} \tag{35}
$$

and the idle time for unsuccessful communication is given as:

$$
\begin{aligned}
T_{idle\_u\_BF} &= T_{idle\_s\_BF} + T_{wfack} \\
T_{idle\_u\_SM} &= T_{idle\_s\_SM} + T_{wfack}
\end{aligned} \tag{36}
$$

where $T_{wfcts}, T_{wfbsdata}$ and T_{wfack} are the waiting for the CTS packet period, waiting for the BSDATA packet period and the waiting period for the ACK packet to arrive. The total time a node spends for sleeping for successful communication for both cooperative schemes is given as:

$$
\begin{aligned}
T_{sleep\_s\_BF} &= 1 - T_{tx\_s\_BF} - T_{rx\_s\_BF} - T_{idle\_s\_BF} - T_{listen} - T_{sensor} \\
T_{sleep\_s\_SM} &= 1 - T_{tx\_s\_SM} - T_{rx\_s\_SM} - T_{idle\_s\_SM} - T_{listen} - T_{sensor}
\end{aligned} \tag{37}
$$

and the sleep time for unsuccessful communication is given as:

$$
\begin{aligned}
T_{sleep\_u\_BF} &= 1 - T_{tx\_u\_BF} - T_{rx\_u\_BF} - T_{idle\_u\_BF} - T_{listen} \\
T_{sleep\_u\_SM} &= 1 - T_{tx\_u\_SM} - T_{rx\_u\_SM} - T_{idle\_u\_SM} - T_{listen}
\end{aligned} \tag{38}
$$

Thus, the total energy consumption for successful cooperative low duty cycle MIMO system communication can be obtained as:

$$
\begin{aligned}
E_{BF\_s} = &\left(P_{pa} + P_{ct}\right) \cdot T_{tx\_s\_control} + \left(P_{paBs} + P_{ct}\right) \cdot T_{tx\_Bsdata} + \\
&\left(P_{paBF} + P_{ct}\right) \cdot T_{tx\_data} + \left(P_r + P_{cr}\right) \cdot T_{rx\_s\_control} + \left(P_{rBs} + P_{cr}\right) \cdot T_{rx\_Bsdata} + \\
&\left(P_{rBF} + P_{cr}\right) \cdot T_{rx\_data} + P_{idle} \cdot T_{idle\_s\_BF} + P_{sleep} \cdot T_{sleep\_s\_BF} + E_{listen}
\end{aligned} \tag{39}
$$

and

$$E_{SM\_s} = (P_{pa} + P_{ct}) \cdot T_{tx\_s\_control} + (P_{paBs} + P_{ct}) \cdot T_{tx\_Bsdata} + (P_{paBF} + P_{ct}) \cdot T_{tx\_data} + (P_{paBr} + P_{ct}) \cdot T_{tx\_Br} + (P_{paBr} + P_{ct}) \cdot T_{tx\_col} + (P_r + P_{cr}) \cdot T_{rx\_s\_control} + (P_{rBs} + P_{cr}) \cdot T_{rx\_Bsdata} + (P_{rSM} + P_{cr}) \cdot T_{rx\_data} + P_{idle} \cdot T_{idle\_s\_SM} + P_{sleep} \cdot T_{sleep\_s\_SM} + E_{listen} \quad (40)$$

and the total energy consumption for unsuccessful cooperative low duty cycle MIMO system communication can be obtained as:

$$E_{BF\_u} = (P_{pa} + P_{ct}) \cdot T_{tx\_u\_control} + (P_{paBs} + P_{ct}) \cdot T_{tx\_Bsdata} + (P_{paBF} + P_{ct}) \cdot T_{tx\_data} + (P_r + P_{cr}) \cdot T_{rx\_u\_control} + (P_{rBs} + P_{cr}) \cdot T_{rx\_Bsdata} + (P_{rBF} + P_{cr}) \cdot T_{rx\_data} + P_{idle} \cdot T_{idle\_u\_BF} + P_{sleep} \cdot T_{sleep\_u\_BF} + E_{listen} \quad (41)$$

and

$$E_{SM\_u} = (P_{pa} + P_{ct}) \cdot T_{tx\_u\_control} + (P_{paBs} + P_{ct}) \cdot T_{tx\_Bsdata} + (P_{paBF} + P_{ct}) \cdot T_{tx\_data} + (P_{paBr} + P_{ct}) \cdot T_{tx\_Br} + (P_{paBr} + P_{ct}) \cdot T_{tx\_col} + (P_r + P_{cr}) \cdot T_{rx\_u\_control} + (P_{rBs} + P_{cr}) \cdot T_{rx\_Bsdata} + (P_{rSM} + P_{cr}) \cdot T_{rx\_data} + P_{idle} \cdot T_{idle\_u\_SM} + P_{sleep} \cdot T_{sleep\_u\_SM} + E_{listen} \quad (42)$$

Therefore, the total energy consumption for the cooperative low duty cycle MIMO system can be modelled as a function of the retransmission rate:

$$E_{BF} = \left(\frac{P_{pBF}}{1 - P_{pBF}} \right) E_{BF\_u} + E_{BF\_s} + E_{sensor} \quad (43)$$

$$E_{SM} = \left(\frac{P_{pSM}}{1 - P_{pSM}} \right) E_{SM\_u} + E_{SM\_s} + E_{sensor} \quad (44)$$

where P_{pBF} and P_{pSM} are the packet error probability of the cooperative BF and SM systems respectively which can be obtained from (Ahmad et al., 2008a).

5. Packet Latency Performance Model

As we noted, each packet transmission in cooperative transmission requires more steps which introduces more overhead. These steps may increase packet delays. However, the reduction of PER as the diversity gain increases from the cooperative MIMO exploitation can reduce the retransmissions rates which in turn can reduce packet latency. Previous work in (Ahmad et al., 2008a) models packet latency performance for the non-cooperative SISO system. Comparison is then made with the models developed for the cooperative MIMO systems as shown in (Ahmad et al., 2008b). In addition to the delay incurred as calculated and analyzed in (Ahmad et al., 2008a & 2008b) for $CMAC_{ON}$ with both BF and SM cooperative schemes, the cyclic RTS-CTS transmission scheme periods which are calculated in Equation (1) are included. Also, the IFS periods for both RTS and CTS packet transmissions as calculated in Equation (2) are included.

6. Performance Analysis and Discussion

All the important parameters for energy consumption modelling are listed in (Mainwaring et al., 2002; Cui et al., 2004; Polastre et al., 2004; Yang et al., 2007) with the times taken to transmit and receive 1 bit, T_{rx_b} and T_{tx_b} fixed at 4 μs corresponding to the bit rate of the system. The values of the system parameters used in Figures 14 and 15 for latency analysis are as follows: T_{rts} = 0.52 ms, T_{cts} = 0.44 ms, T_{ack} = 0.432 ms, T_{Bs} = 4.528 ms, T_{Br} = 0.432 ms, T_{data} = 4.096 ms, T_{col} = 32.8 ms, and T_{wfack} for SM scheme = 70 ms (Yang et al., 2007) and T_{wfack} for BF scheme = 0.864 ms (Kohvakka et al., 2006).

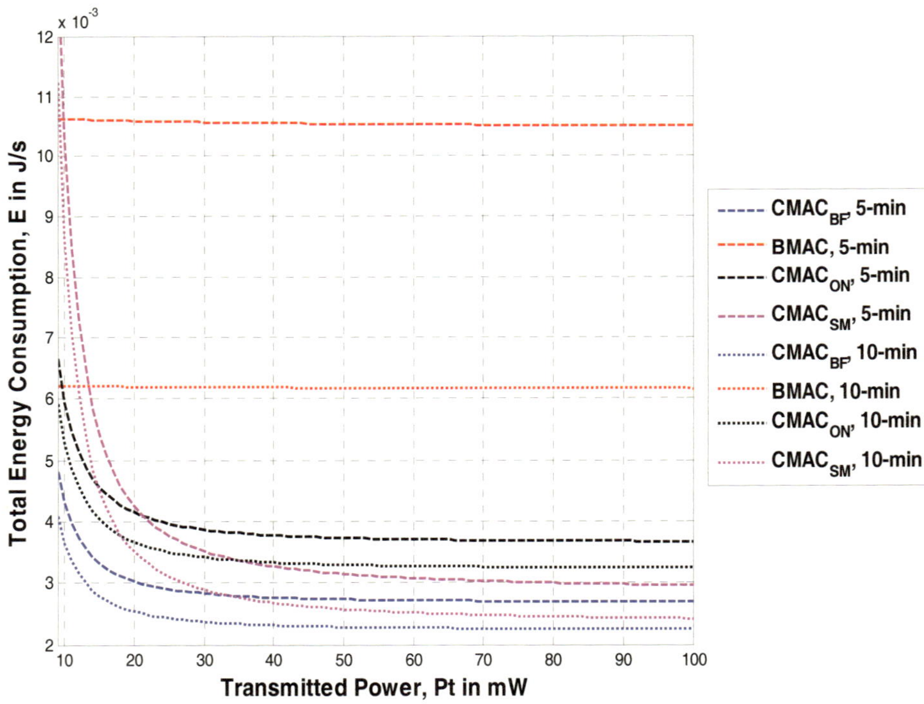

Fig. 5. Total energy consumption vs. transmission power of various MAC protocols with M = 2 and N = 1 (Cooperative BF) and M = N = 2 (Cooperative SM) for 5-min and 10-min sample periods

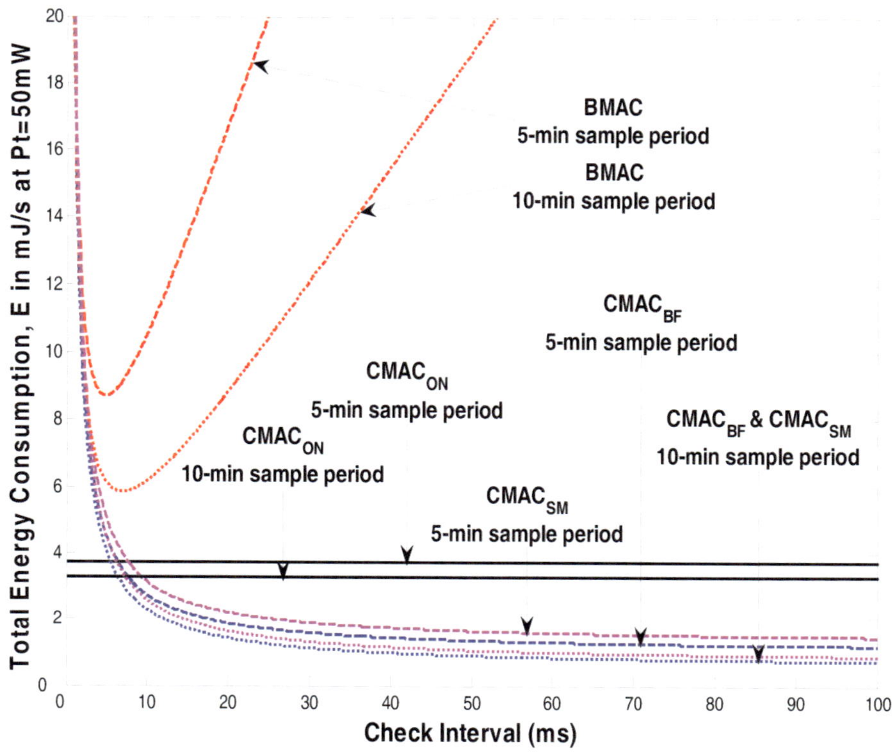

Fig. 6. Total energy consumption vs. check interval of various MAC protocols with $M = 2$ and $N = 1$ (Cooperative BF) and $M = N = 2$ (Cooperative SM) for 5-min and 10-min sample periods

We can see in Figure 5 that both CMAC and CMAC$_{ON}$ outperform B-MAC and that the CMAC$_{BF}$ is more energy efficient than CMAC$_{SM}$ with two transmitting nodes for all the sampling periods. If we let the sampling period be long enough, the performance difference between CMAC and B-MAC should be reduced at the same check interval. Thus, we can deduce that CMAC is more energy efficient than B-MAC at shorter sampling periods which makes CMAC more practical for applications with frequent sampling periods.

As shown in Figure 6, B-MAC has the optimal check interval at 5 ms for the 5 minutes sampling period. We can expect that the optimal check interval gets higher when the sampling period gets higher. As measured at 10 minutes sampling period, the optimal check interval is 7 ms with 2 ms increase. The same observation is applied for CMAC as shown in Figure 7. Furthermore from Figure 6, we can observe that below 3 ms, both B-MAC and CMAC suffer higher transient energy which puts the lower bound or lower constraint on the operating check interval. Clearly, above 7 ms, CMAC outperforms both CMAC$_{ON}$ and B-MAC. B-MAC may suffer from higher transmission power due to a longer preamble packet as the check interval gets higher. Interestingly, CMAC$_{SM}$ has the same optimal check interval with CMAC$_{BF}$ for various sampling periods as shown in Figure 7.

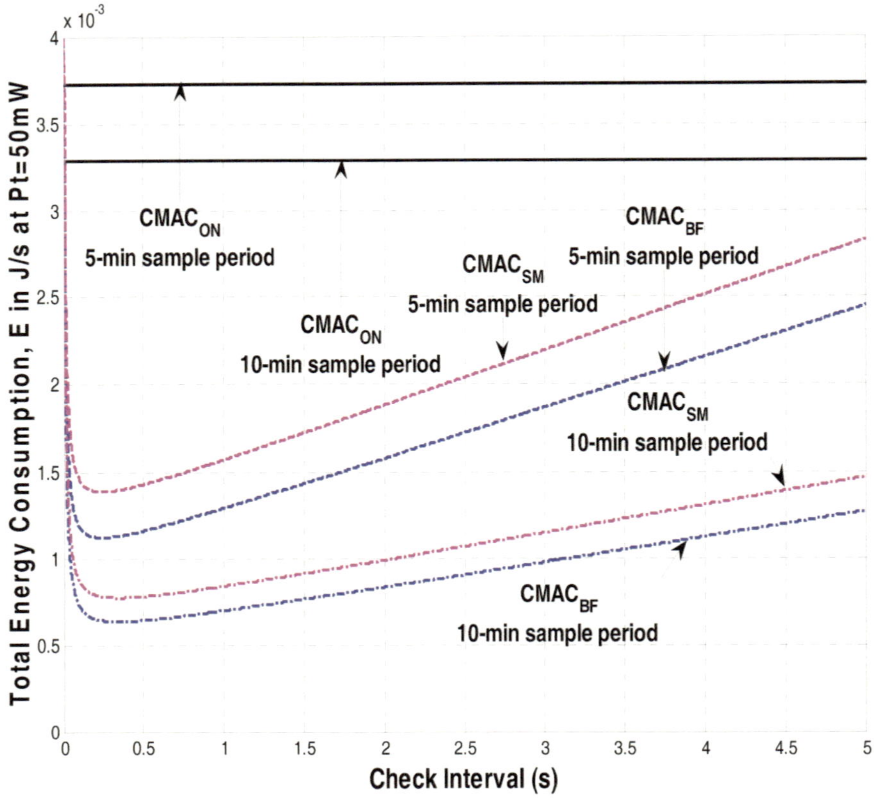

Fig. 7. Total energy consumption vs. check interval of CMAC protocols when $M = 2$ and $N = 1$ (Cooperative BF) and $M = N = 2$ (Cooperative SM) for 5-min and 10-min sample periods

Figure 8 shows the impact of M on the energy consumption of CMAC and $CMAC_{ON}$. We can observe that the increase of energy consumption is small as M increases even when we increase M from 2 to 10 nodes. As long as the nodes are operating within an optimal range during cooperative communication (Nguyen et al., 2007), the small circuit energy can be tolerated in a cooperative low duty cycle MIMO system. The impact of N is shown in Figure 9. As we observed earlier, increasing M does not have a significant impact on the total energy consumption for both schemes. Interestingly, we also observe that N does not have a significant impact on the total energy consumption. Therefore, as long as we can tolerate a little increase of circuit energy by increasing the number of M and N, then we can choose to use either the BF or SM scheme in a cooperative low duty cycle MIMO system. However, the optimal choice is still to use $CMAC_{BF}$ and to set $M = 2$ and this result agrees with the previous results in (Ahmad et al., 2008a). On the other hand, when we consider high-speed WSNs, obviously $CMAC_{SM}$ is the optimal choice.

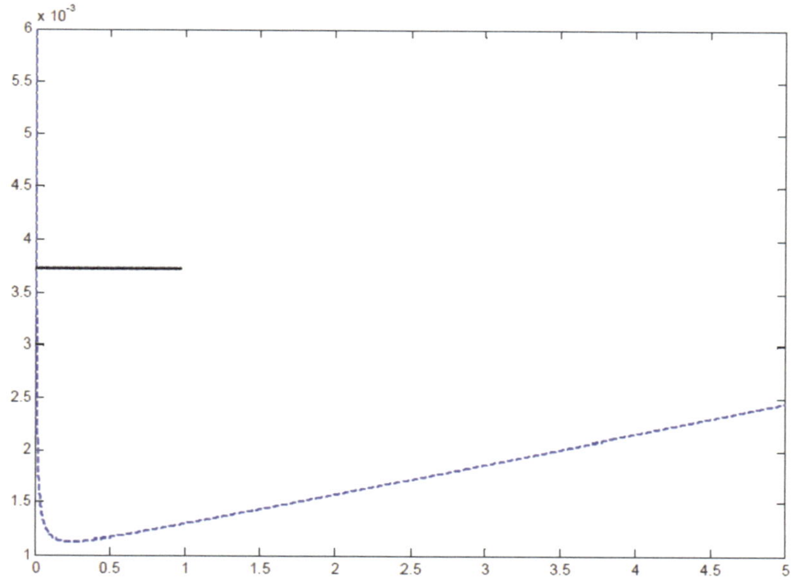

Fig. 8. Total energy consumption vs. check interval of CMAC protocols for various M with $N = 1$ (Cooperative BF) and $N = 2$ (Cooperative SM)

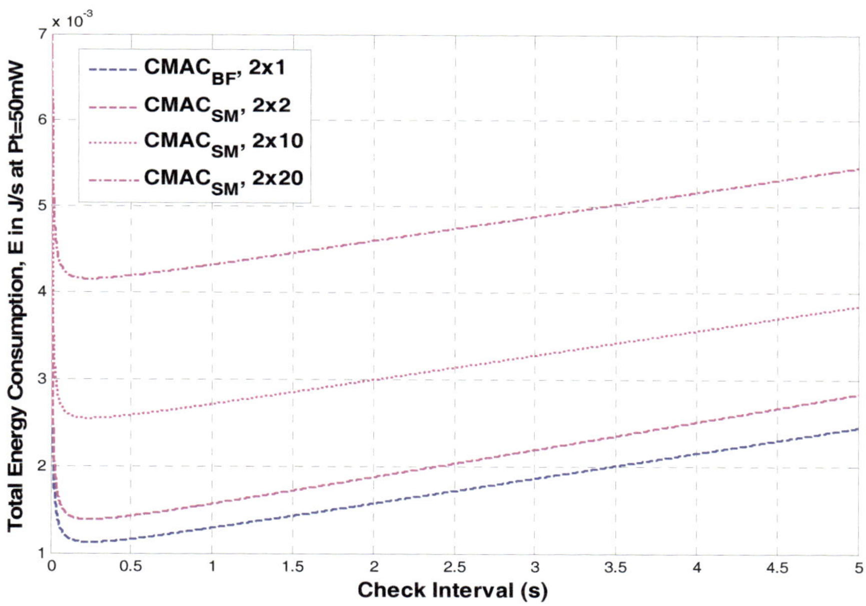

Fig. 9. Total energy consumption vs. check interval of CMAC protocols for various N (Cooperative SM) with fixed $M = 2$ for all cooperative schemes

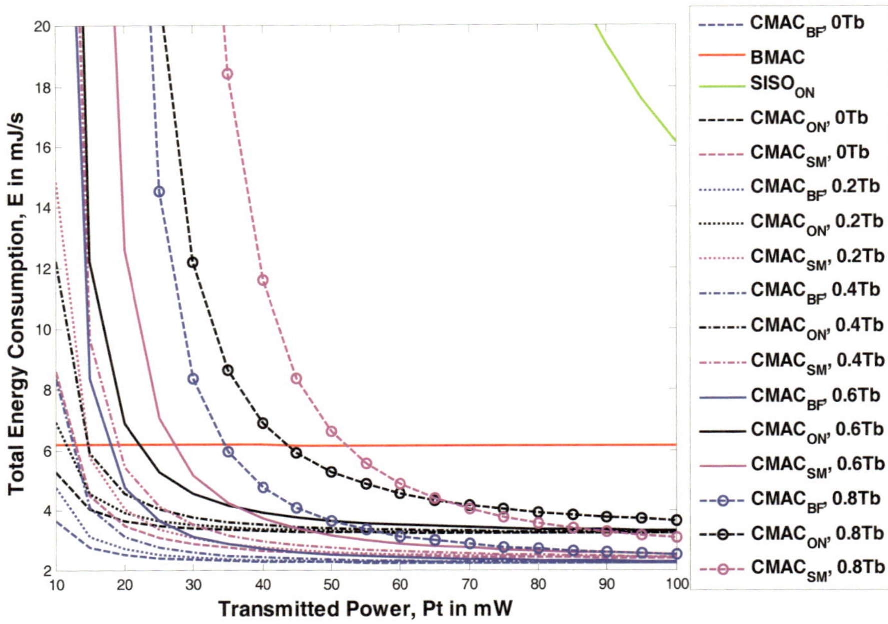

Fig. 10. Total energy consumption vs. transmission power for various imperfect synchronization cooperative schemes with $M = 2$ and $N = 1$ (Cooperative BF) and $M = N = 2$ (Cooperative SM)

Figure 10 shows that $CMAC_{BF}$ outperforms the other schemes below $0.8T_b$ at common transmission power above 40mW. Figure 11 shows the $CMAC_{SM}$ suffers the timing error effect at above $0.9T_b$ where $SISO_{ON}$ outperforms $CMAC_{BF}$. Also we observe that B-MAC outperforms both $CMAC_{BF}$ and $CMAC_{ON}$ utilizing the BF scheme with $0.9T_b$ at a lower check interval below 200ms. A closer look at all the cooperative MAC schemes is shown in Figure 12 where the jitter difference is varied from $0T_b$ to $0.8T_b$. $CMAC_{BF}$ experiences 1.3mJ/s increases between $0T_b$ and $0.8T_b$. The increase is still small when we compare it to $CMAC_{SM}$ and $CMAC_{ON}$ utilising the BF scheme with 4.6mJ/s and 3.5mJ/s increases, respectively.

The impact of the number of cooperative receiving nodes, N, in the cooperative SM scheme is shown in Figure 13. We can reduce the energy cost from 4.6mJ/s increase to 0.2mJ/s increase when $N = 6$. As N gets higher, the circuit energy gets higher and thus the total energy consumption also gets higher. However, we can tolerate the small circuit energy at higher jitter differences as shown since $CMAC_{ON}$ utilising the BF scheme with $N = 20$ at $0.8T_b$ has lower energy than $CMAC_{ON}$ utilizing the BF scheme with $N = 2$ at $0.8T_b$. From all the observations, we suggest that $CMAC_{BF}$ is the optimal choice below $0.9T_b$ jitter difference. As shown in Figures 14 and 15, B-MAC enjoys lower packet latency and outperforms the other schemes even when the diversity gain of the cooperative SM scheme is increased. $CMAC_{ON}$ utilising the BF scheme outperforms B-MAC when the transmission power is higher than 50mW below $0.4T_b$. $CMAC_{BF}$ with $0T_b$ suffers a slightly higher delay compared to B-MAC when the transmission power is 50mW. In order to maintain lower latency, as low as 50 ms, $CMAC_{BF}$ must operate below $0.6T_b$ jitter difference.

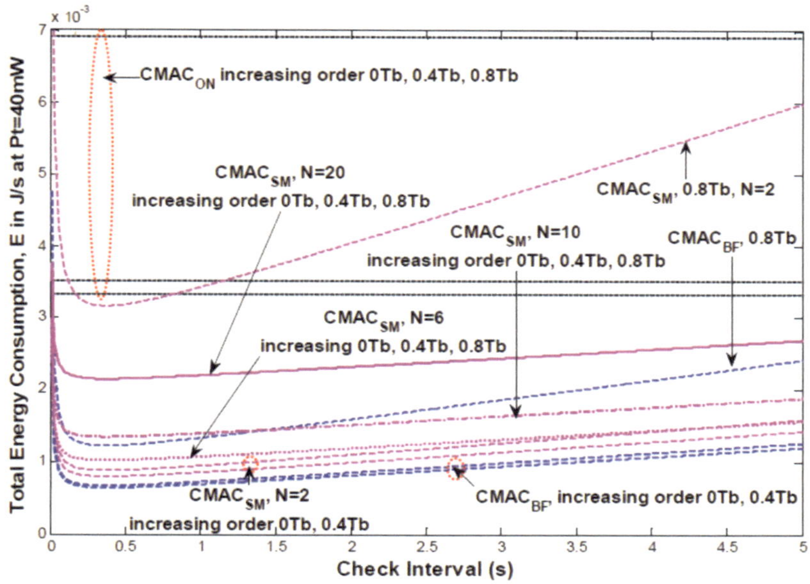

Fig. 11. Total energy consumption vs. check interval for various imperfect synchronisation cooperative schemes with M = 2 and N = 1 (Cooperative BF) and M = N = 2 (Cooperative SM) at clock jitter = $0.9T_b$

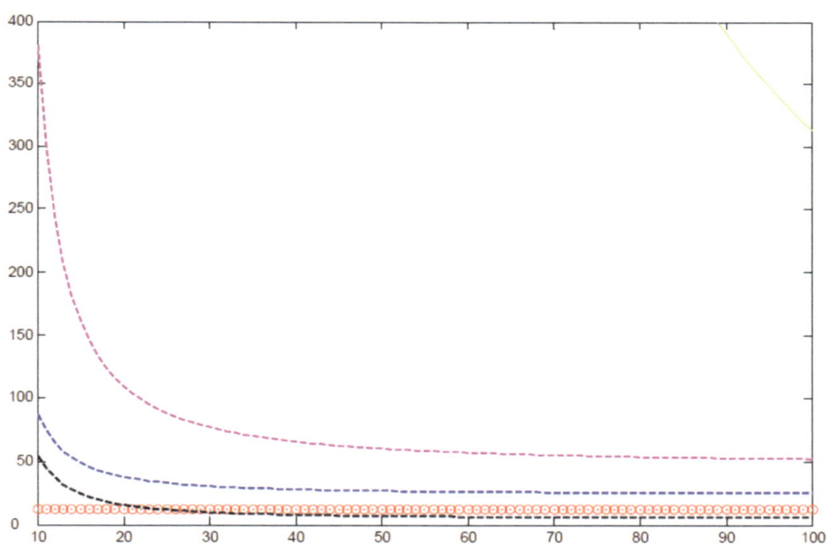

Fig. 12. Total energy consumption vs. check interval for various imperfect synchronisation schemes with M = 2 and N = 1 (Cooperative BF) and M = N = 2 (Cooperative SM) with clock jitter ≤ $0.8T_b$

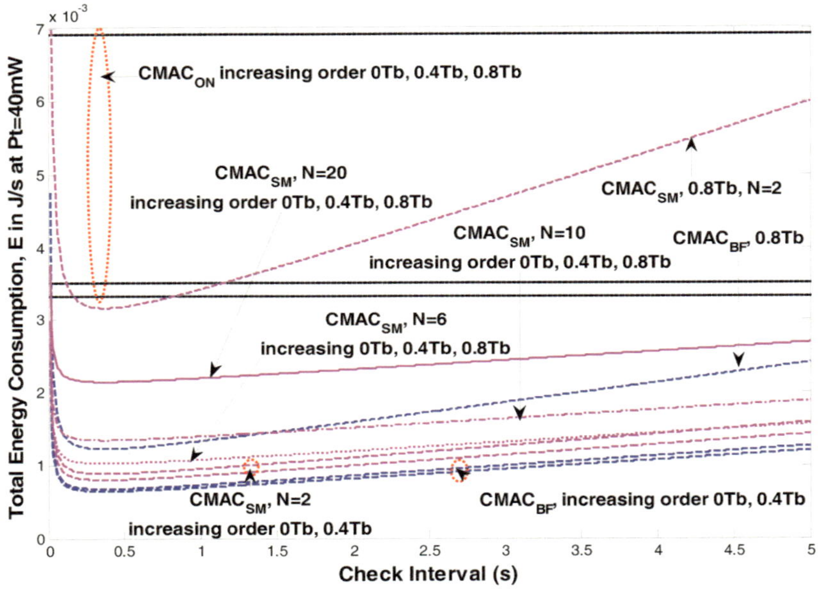

Fig. 13. Total energy consumption vs. check interval for various imperfect synchronization schemes with $M = 2$ and $N = 1$ (Cooperative BF) and with $M = 2$ and various $N = 2, 6, 10,$ and 20 (Cooperative SM) with clock jitter $\leq 0.8T_b$

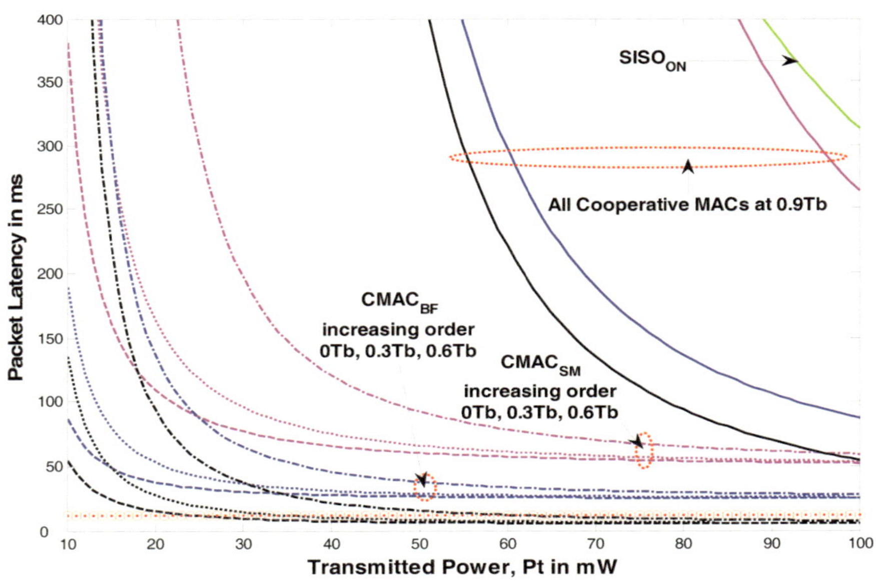

Fig. 14. Packet latency vs. transmission power of various imperfect synchronization schemes with $M = 2$ and $N = 1$ (Cooperative BF) and $M = N = 2$ (Cooperative SM) for $0T_b$, $0.3T_b$, $0.6T_b$ and $0.9T_b$

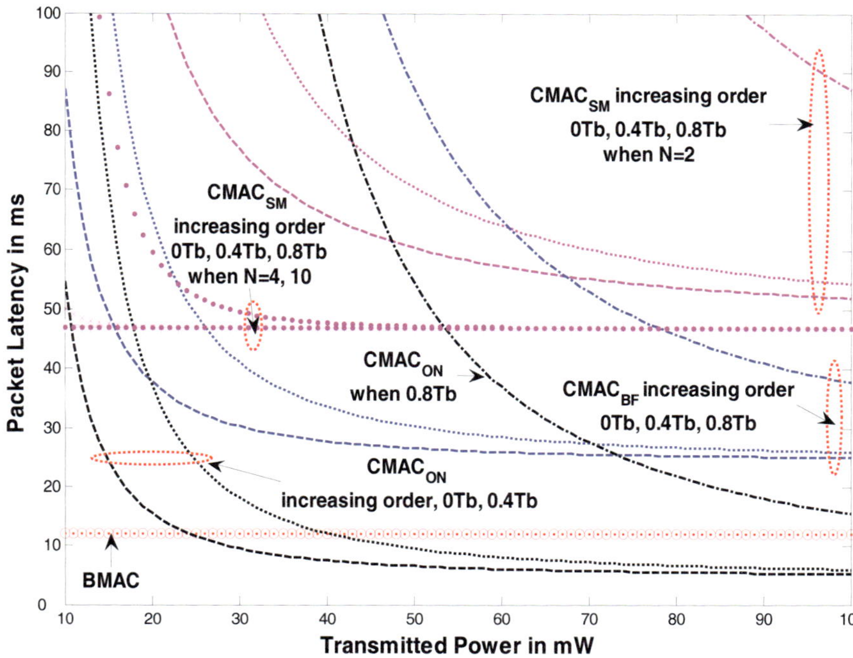

Fig. 15. Packet latency vs. transmission power of various imperfect synchronization schemes with M = 2 and N = 1 (Cooperative BF) and with M = 2 and various N = 2, 4, and 10 (Cooperative SM) for $0T_b$, $0.4T_b$ and $0.8T_b$

7. Conclusions

In order to address the idle listening and overhearing problems in a system with the CMAC$_{ON}$ protocol, we have proposed a new Cooperative low duty cycle MAC protocol (CMAC) for two cooperative MIMO schemes: optimal Beamforming (CMAC$_{BF}$) and Spatial Multiplexing (CMAC$_{SM}$). We have developed analytical models to evaluate total energy consumption and packet latency for both schemes. We have considered both synchronous and asynchronous scenarios. We have taken into consideration all the related energy costs (transmission, reception, idle listening, establishing cooperative mechanism, sleep, etc.) in the system performance modeling. We have applied the models for periodic monitoring applications.

We conclude that the new cooperative low duty cycle MAC with the optimal Beamforming scheme (CMAC$_{BF}$) outperforms the other cooperative and SISO schemes in terms of total energy consumption with the number of cooperating nodes set to M = 2. In order to achieve both lower energy and lower latency, CMAC$_{BF}$ must operate at M = 2 and with the clock jitter difference below $0.6Tb$. These results can be used to assist with the design of CMAC for multi-hop communication. Moreover, the trade-off relationship between energy efficient operation and latency can be utilized to find the optimal number of hops and the optimal number of cooperating nodes that should be involved in the transmission.

8. References

Ahmad, M.R.; Dutkiewicz, E. & Huang, X. (2008a). Performance Analysis of Cooperative MIMO Transmission Schemes in WSN, *Proceedings of IEEE International Symposium on Personal, Indoor and Mobile Radio Communications (PIMRC)*, Cannes, France, 15-18 September 2008.

Ahmad, M.R.; Dutkiewicz, E. & Huang, X. (2008b). Performance Evaluation of MAC Protocols for Cooperative MIMO Transmissions in Sensor Networks, *Proceedings of ACM International Workshop on Performance Evaluation of Wireless Ad-hoc, Sensor, and Ubiquitous Networks,* Vancouver, Canada, 2008.

Buettner, M.; Yee, G.; Anderson, E. & Han, R. (2006). X-MAC: A Short Preamble MAC Protocol for Duty-Cycled Wireless Sensor Networks, *Proceedings of ACM Conference on Embedded Networked Sensor Systems (SENSYS)*, Baltimore, Maryland, USA, 2006.

Cui, S.; Goldsmith, A.J. & Bahai, A. (2004). Energy-efficiency of MIMO and Cooperative MIMO Techniques in Sensor Networks. *IEEE Journal on Selected Areas in Communications,* Vol. 22, No. 6, pp. 1089-1098.

Jagannathan, S.; Aghajan, H. & Goldsmith, A.J. (2004). The Effect of Time Synchronization Errors on the Performance of Cooperative MISO Systems, *Proceedings of IEEE Global Telecommunications Conference (GLOBECOM)*, Dallas, Texas, USA, 2004.

Kohvakka, M.; Kuorilehto, M.; Hannikainen, M. & Hamalainen, T.D. (2006). Performance Analysis of IEEE 802.15.4 and Zigbee for large-Scale Wireless Sensor Network Applications, *Proceedings of ACM International Workshop on Performance Evaluation of Wireless Ad-hoc, Sensor, and Ubiquitous Networks,* Malaga, Spain, 2006.

Kuorilehto, M.; Kohvakka, M.; Suhonen, J.; Hamalainen, P.; Hannikainen, M. & Hamalainen, T.D. (2007). MAC Protocols, In: *Ultra-Low Energy Wireless Sensor Networks in Practice,* pp. 73-88, John Wiley, 978-0-470-05786-5, West Sussex, England.

Mainwaring, A.; Polastre, J.; Szewczyk, R.; Culler, D. & Anderson, J. (2002). Wireless Sensor Networks for Habitat Monitoring, *Proceedings of ACM International Workshop on Wireless Sensor Networks and Applications*, 2002.

Nguyen, T.-D.; Berder, O. & Sentieys, O. (2007). Cooperative MIMO Schemes Optimal Selection for Wireless Sensor Networks, *Proceedings of IEEE Vehicular Technology Conference (VTC)*, Baltimore, Maryland, USA, 2007.

Polastre, J.; Hill, J. & Culler, D. (2004). Versatile Low Power Media Access for Wireless Sensor Networks, *Proceedings of ACM Conference on Embedded Networked Sensor Systems (SENSYS)*, pp. 95-107, Baltimore, Maryland, USA, November 2004.

Wong, K.J. & Arvind, D. (2006). Low Power Decentralized MAC Protocols for Low Data Rate Transmissions in Specknets, *Proceedings of ACM International Workshop on Multihop Ad-hoc Networks: From Theory to Reality*, Florence, Italy, 2006.

Yang, H.; Shen, H.-Y. & Sikdar, B. (2007). A MAC Protocol for Cooperative MIMO Transmissions in Sensor Networks, *Proceedings of IEEE Global Telecommunications Conference (GLOBECOM)*, pp. 636-640, ISBN 1-4244-1042-8, Washington DC, USA, 26-30 November 2007.

6

Energy Efficient and Secured Cluster Based Routing Protocol for Wireless Sensor Networks

Dananjayan P, Samundiswary P and Vidhya J
Pondicherry Engineering College
Pondicherry,
India

1. Introduction

Recent advances in wireless and ubiquitous computing have prompted much research attention in the area of wireless sensor network (WSN). Sensor network consists of hundreds to thousands of low power multifunctioning sensor nodes operating in hostile environment with limited computational and sensing capabilities. They represent a new paradigm to support a wide variety of data gathering applications such as military, environmental monitoring and other fields ranging from traffic management to high secured monitoring of physical phenomenon (Akyildiz et al., 2002, Kazem et al., 2007). The main task of sensor nodes is to sense and collect data from a target domain, process the data and route the information to the specific sites where the underlying application resides. To achieve these potential, WSNs require novel routing techniques that take into consideration the immense scalability and inaccessibility of sensor devices with limited resources deployed in a harsh environment (Ilyas & Mahgoub, 2005). Moreover, sensor devices are subjected to severe fading, interference and susceptible to various attacks when operated in wireless medium. These constraints present unique design challenges. One of the challenges considered in the chapter is interconnecting a large group of sensors in a viable and secure network. This involves the need of designing a routing protocol which prolongs the network lifetime.

Routing of sensed data from sensor nodes to base station in a wireless sensor network occurs in different methods (Karki & Kamal, 2004). The classical approaches like direct transmission (DT) and multihop routing do not guarantee well balanced distribution of energy among the sensor nodes and are vulnerable to severe attacks. Using DT, each sensor directly sends the gathered information to remote receiver (sink) independent of each other. This approach has an inherent scalability problem and is prone to channel fading. With multihop routing, data is routed over minimum cost routes, and the nodes near the sink tend to die faster (Heinzelman et al., 2000). It can be easily compromised by attackers.

Clustering is the most promising technique that can significantly save the energy of sensor nodes and improve the scalability of the network. In clustering approach, sensors group together to form clusters. One of the sensors in each of the cluster will be elected as cluster head. The elected cluster head will be responsible for relaying data from each sensor in the cluster to the remote receiver. In addition, data fusion and data compression can occur in the cluster head by considering the potential correlation among data from neighbouring sensors (Do hyun mam & Hong-Ki-Min, 2007, Muruganathan et al., 2005).

Clustered sensor networks can be classified into two broad types: homogenous and heterogeneous sensor networks (Vivek & Catherine, 2004). In homogeneous sensor network, all the sensor nodes are identical in terms of energy and hardware complexity. This type of network consists of purely static clustering (cluster heads once elected, serve for the entire lifetime of the network) and the head node can be easily compromised. It is evident that the cluster head nodes will be over-loaded with the long range transmissions to the remote base station. And also, the extra processing is necessary for the cluster head for data aggregation and protocol co-ordination. As a result, the cluster head nodes expire before other nodes. It is desirable to ensure that all the nodes run out of their battery at about the same time.

One way to ensure this is to rotate the role of a cluster head randomly and periodically over all the nodes as proposed in low energy adaptive clustering hierarchy (LEACH) (Heinzelman et al., 2002, Yu et al., 2007). Since, all the nodes should be capable of acting as cluster heads; the network should possess the necessary hardware capabilities. Hence, the homogenous network requires high hardware cost. It also suffers from poor performance and scalability. To improve the network performance, heterogeneous sensor network (HSN) is formed by deploying a small number of high-end sensors (H-sensors) in addition to a large number of low-end sensors (L-sensors). Compared to an L-sensor, an H-sensor has better computation capability, larger storage and better reliability. However, the performance of HSN will be degraded when sensor nodes are distributed in an insecure and wireless environment. Hence this chapter considers two routing protocols to forward the data packets from source to remote receiver using the cluster based heterogeneous sensor network to overcome fading and defend against attacks such as selective forwarding and sinkhole attacks.

To reduce the fading effects in wireless channel, muti-input muti-output (MIMO) scheme is implemented for sensor network to save energy consumption at sensor nodes (Cui et al., 2004, Bravos & Efthymoglou et al., 2007). Applying multiple antenna technique directly to sensor network is impractical because, the limited size of sensor node usually supports a single antenna. If cooperative transmission and reception from antennas in a group of sensor nodes is used, an equivalent MIMO system for WSN can be realised. Normally, a MIMO system needs to estimate all channels between source and destination. If cooperative transmissions from multiple sensor nodes are allowed, the amount of channel estimation at the receiver can be reduced and hence can save the energy of sensor nodes (Cheng et al., 2006, Jayaweera, 2004).

Li et al., 2005 analysed cooperative multi input single output (MISO) transmission scheme based on LEACH protocol. However cooperative MISO system performs only single hop transmission and does not prolong the network lifetime. To overcome these drawbacks, the proposed model modifies the LEACH routing scheme using HSN architecture and suggests two solutions such as cooperative LEACH (C-LEACH) and cluster head cooperative LEACH (CH-C-LEACH) scheme to maximise the network lifetime. The proposed C-LEACH scheme allows cluster heads to form a multihop backbone and incorporates cooperative MIMO on each single hop transmission by utilising a set of sending and receiving cooperative nodes in each cluster. In CH-C-LEACH scheme, the cluster heads gets paired with other cluster head. They intelligently exchange data and balance communication load and transmit data cooperatively to the base station. To enhance the performance of the proposed routing scheme, cooperative MIMO utilises space time block code (STBC) to

provide significant diversity gain (Tarokh et al., 1999). For the proposed cooperative heterogeneous MIMO routing scheme, the energy consumed and the number of nodes alive for each round of data transmission is evaluated to reduce the channel fading effects and is compared with the traditional LEACH protocol.
Moreover, the lifetime of the network can be enhanced by providing security and privacy against network layer attacks when the nodes are scattered in an unsupervised environment. In order to protect network, few of the routing protocols such as sensor protocols for information via negotiation(SPINS) (Adrian Perrig et al., 2001) and path redundancy based security algorithm (PRSA) for homogeneous based wireless sensor networks (Sami et al., 2007) address the security mechanism and authentication against the various attacks.
Some of the secured routing protocols of heterogeneous sensor networks (Xiaojiang et al., 2006) can detect the malicious nodes and deliver the packets to the sink successfully. But these routing protocols increase the buffering requirements, overheads and delay. Hence, PRSA is extended for heterogeneous sensor networks by including alternate path mechanism to protect the nodes from selective forwarding (Jeremy & Xiaojiang, 2008) and sinkhole attacks in HSN. For the proposed secured routing mechanism, the network performance in terms of energy, delay and delivery ratio in the presence of compromised nodes is evaluated and compared with the heterogeneous network model.
The chapter is organised as follows: section 2 describes the heterogeneous sensor network model. The proposed cluster based cooperative MIMO routing protocols such as C-LEACH and CH-C-LEACH are discussed to minimise the channel fading effects in section 3. The energy consumption model of proposed scheme is analysed in section 4. Simulation results of the cooperative MIMO scheme in terms of energy and delay are discussed to minimize the channel fading in section 5. The various network layer attacks that exist in the sensor network are outlined in section 6. To defend against these attacks section 7 describes the proposed secured path redundancy algorithm in heterogeneous sensor network. Simulation results of the proposed algorithm are discussed in section 8 in terms of energy consumption, delay and delivery ratio and conclusion are drawn in section 9.

2. Heterogeneous sensor network

In the HSN model, H-sensors and L-sensors are randomly distributed in the field and clusters are formed. The cluster formation is shown in Fig.1, where L-sensors are the small squares, H-sensors are large rectangles, and the large square at the top-right corner is the base station (BS). H-sensors serve as cluster heads. The H-sensors have more energy resource, longer transmission range and can handle higher data rate than L-sensors. All the H-sensors form a backbone in the network. H-sensors use multi-hop communications to reach the BS. L-sensors can use single-hop or multi-hop communications to reach H-sensors (Xiaojiang et al., 2006, 2007).

2.1 Routing in heterogeneous sensor network

The primary functionality of a wireless sensor network is to sense the environment and transmit the acquired information to the BS for further processing. Since sensor nodes are small and unreliable devices, they are prone to failures. The routing protocol designed for the network has to be robust to sensor failures by providing new paths. The basic idea of routing in HSN is to let each non-cluster head (L-sensor) to send data to its cluster head (an H-sensor).

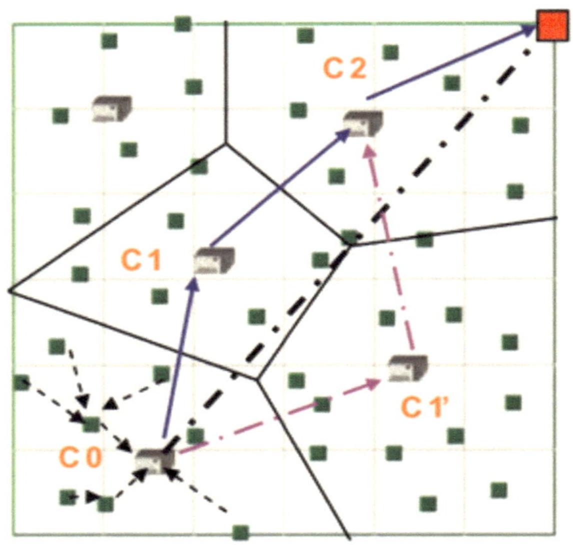

Fig. 1. Heterogeneous sensor network model

i. *Intra- cluster routing*

Routing within a cluster (from an L-sensor to its cluster head) is referred to as intra-cluster routing which is illustrated in Fig.1. L-sensor sends its location information to the cluster head during the cluster formation. The location of H is broadcasted to all L-sensors in the cluster. All the L-sensors in a cluster form a tree, rooted at the cluster head (denoted as H) so that each L-sensor sends packets to its H-sensor, when it generates packets. If data from nearby L-sensor nodes are highly correlated, then a minimum spanning tree (MST) can be adopted to approximate the least energy consumption case.

A centralised algorithm created by H-sensor can be used to construct an MST. Then H disseminates the MST structure information to L-sensors, i.e., informing each L-sensor which node its parent is. If a data fusion is conducted at intermediate L-sensors nodes, then MST consumes the least total energy in the cluster. If there is few or no data fusion among L-sensors in a cluster, a shortest-path tree (SPT) should be used to approximate the least total energy consumption.

Similarly, the cluster head (H-sensor) can construct an SPT by using a centralised algorithm and the locations of L-sensors (Xiaojiang et al., 2006, 2007). In the above route setup, each L-sensor may record two or more parent nodes. One parent node serves as the primary parent, and other parent nodes serve as backup parent. If the primary parent node fails, an L-sensor can use a backup parent for data forwarding. Further each L-sensor records one or more backup cluster heads during cluster formation. When a cluster head fails, L-sensors in the cluster send their packets to a backup cluster head.

ii. *Inter-cluster routing*

Routing across clusters (from an H-sensor to the BS) is referred to as inter-cluster routing which is shown in Fig.1. After receiving data from L-sensors, cluster heads may perform data aggregation via the H-sensor backbone. Each cluster head exchanges location information with neighbor cluster heads. During route discovery, a cluster head draws a straight line L between itself and the BS, based on the location of the BS and itself which is shown in Fig.1. Line L intersects with a serial of clusters, and these clusters are denoted as $C_0, C_1, \ldots, C_k$,which are referred to as relay cells.

The packet is forwarded from the source cluster head to the BS via cluster heads in the relay cells. H-sensors are more reliable nodes than L-sensors. However, an H-sensor may also fail because of various reasons, such as harsh environment, or may be destroyed by an adversary. If any cluster head in the relay cells is unavailable, then a backup path is used. A backup path is set up as follows: The current cluster head (say R1) draws a straight line L′ between itself and the BS, and line L intersects with several cells $C'_1,...,C'_{k-1},C'_k$. If the next cell is the cell having the failed cluster head, R1 will use a detoured path to avoid the cell. The sequence cells $C'_1,.....,C'_{k-1},C'_k$ will be the new relay cell and are used to forward the packet to the BS.

3. Proposed cluster-based cooperative MIMO routing scheme

A heterogeneous cluster based sensor network model is considered as discussed in section 2. The base station for the network model is assumed to have no energy constraints and is equipped with one or more receiving antennas. The sensor nodes are geographically grouped into clusters consisting of H-sensors, L-sensors, cooperative sending and receiving nodes that sense the data from the sensing field. The H-sensors are reelected after each round of data transmission as in LEACH protocol (Xiangnin & Song Yulin, 2007, Vidhya & Dananjayan, 2009).

3.1 Cooperative heterogeneous MIMO LEACH scheme

The proposed multihop cooperative MIMO LEACH transmission model is illustrated in Fig.2. The transmission procedure of the proposed scheme is divided into multiple rounds. Each round has three phases:

i. *Cluster formation phase*

In this phase, clusters are organised and cooperative MIMO nodes (Yuan et al, 2006) are selected according to the steps described below:

a. *Cluster head advertisement*

Initially, when clusters are being created, each node decides whether or not to become a cluster head for each round as specified by the original LEACH protocol. Each self-selected cluster head, then broadcasts an advertisement (ADV) message using non-persistent carrier sense multiple access (CSMA) MAC protocol. The message contains header identifier (ID).

b. *Cluster set up*

Each non-cluster head node i.e L-sensor node chooses one of the strongest received signal strength (RSS) of the advertisement as its cluster head, and transmits a join-request (Join-REQ) message back to the chosen cluster head i.e H-sensor. The information about the node's capability of being a cooperative node, i.e., its current energy status is added into the message.

If H-sensor receives advertisement message from another H-sensor y, and if the received RSS exceeds a threshold, it will mark H-sensor y as the neighbouring H-sensor and it records y's ID. If the base station receives the advertisement message, it will find the cluster head with the maximum RSS, and sends the base station position message to that cluster head marking it as the target cluster head (TCH).

c. *Schedule creation*

After all the H-sensors have received the join-REQ message, each cluster head creates a time division multiple access(TDMA) schedule and broadcasts the schedule to its cluster members as in original LEACH protocol (Vidhya & Dananjayan, 2010). This prevents collision among data messages and allows the radio of each L-sensor node to be turned off until its allocated transmission time to save energy.

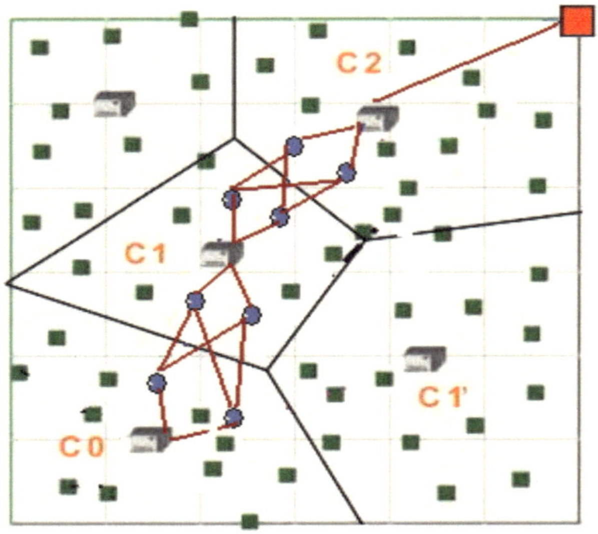

Fig. 2. C-LEACH transmission model

d. *Cooperative node selection*

After the cluster formation, each H-sensor will select J cooperative sending and receiving nodes for cooperative MIMO communication with each of its neighbouring cluster head. Nodes with higher energy close to the H-sensor will be elected as sending and receiving cooperative nodes for the cluster. At the end of the phase, the cluster head will broadcast a cooperative request (COOPERATE-REQ) message, to each cooperative node which contains the ID of the cluster itself, the ID of the neighbouring H-sensor y, the ID of the transmitting and receiving cooperative nodes and the index of cooperative nodes in the cooperative node set for each cluster head to each cooperative node. Each cooperative node on receiving the COOPERATE- REQ message, stores the cluster head ID, the required transmitted power and sends back a cooperate-acknowledgement (ACK) message to the H-Sensor.

ii. *Routing table construction*

Each H-sensor will maintain a routing table which contains the destination cluster ID, next hop cluster ID, IDs of cooperative sending and receiving nodes. Each cluster head will simply inform its neighbouring cluster heads of its routing table. After receiving route advertisements from neighbouring cluster heads, the cluster heads will update the route cost and advertise to their neighbouring cluster heads about the modified routes. Then the TCH will flood a target announcement message containing its ID to each H-sensor to enable transmission paths to the base station.

iii. *Data transmission phase*

In this phase, the L-sensors will transmit their data frames to the H-sensor as in LEACH protocol during their allocated time slot. Each cluster member will transmit its data as specified by TDMA schedule in cluster formation phase, and will sleep in other slots to save energy. The duration and the number of frames are same for all clusters and depend on the number of L-sensor nodes in the cluster. After a cluster head receives data frames from its cluster members as shown in Fig.2, it performs data aggregation to remove redundant data and broadcasts the data to J cooperative MIMO sending nodes. When each cooperative sending node receives the data packet, they encode the data using STBC (Tarokh et al.,1999) and transmit the data cooperatively. The receiving cooperative nodes use channel state information to decode the space time coded data. The cooperative node relays the decoded data to the neighbouring cluster head node and forwards the data packet to the TCH by multihop routing.

3.2 Cluster head cooperative heterogeneous MIMO LEACH scheme

To further prolong the network lifetime a CH-C-LEACH scheme is proposed and is illustrated in Fig.3. In this scheme the cluster head nodes cooperate and pair among themselves to transmit data cooperatively rather than selecting the cooperative sending and receiving groups in each cluster as specified in section 3.1. The transmission procedure of the proposed scheme split into different rounds and each round has four phases:

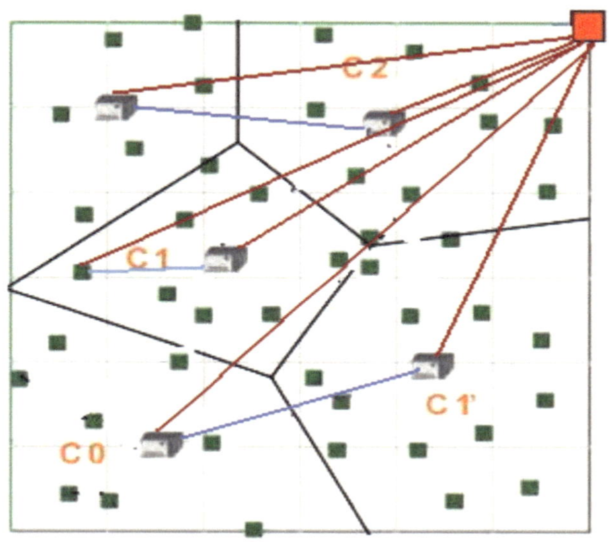

Fig. 3. CH-C-LEACH transmission model

i. *Cluster formation phase*

During this phase, clusters are organised following the same procedure of C-LEACH scheme as described in section 3.1.

ii. *Intra-cluster transmission and data aggregation*

In this phase, the L-sensor sends its packets to the H-sensor. The cluster head then performs data aggregation. At this point, each cluster head knows the volume of data it needs to transmit to the base station.

iii. *Data volume advertisement*

In this phase, the H-sensors inform each other about their data volume by broadcasting a short message that contains the node's ID and the volume of data it needs to transmit. All messages are recorded by each H-sensor. Besides, according to the received signal strength of the advertisement, each cluster head estimates the distances to all other cluster heads and records the information.

iv. *Data exchange and cooperative transmission*

In this phase each H-sensor gets paired with other H-sensor and transmits data cooperatively. The data transmission in CH-C-LEACH scheme is shown in Fig.4 and is described below:

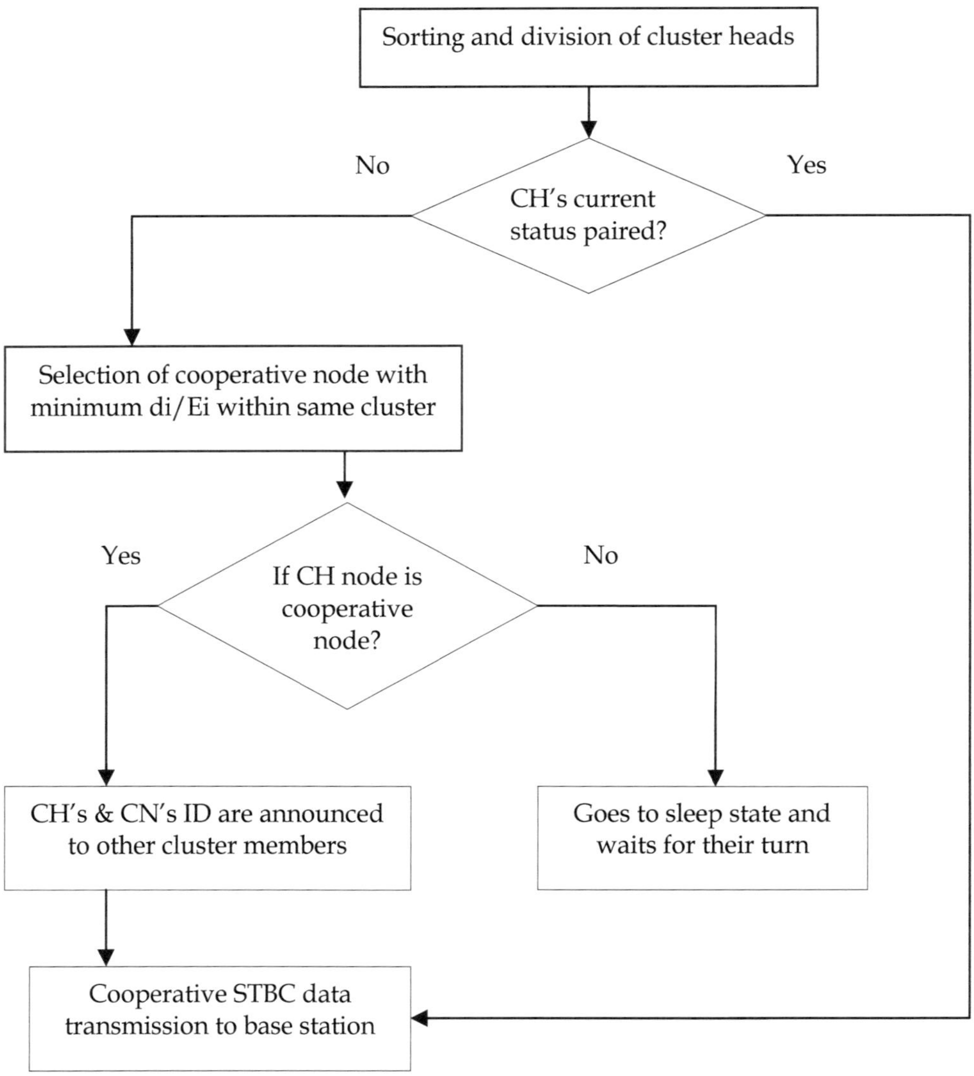

Fig. 4. Flow chart of data transmission in CH-C-LEACH scheme

a. *Sorting and division*

Based on the volume of data available at cluster head, each CH sorts the data and gets the reordered sequence for pairing to enable cooperative MIMO data transmission.

b. *Cooperative node selection and transmission*

If the number of H-sensors is odd, one of the H-sensor selects a cooperative node with minimal di/ Ei within its own cluster, where Ei is the energy status reported by node i and di is the distance between node i and the cluster head. This H-sensor informs the selected cooperative node by broadcasting a short message containing the cluster head's ID, the selected node's ID and an appropriate transmission time T that this pair needs to transmit data to base station. Upon receiving the message, all nodes except this pair of nodes can turn off their radio components to save energy. The cluster heads should wake up at time T, and other L–sensor nodes can remain in the sleep state till the next round. On the other hand, the

H-sensor node sends its data to the selected cooperative node, and they encode the transmission data according to STBC and transmit the data to the base station cooperatively. Once the transmission ends, these two nodes go into the sleep state till the next round.

4. Energy consumption model of the proposed scheme

The energy consumed during each round of data transmission using C-LEACH scheme results from the following sources such as: L-sensor transmitting their data to the H-sensor, routing table constructed by the H-sensor, cluster head transmitting the aggregated data to the cooperative nodes, cooperative node transmitting the data to the receiving cooperative nodes and to the receiving H-sensor. The energy consumed using CH-C-LEACH is due to cluster members transmitting their data to the H-sensor, cluster head transmitting the aggregated data to the cooperative cluster head and H-sensor nodes cooperate to transmit the data to the base station.

i. *Energy consumption of cluster member*

The energy consumed by the source nodes i.e L-sensor to transmit one bit data to the cluster head node for C-LEACH and CH-C-LEACH scheme is given by

$$E_{bs}(k_c) = -\frac{1}{\pi k_c}(1+\alpha)N_f\sigma^2 \ln(P_b)G_1M^2M_1 + \frac{P_{ct}+P_{cr}}{B} \tag{1}$$

where k_c is the number of clusters, α is the efficiency of radio frequency (RF) power amplifier, N_f is the receiver noise figure, $\sigma^2=N_o/2$ is the power density of additive white Gaussian noise (AWGN) channel, P_b is the bit error rate (BER) obtained while using phase shift keying, G_1 is the gain factor, M is the network diameter, M_1 is the gain margin, B is the bandwidth, P_{ct} is the circuit power consumption of the transmitter and P_{cr} is the circuit power consumption of the receiver.

The total number of bits transmitted to cluster head of each cluster in each round is given by

$$S_1(k_c) = \left\lceil \frac{N}{k_c} \right\rceil F_n Ps \tag{2}$$

where N is the number of sensor nodes, F_n is the number of symbols in a frame, P is the transmit probability of each node and s is the packet size.

The energy consumed by a cluster member to transmit data to the cluster head is given by

$$E_s(k_c) = k_c S_1(k_c) E_{bs}(k_c) \tag{3}$$

ii. *Energy consumption of cluster heads*

To construct routing table, the energy consumed by H-sensor node for C-LEACH scheme is

$$E_r(k_c) = k_c R_{ts} R_{bt}\left((1+\alpha)M_1N_f \frac{N_0(4\pi)^2(2M)^k}{P_bG_tG_r\lambda^2(\pi k_c)^{k_c/2}} + \frac{P_{ct}+4P_{cr}}{B} \right) \tag{4}$$

where R_{bt} is the time required for exchanging routing information, R_{ts} is the routing table size, k is the path loss factor, G_t is the gain of transmitting antenna, G_r is the gain of receiving antenna and λ is the wavelength of transmission.
The energy per bit consumed by the cluster head node to transmit the aggregated data to J cooperative nodes for C-LEACH and CH-C-LEACH scheme is given by

$$E_{bc0}(k_c, J) = -\frac{1}{\pi k_c}(1+\alpha)N_f\sigma^2 \ln(P_b)G_1 M^2 M_l + \frac{P_{ct} + JP_{cr}}{B} \tag{5}$$

The amount of data after aggregation for each round by H-sensor node is given by

$$S_2(k_c) = \frac{S_1(k_c)}{([N/k_c]Pagg - agg + 1)} \tag{6}$$

where agg is the aggregation factor.
The energy consumed by cluster head node to transmit the aggregated data to J cooperative nodes is given by

$$E_{c0}(k_c, J) = k_c S_2(k_c) E_{bc0}(k_c, J) \tag{7}$$

iii. *Energy consumption of cooperative nodes*
The transmitter cooperative nodes of the cluster will encode and transmit the sequence according to orthogonal STBC to the H-sensor node. Consider the block size of the STBC code is F symbols and in each block pJ training symbols are included and are transmitted in L symbol duration. The actual amount of data required to transmit the $S_2(k_c)$ bits is given by

$$S_e(k_c, J) = FS_2(k_c)/R(F - pJ) \tag{8}$$

where R is the transmission rate.
The energy consumed by J cooperative sending nodes to transmit MIMO data to the J cooperative receiving nodes for C-LEACH scheme is given by

$$E_{cs}(k_c, J) = S_e(k_c, J)\left((1+\alpha)M_l N_f \frac{JN_0 (4\pi)^2 (2M)^k}{P_b^{1/J} G_t G_r \lambda^2 (\pi k_c)^{k_c/2}} + \frac{JP_{ct} + JP_{cr}}{B} \right) \tag{9}$$

Similarly, the energy consumed by J receiving cooperative nodes/cluster head cooperative nodes to transmit data to the neighbouring cluster head/base station respectively for C-LEACH and CH-C-LEACH scheme is given by

$$E_{cr}(k_c, J) = S_e(k_c, J)\left((1+\alpha)M_l N_f \frac{JN_0 (4\pi)^2 (2M)^k}{P_b^{1/J} G_t G_r \lambda^2 (\pi k_c)^{k_c/2}} + \frac{JP_{ct} + P_{cr}}{B} \right) \tag{10}$$

iv. *Over all energy consumption for a round*
The energy consumption for each round of cooperative multihop MIMO data transmission for C-LEACH scheme can be obtained from Equations (3), (4), (7), (9) and (10) and it is given by

$$E(k_c, J) = E_s(k_c) + E_r(k_c) + n_k E_{c0}(k_c, J) + n_k E_{cs}(k_c, J) + n_k E_{cr}(k_c, J) \quad (11)$$

where n_k is the average number of hops.
The energy consumption for each round of data transmission for CH-C-LEACH scheme is given by

$$E(k_c, J) = E_s(k_c) + n_k E_{c0}(k_c, J) + n_k E_{cr}(k_c, J) \quad (12)$$

5. Simulation results

The analysis of the proposed cooperative heterogeneous MIMO schemes discussed in section 4 is carried out using MATLAB to evaluate the energy consumption and maximise the lifetime of the sensor network. A sensing field with a population of N= 100 nodes is considered for simulation with 80 normal nodes and 20 advanced nodes deployed over the region randomly. The initial energy of a normal node is set to 0.5 J and the energy of the advanced node is 2 J.

5.1 Energy consumption analysis

The performance of the proposed C-LEACH scheme is compared with that of the original LEACH scheme in terms of energy and is shown in Fig.5.

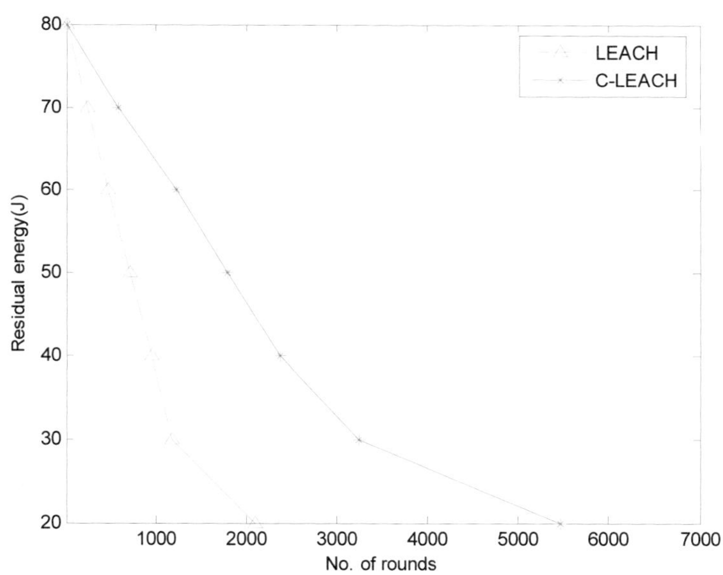

Fig. 5. Energy analysis of C-LEACH scheme

With the use of two cooperative nodes for data transmission, the energy consumption of the network is decreased. This is due to the diversity gain of the MIMO STBC encoded system. From the graph it is clear that the proposed scheme utilising two cooperative sending and receiving nodes can achieve twice the energy savings than LEACH protocol. Fig.6 illustrates the energy performance of proposed CH-C-LEACH scheme. When the cluster head nodes are paired and involved in MIMO data transmission the residual energy of the network for

1000 rounds is 30% more than the LEACH protocol. This is due to the diversity gain of MIMO system.

The performance comparison of proposed C-LEACH and CH-C-LEACH scheme is plotted in Fig.7. The proposed CH-C-LEACH scheme performs better than the proposed C-LEACH scheme by approximately 150 rounds. This is because C-LEACH contributes additional energy consumption in selection of cooperative nodes within a cluster during the cluster setup process.

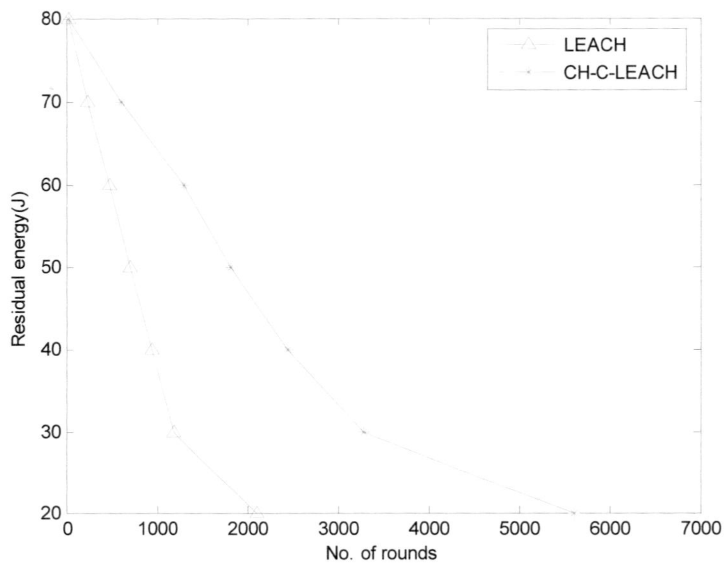

Fig. 6. Energy analysis of CH-C-LEACH scheme

5.2 Network lifetime

The number of nodes alive for each round of data transmission is observed for the proposed scheme to evaluate the lifetime of the network. Fig.8 shows the performance of the system for the LEACH and proposed C-LEACH scheme. It is observed that the proposed C-LEACH scheme outperforms LEACH scheme due to balanced energy dissipation of individual node through out the network.

Similar performance is observed for the proposed CH-C-LEACH scheme in Fig.9. The number of nodes alive after each round of data transmission is greater than LEACH scheme. It is vivid from the graph that 70% of nodes in the LEACH network die in 1250 rounds whereas the proposed CH-C-LEACH scheme prolongs the life time up to 4250 rounds. The performance comparison of proposed C-LEACH and CH-C-LEACH scheme is plotted in Fig.10. The proposed CH-C-LEACH scheme performs better than the proposed C-LEACH scheme by approximately 250 rounds. This is because, the larger energy consumption involved in the data transmission process for C-LEACH scheme reduces the number of alive nodes in the network.

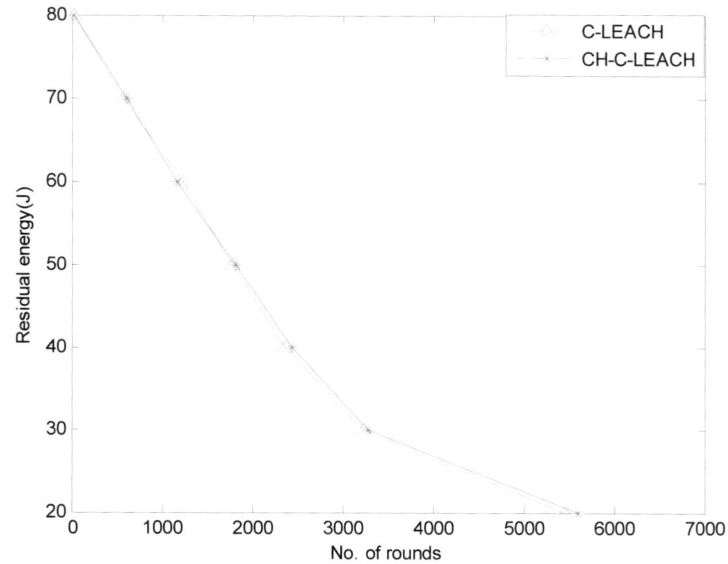

Fig. 7. Energy analysis comparison of C-LEACH and CH-C-LEACH scheme

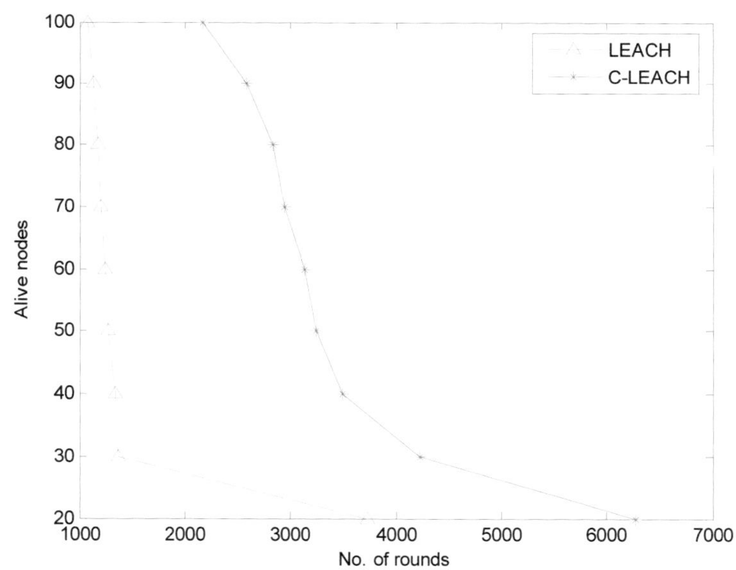

Fig. 8. Network lifetime of C-LEACH scheme

5.3 Percentage of Node death

The number of rounds for every 10% of node death is observed for LEACH and the proposed C-LEACH scheme in Fig.11. From the results it is evident that the lifetime of LEACH protocol is limited to 3750 rounds and the proposed MIMO scheme extents up to 6250 rounds. The proposed C-LEACH scheme provides an extended lifetime of approximately twice LEACH protocol. Similar performance can be observed with CH-C-LEACH scheme and is shown in Fig.12. The proposed CH-C-LEACH scheme has longer life time than LEACH scheme. Also, the proposed CH-C-LEACH scheme performs better than the proposed C-LEACH scheme by extending the lifetime of approximately 500 rounds as shown in Fig.13.

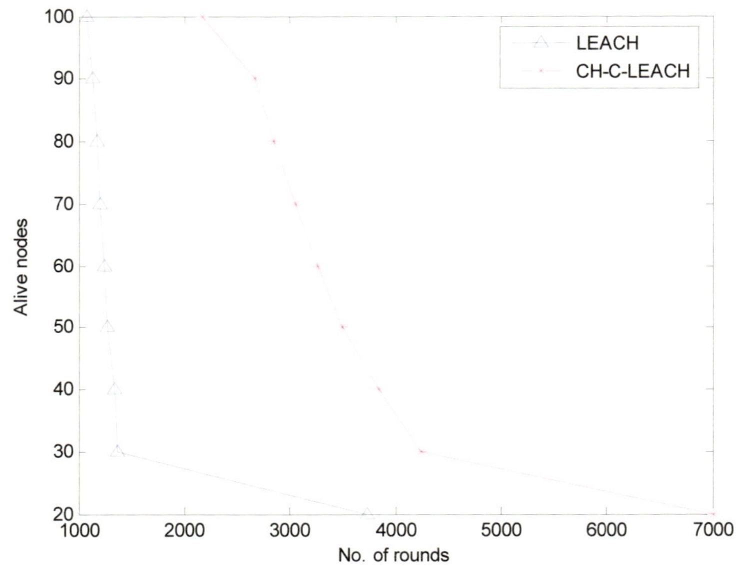

Fig. 9. Network lifetime of CH-C-LEACH scheme

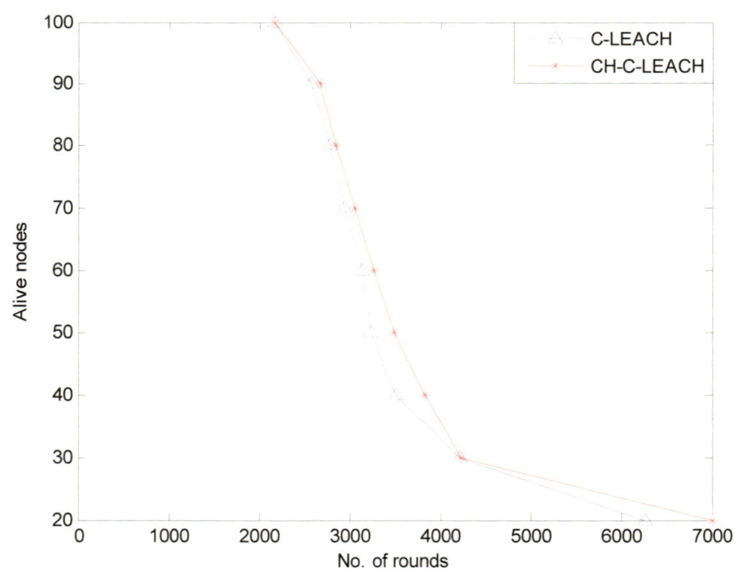

Fig. 10. Comparison of network lifetime for C-LEACH and CH-C-LEACH scheme

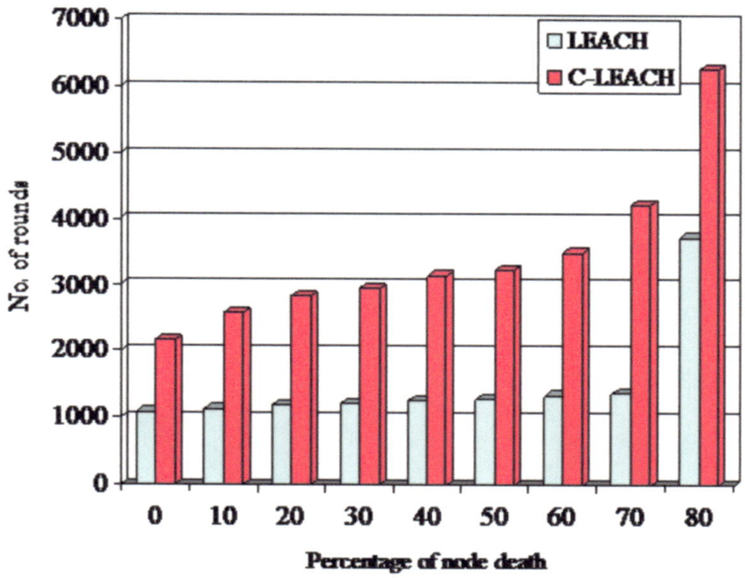

Fig. 11. Percentage of node death with C-LEACH scheme

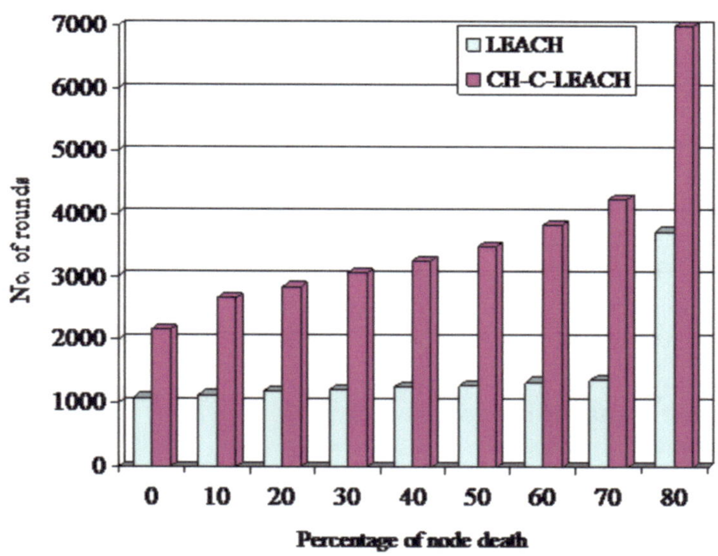

Fig. 12. Percentage of node death with CH-C-LEACH scheme

6. Attacks in wireless sensor network

Security plays an important role in WSN since the nodes are exposed to attacks in ruthless environment. Due to the unattended deployment of the sensor nodes, the attackers can easily capture and convert them as malicious nodes. Routing protocols are common target of these compromised nodes. So the capability of avoiding compromised nodes is quite weak.

The adversary can damage the nodes in physical layer or manipulate data in the data link layer and choose incorrect routing path to destroy the network. The malicious nodes can either join the network externally or may originate internally by compromising an existing benevolent node (Le et al., 2008).

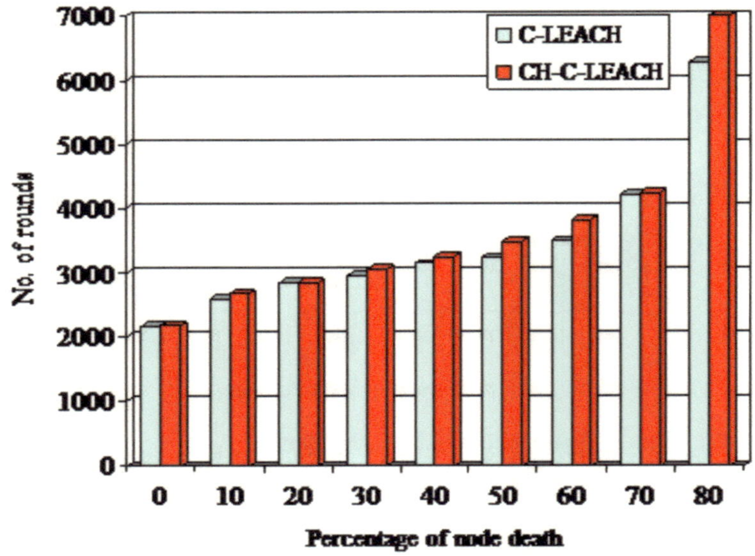

Fig. 13. Percentage of node death with C-LEACH and CH-C-LEACH scheme

These nodes can carry out both passive and active attacks. In passive attacks a malicious node only eavesdrop upon the packet contents, while in active attacks it may imitate, drop or modify legitimate packets. The main active attacks are as follows: spoofed, altered, or replayed routing information, selective forwarding attacks, sinkhole attacks, wormholes, sybil attacks and HELLO flood attacks which are applied to compromise the routing protocols of wireless sensor network. The various types of attacks that occur in sensor networks are shown in Fig.14.

6.1 Attacks in heterogeneous sensor network

i. *Selective Forwarding*

In a selective forwarding attack, malicious nodes may refuse to forward certain messages and simply drop them, ensuring that they are not propagated any further. A simple form of this attack is when a malicious node behaves like a black hole; it refuses to forward every packet it sees. However, such an attacker runs the risk that neighboring nodes will conclude that it has failed and decided to seek another route. A more subtle form of this attack is when an adversary selectively forwards packets. An adversary interested in suppressing or modifying packets originating from a selected set of nodes can reliably forward the remaining traffic and limit suspicion of its wrong doing (Xiaojiang et al., 2006, 2007).

Selective forwarding attacks are typically most effective when the attacker is explicitly included on the path of a data flow. However, it is conceivable that an adversary overhearing a flow passing through neighboring nodes might be able to emulate selective forwarding by jamming or causing a collision on each forwarded packet of interest. The mechanics of such an effort are tricky at best, and may border on impossible. Thus, an adversary launching a selective forwarding attack will likely follow the path of least resistance and attempt to include itself on the actual path of the data flow.

ii. *Sinkhole attack*

In a sinkhole attack, a malicious node uses the faults in a routing protocol to attract much traffic from a particular area, thus creating a sinkhole (Karlof et al., 2003). The adversary's goal of this attack is to lure nearly all the traffic from a particular area through a compromised node, creating a metaphorical sinkhole with the adversary at the center. Because, sinkhole attacks can enable many other attacks (selective forwarding, for example). Sinkhole attacks typically work by making a compromised node look especially attractive to surrounding nodes with respect to the routing algorithm (Xiaojiang, 2008). For instance, an adversary could spoof or replay an advertisement for an extremely high quality route to a base station. Some protocols might actually try to verify the quality of route with end-to-end acknowledgements containing reliability or latency information.

In this case, a laptop-class adversary with a powerful transmitter can actually provide a high quality route by transmitting with enough power to reach the base station in a single hop. Due to either the real or imagined high quality route through the compromised node, it is likely each neighboring node of the adversary will forward packets destined for a base station through the adversary, and also propagate the attractiveness of the route to its neighbours. Effectively, the adversary creates a large "sphere of influence", attracting all traffic destined for a base station from nodes several hops away from the compromised node. Since all packets share the same ultimate destination, a compromised node needs only to provide a single high quality route to the base station in order to influence a potentially large number of nodes.

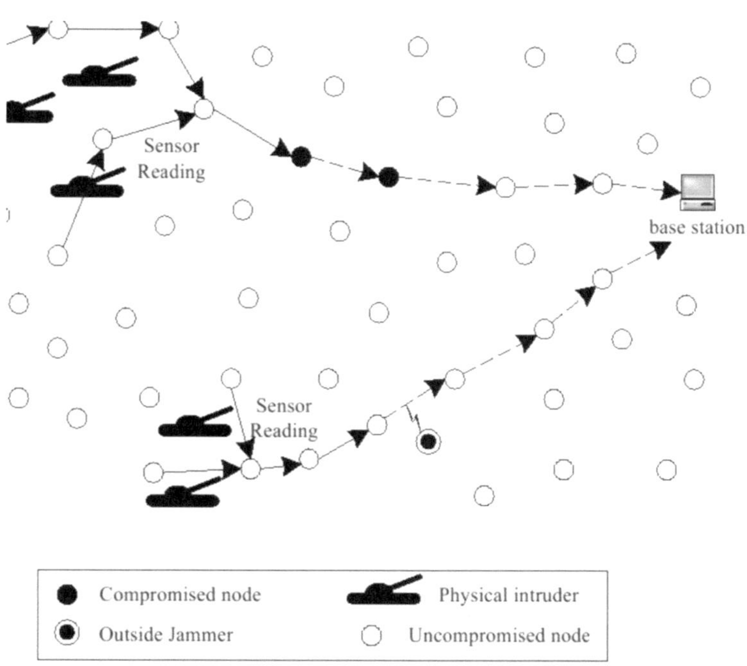

Fig. 14. Attacks in sensor network

7. Secured path redundancy algorithm in heterogeneous sensor network

The alternate path redundancy algorithm is used to find secure multiple paths between the source and destination nodes in the presence of attackers. Selective forwarding and sinkhole attacks are types of attackers that make a compromised node look more attractive to surrounding L-sensor nodes of HSN by forging routing information (Samundiswary & Dananjayan, 2010). The end result is that surrounding L-sensor nodes of HSN will choose the compromised node as the next node to route the data through. This is achieved by removing one or more L-sensor nodes that is suspected to be an active adversary node from the routing path. Such nodes are identified by algorithm using a set of parameters that is usually reflecting the presence of adversary nodes. The parameters used are packet ID, number of hop counts and delay to reach the destination. This secured path redundancy algorithm mechanism can defend against the above mentioned attacks (Xiaojiang, 2008).
Further more, sink mobility brings new challenges to data dissemination in large sensor networks. Sink mobility suggests that information about each mobile sink's location be continuously propagated throughout the sensor field in order to keep all sensor nodes informed about the direction of forwarding future data reports. Unfortunately, frequent location updates from multiple sinks can lead to both excessive drain of sensors' battery resources and increased collisions in wireless transmissions. To avoid these limitations, the same secured path redundancy algorithm for HSN approach is extended for mobile sinks as shown in Fig. 15.

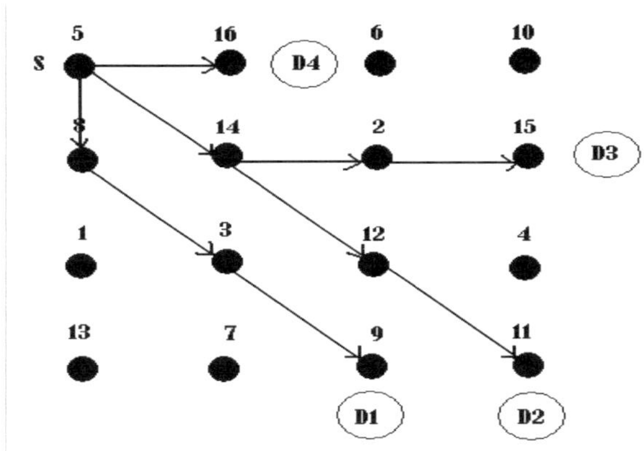

Fig. 15. Mobile sink

8. Simulation results

The secured path redundancy algorithm for static nodes with sink mobility in heterogeneous sensor network is simulated by varying the number of nodes from 25 to 500 with 30 and 50 numbers of malicious nodes for different coverage area in Glomosim. The energy consumption, delivery ratio and delay are calculated for proposed algorithm considering constant bit rate (CBR) traffic in the network.

8.1 Energy consumption

The simulation results shown in Fig.16, Fig.17 and Fig.18 prove that there is a significant reduction in the energy consumption of secured heterogeneous sensor networks by increasing the numbers of nodes and number of mobile sinks for different coverage area and different values of malicious nodes. Fig.16 shows that there is increment in the energy consumption of secured heterogeneous sensor networks for increased coverage area. When the number of nodes increases, the energy consumption of secured heterogeneous sensor networks reduces from 57% to 81.5% compared to heterogeneous sensor networks with 30 malicious nodes and the coverage area of 300m×300m.

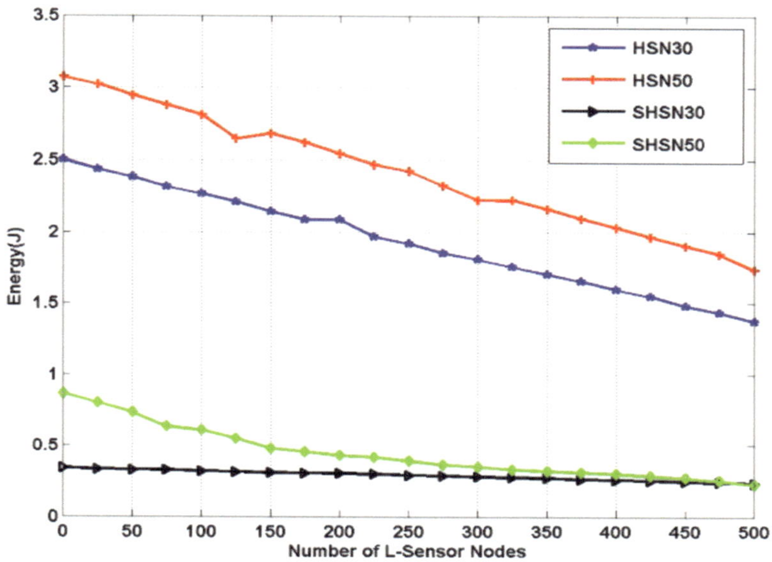

Fig. 16. Energy consumption with number of L-sensor nodes for coverage area 300m×300m

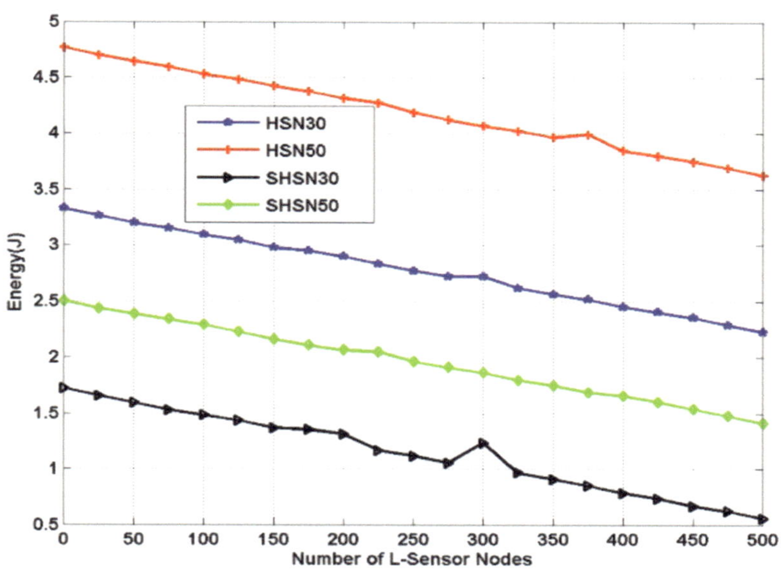

Fig. 17. Energy consumption with number of L- sensor nodes for coverage area 500m×500m

Even if the number of malicious nodes and coverage area increases, the energy consumption reduces by 49% to 67% with respect to heterogeneous sensor networks. Energy consumption of secured heterogeneous sensor networks is lesser than heterogeneous sensor networks because nodes involve alternate shortest secured path and less number of broken paths by using H-sensors even in the presence of malicious nodes.

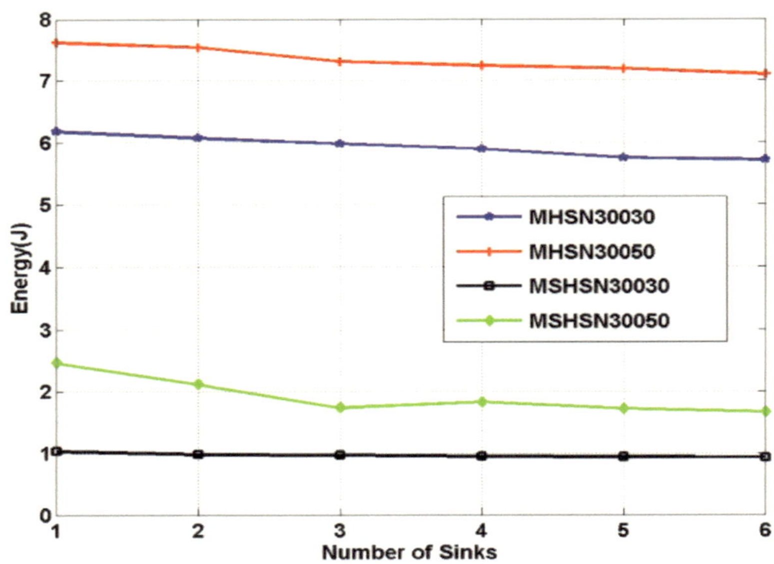

Fig. 18. Energy consumption with number of mobile sinks for coverage area 300m×300m

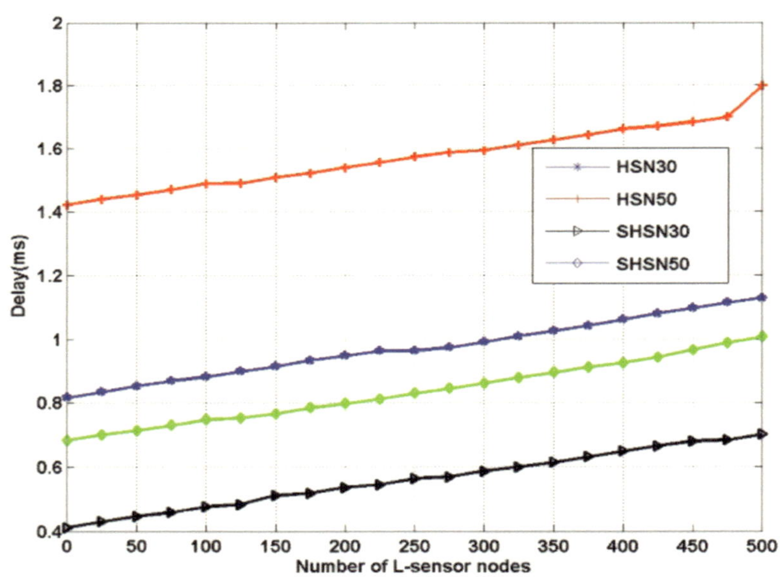

Fig. 19. Delay with respect to number of L-sensor nodes for coverage area 300m×300m

8.2 Delay

The delay graph is illustrated in Fig.19, Fig.20 and Fig.21 considering different coverage areas and various values of mobile sinks with 30 and 50 malicious nodes. The results prove that secured path redundancy algorithm (SPRA) for heterogeneous sensor network nodes is lower than that of HSN by 50% to 55% in case of 30 malicious nodes for network coverage area of 300m×300m and 500m×500m. Since proposed security algorithm for heterogeneous sensor network uses a secured path, packets require less hop count and link failures to reach the mobile sinks from the source even in the presence of malicious nodes.

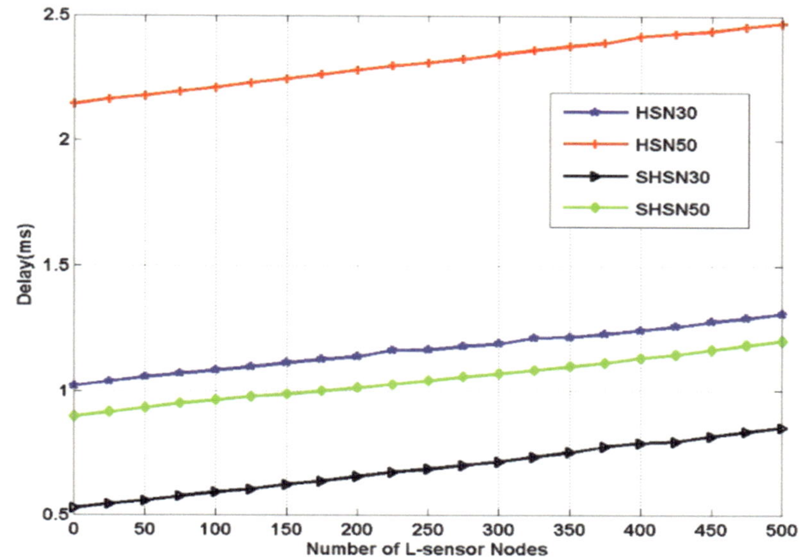

Fig. 20. Delay with respect to number of L-sensor nodes for coverage area 500m×500m

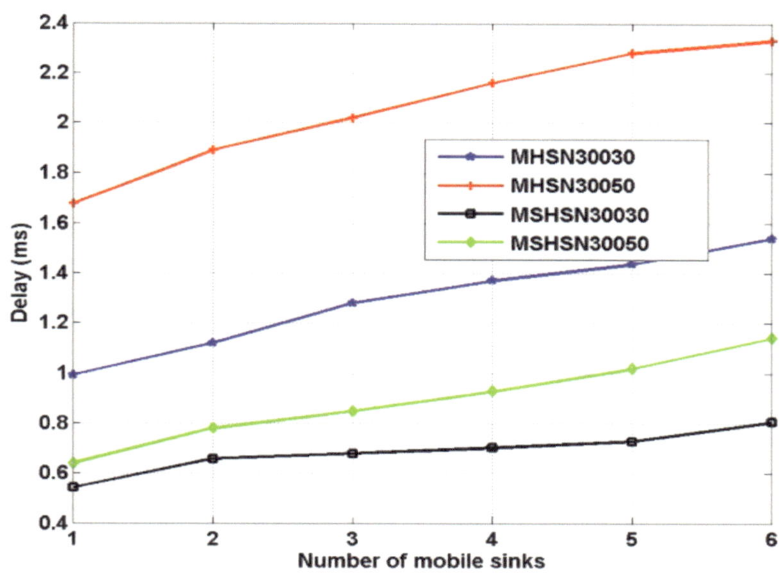

Fig. 21. Delay with respect to number of mobile sinks for coverage area 300m×300m

8.3 Delivery ratio

Delivery ratio of proposed SPRA for heterogeneous sensor networks is higher than conventional HSN which is shown in Fig.22, Fig.23 and Fig.24 for different values of malicious nodes and different coverage area.

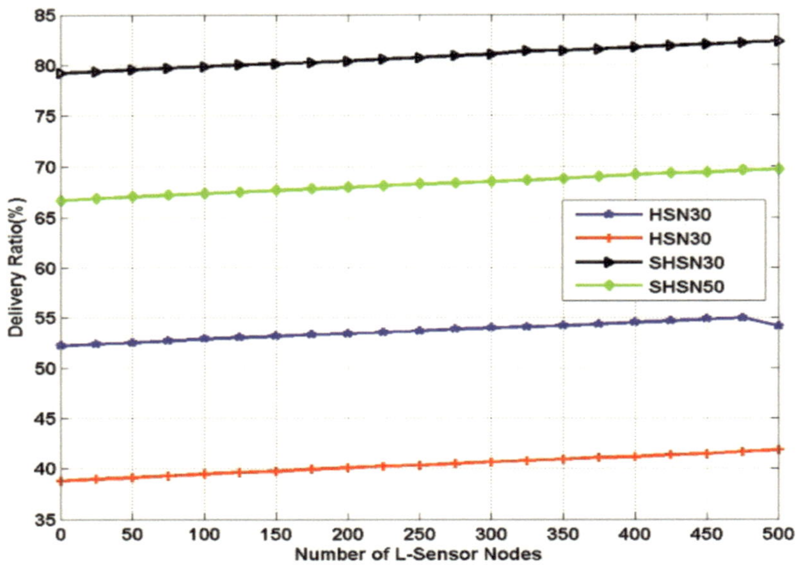

Fig. 22. Delivery ratio with respect to number of L-sensor nodes for coverage area 300m×300m

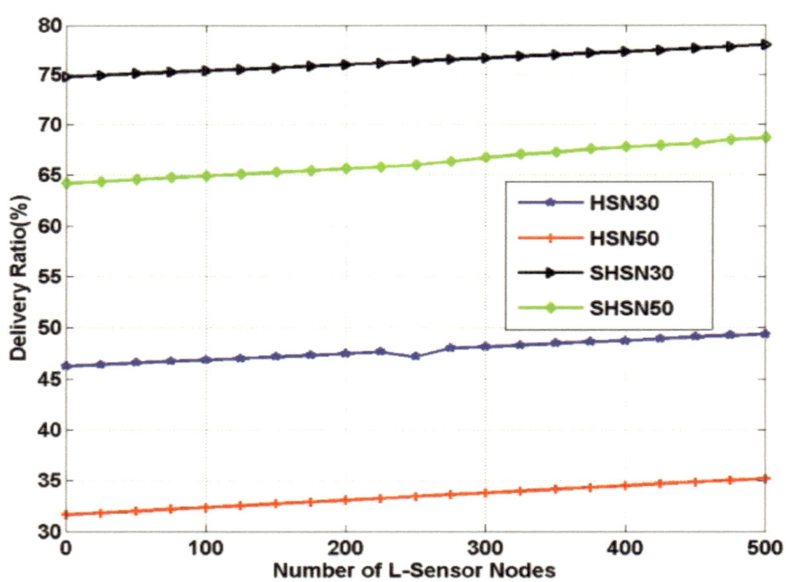

Fig. 23. Delivery ratio with respect to number of L-sensor nodes for coverage area 500m×500m

In Fig.22, the delivery ratio of proposed security algorithm (SPRA) of heterogeneous sensor network is higher than that of heterogeneous sensor network in the presence of malicious nodes by 60%-70%. The fact is that secured HSN packets require less number of hops from the L-sensors to the cluster head than HSN. Moreover, the packet loss is reduced due to secured path from source to sink in secured heterogeneous sensor network.

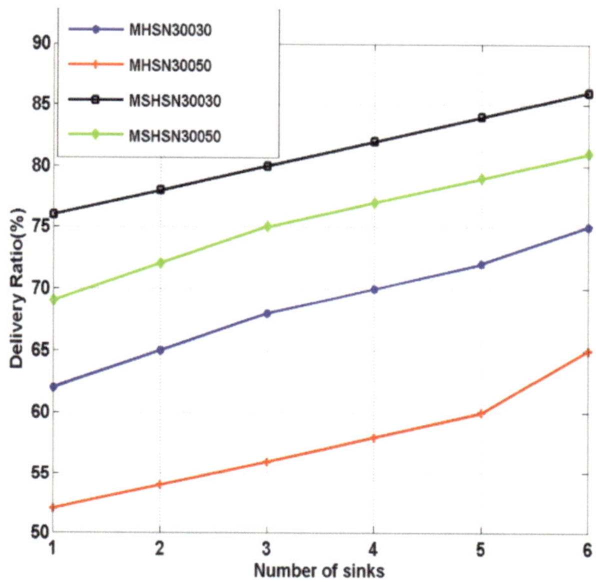

Fig. 24. Delivery ratio with respect to number of mobile sinks for coverage area 300m×300m.

9. Conclusion

This chapter proposed two routing mechanisms to reduce the fading effects and defend against network layer attacks by incorporating cooperative MIMO routing scheme and SPRA in heterogeneous sensor networks.

A cluster-based cooperative heterogeneous MIMO routing scheme using STBC for WSN has been explored for 100 sensor nodes with initial energy of 0.5J for normal nodes and 2J for advanced nodes. The secured path redundancy algorithm for heterogeneous sensor networks is simulated by varying the number of nodes from 100 to 500 and malicious nodes (30 and 50) with mobile sinks (1 to 6).

The performance of the proposed cooperative heterogeneous MIMO system is evaluated to minimise the energy consumption and increase the lifetime of sensor nodes. The simulation results reveal that the LEACH protocol consumes more energy and has shorter lifetime of 3750 rounds due to the adverse channel fading effects. The proposed cooperative heterogeneous MIMO CH-C-LEACH performs better and extends 3250 rounds and 750 rounds more than the LEACH scheme and C-LEACH scheme respectively for data transmission. The proposed scheme saves up to 50% energy compared to LEACH by the exploitation of the diversity gain of MIMO systems.

The performance of the proposed SPRA of heterogeneous sensor network is verified through simulation by evaluating energy consumption, delay and the delivery ratio in the presence of selective forwarding and sink hole attacks. The simulation results prove that secured path redundancy algorithm in heterogeneous sensor networks has better network performance than that of conventional heterogeneous sensor networks. The reduction in the energy consumption of 60% is achieved by using this algorithm compared to that of conventional heterogeneous sensor networks. The results also demonstrate that the enhancement in delivery ratio of approximately 65% and end to end delay of roughly 52% is achieved through secured heterogeneous sensor network. The improved performance of this algorithm is due to the usage of a secured alternate path which involves less number of broken paths, hop count and less packet loss to reach the destination node.

The further enhancement of the work is to extend the routing scheme taking into account mobile H- sensors. To reduce the energy consumption further due to fading effects other space time encoding schemes and modulation levels of PSK can be implemented to improve network lifetime. For enhancing the security of the system a modified algorithm can be suggested to defend against other network layer attacks such as worm hole and sybil attack.

10. References

Adrian Perrig,; Robert Szewczyk,; Victor Wen,; David Culler & Tygar, J.D. (2001). SPINS: Security protocols for sensor networks, *Proceedings of ACM Annual International Conference on Mobile Computing and Networking*, pp.189-199, ISBN: 1-58113-422-3, Rome, Italy, July, 2001, ACM, New York.

Akyildiz, L.; Su, W.; Sankarasubramanian, Y. & Cayirci, E. (2002). A survey on sensor networks. *IEEE Communications Magazine*, Vol.40, No.8, August 2002, pp.102-114.

Bravos, G.N. & Efthymoglou, G. (2007). MIMO-based and SISO multihop sensor network: Energy efficiency evaluation. *Proceedings of 3rd IEEE International Conference on Wireless and Mobile Computing, Networking and Communications*. pp.13, ISBN: 0-7695-2889-9, NewYork, USA, October 2007.

Cheng, W.; Xu, K.; Yang, Z & Feng, Z. (2006). An energy-efficient cooperative MIMO transmission scheme for wireless sensor networks. *Proceedings of International Conference on Wireless Communication, Networking and Mobile Computing*, pp.1-4, ISBN: 1-4244-0517-3, Wuhan, September 2006.

Cui, S.; Goldsmith, A.J. & Bahai, A. (2004). Energy-efficiency of MIMO and cooperative techniques in sensor networks, *IEEE Journal on Selected Areas in Communications*, Vol.22, No.6, August 2004, pp.1089-1098.

Do hyun mam & Hong-Ki-Min. (2007). An Efficient Ad hoc routing using a hybrid clustering method in a wireless sensor network. *Proceedings of 3rd IEEE International Conference on Wireless and Mobile Computing, Networking and Communications*, pp.60, ISBN: 0-7695-2889-9, NewYork, USA, October 2007.

Heinzelman, W.B.; Chandrakasan, A.P. & Balakrishnan, H. (2000). Energy –efficient communication protocol for wireless micro sensor networks, *Proceedings of the 33rd Hawaii International Conference on System Science*, pp.3005-3014, Mauri, Hawaii, January 2000.

Heinzelman, W.B.; Chandrakasan, A.P. & Balakrishnan, H. (2002) . An application-specific protocol architecture for wireless microsensor networks, *IEEE Transactions on Wireless Communications*, Vol.1, No.4, October 2002, pp.660 - 670.

Ilyas, M. & Mahgoub, I. (2005). *Handbook of sensor networks: Compact wired and wireless sensing systems*. Boca Raton, FL.

Jayaweera, S.K. (2004). Energy analysis of MIMO techniques in wireless sensor networks. (2004). *Proceedings of 36th Annual Conference on Information Sciences and Systems*, Princeton, NJ, March 2008.

Jeremy Brown & Xiaojiang Du. (2008). Detection of selective forwarding attacks in heterogeneous sensor networks. *Proceedings of IEEE International Conference on Communications*, pp.1583-1587, ISBN: 978-1-4244-2075-9, Beijing, China, May 2008, IEEE.

Karki, J.A. & Kamal, A.E. (2004). Routing techniques in wireless sensor networks: A survey. *IEEE Personal Communication*, Vol.11, No.6, December 2004, pp.6-28.

Karlof.C. & Wagner.D (2003).Secure routing in wireless sensor networks: attacks and countermeasures. *Proceedings of 1st International Workshop on Sensor Protocols and Applications*, pp.113-127, ISBN: 0-7803-7879-2, Anchorage, AK, May 2003, IEEE.

Kazem Sohraby,; Daniel Minoli & Taieb Zanti. (2007). *Wireless sensor network technology, protocols and applications*. John Wiley and Sons Inc., ISBN: 978-0-471-74300-2.

Li, X.; Chen, M. & Liu, W. (2005). Application of STBC-encoded cooperative transmissions in wireless sensor networks. *IEEE Signal Processing Letters*, Vol.22, No.2, February 2005, pp.134-137.

Le Xuan Hung,; Ngo Trong Canh,; Sungyoung Lee,; Young-Koo Lee & Heejo Lee. (2008). An energy-efficient secure routing and key management scheme for mobile sinks in wireless sensor networks using deployment knowledge. *Journal on Sensors*, Vol.8, December 2008, pp.7753-7782.

Muruganathan, S.D.; Ma, D.C.F.; Bhasin, R.I & Fapojuwo, A.O. (2005). A centralized energy-efficient routing protocol for wireless sensor networks. *IEEE Radio Communications*, March 2005, pp. s8-s13.

Sami,; Al-Wakeel, S. & Al-Swailem, A. (2007). PRSA: A path redundancy based security algorithm for wireless sensor networks. *Proceedings of IEEE Wireless Communication and Networking Conference*, pp.4156-4160, ISBN: 1-4244-0658-7, Kowloon, China, March, IEEE.

Samundiswary, P. & Dananjayan, P. (2010). Detection of Sinkhole attacks for mobile nodes in heterogeneous sensor networks with mobile sinks. *International Journal on Computer and Electrical Engineering*, Vol.2, No.1, February 2010, pp.127-133, ISSN online: 1793-8198.

Tarokh,V.; Jafarkhani, H. & Calderbank, A.R. (1999). Space-time block codes from orthogonal designs. *IEEE Transactions on Information Theory*, Vol.45, No.5, July 1999, pp. 1456-1467.

Vidhya, J. & Dananjayan, P. (2009). A hybrid clustering protocol for energy-efficient routing in wireless sensor networks. *International Journal of Electronics Engineering*, Vol.1, No.1, January 2009, pp.7-12, ISSN: 0973-7383.

Vidhya, J. & Dananjayan, P. (2010). Life time maximization of multihop WSN protocol using cluster based cooperative MIMO scheme. *International Journal of Computer Theory and Electrical Engineering*, Vol.2, No.1, February 2010, pp. 1793-8201, ISSN online: 1793-8201.

Vivek Mhatre & Catherine Rosenberg. (2004). Homogeneous Vs Heterogeneous clustered sensor networks: A comparative study. *Proceedings of IEEE International Conference on Communications*, Vol.6, pp.3646-3651, ISBN: 0-7803-8533-0, Paris, France, June 2004, IEEE.

Xiaojiang Du, ; Sghaier Guizani,; Yang Xiao & Hsiao-Hwa Chen. (2006). A secure routing protocol for heterogeneous sensor networks. *Proceedings of IEEE Global Telecommunication Conference (GLOBECOM'06)*, pp.1-5, ISBN: 1-4244-0356-1, San Francisco, California, December, IEEE.

Xiaojiang Du,; Mohsen Guizani,; Yang Xiao & Hsiao-Hwa Chen. (2007). Two tier secure routing protocol for heterogeneous sensor networks. *IEEE Transactions on Wireless Communications*, Vol.6, No.9, September 2007, pp.3395-3407, ISSN: 1536-1276.

Xiaojiang Du. (2008). Detection of compromised sensor nodes in heterogeneous sensor networks. *Proceedings of IEEE International Conference on Communications*, pp.1446-1450, ISBN: 978-1-4244-2075-9, Beijing, China, May, IEEE.

Xiangning, F. & SongYulin. (2007). Improvement on LEACH protocol of wireless sensor network. *Proceedings of International Conference on Sensor Technologies and Applications*, pp. 260-264, Valencia, Spain, October 2007.

Xu, K.; Hong, X. & Gerla, M. (2005). Improving routing in sensor networks with heterogeneous sensor nodes. *Proceedings of IEEE 61st Vehicular Technology Conference*, pp.2528-2532, ISBN: 0-7803-8887-9, Stockholm, Sweden, Vol.4, May 2005, IEEE.

Yuan, Y.; He, Z. & Chen, M. (2006). Virtual MIMO- based cross-layer design for wireless sensor networks. *IEEE Transactions on Vehicular Technology*, Vol.55, No.3, May 2006, pp.856 -864.

Yu, M.; Leung, K. & Malvankar, A. (2007). A dynamic clustering and energy efficient routing technique for sensor networks. *IEEE Transactions on Wireless Communication*, Vol.6, No.8, August 2007, pp.3069-3078.

7

Data Aggregation Tree Construction: Algorithms and Challenges [1]

Zahra Eskandari[2] and Fatemeh Ayughi
Islamic Azad University of Kashmar, Islamic Azad University of Qazvin
Iran

1. Introduction

Wireless Sensor Networks (WSNs) are a set of communication networks which consist of sensor nodes. These nodes sense the events occurred in their own area and transmit the data related to these events to the sink node. The sink node is considered as the gateway. The sink is in relation with the end user and disseminates the queries requested by the user (Akyildiz et al, 2002) in the network. Receiving the query, the nodes of the network should sense the query's required data from the environment and send them to the sink.

The first advantage of wireless technology is easy deployment of sensors, so that outdoor environments like forests, deserts and the wildness in general can be covered. The second advantage is the possibility of networking mobile nodes. The application scenarios are various, ranging from the obvious military applications, such as distributed battlefield sensing or frontier control, to peaceful and civilian uses. Examples are: habitat monitoring (birds, whales), home intelligence (e.g. local climate control and smart appliances), biomedical, patient tracking, disaster relief, surveillance, fire control, agricultural, and industrial control (Cantoni et al, 2006).

WSNs have specific characteristics. In these networks, the nodes are randomly deployed in the environment, i.e. the geographical locations of these nodes are undetermined (Eskandari et al a, 2008) and these nodes are inaccessible. Furthermore, the nodes are deployed in the environment densely. These nodes have generally low capability for processing and storing. So the tasks that the nodes perform should not be computationally complex.

Furthermore, one of the main constraints in these networks is energy resource due to size and cost limitation in their nodes (Lee & Wong c, 2006), so, the tasks should be energy efficient. Up to now, many attempts have been made to minimize energy consumption (Chlamtac & Kutten, 1987; Chlamtac & Weinstein, 1991; Heinzelman et al, 2000; Min & Chandrakasan, 2001; Upadhyayula et al, 2003; Krishnamachari et al, 2002; Intanagonwiwat et al, 2004).

In monitoring application, the sensor nodes sense data from the environment periodically and transmit these data to the sink node. The nodes in the network are densely developed,

---

1 This work is supported by Islamic Azad University of Kashmar.

2 Professor of computer engineering department, Islamic Azad University of Kashmar.

so the information associated with an event is sensed by more than one sensor node. The nodes transmit the redundant information to the sink (Liang & Liu, 2006).

Transmission of these redundant data wastes energy. As energy resources are the most important limitation of WSNs and data transmission is the most costly function in the network. This leads to decrease in the node's power, quickly (Akyildiz et al, 2002). After some rounds, network nodes energy is finished and this leads to cases in which the network can not work anymore. Regarding the above mentioned points, in order to increase the network's lifetime, the number of transmitted data packets should be minimized (Akyildiz et al, 2002; Eskandari et al a, 2008).

As described in (Upadhyayula & Gupta, 2006), a round is defined as the collection of one data unit from every node in the network and delivering the resulting aggregated data to the sink node. And, also based on this work, the lifetime of a tree is defined as the number of rounds that can be performed before the failure of certain percentage of total nodes.

Based on energy model described in (Kamimura et al, 2004), a sensor node consumes E_{elec} (J/bit) in transmitter or receiver circuitry and Eamp (J/bit/m^2) in transmitter amplifier to achieve an acceptable signal noise ratio. A sensor node expends energy ETij (k) or ERi(k) in transmitting or receiving a k-bit packet to or from distance distij, given by the following equations:

$$ET_{ij}(k) = E_{elec} * k + E_{amp} * k * dist_{ij}^{\lambda} \tag{1}$$

$$ER_i(k) = E_{elec} * k \tag{2}$$

The exponent λ heavily depends on the communication medium (Upadhyayula & Gupta, 2006). As described in (Younis & Fahmy, 2004) if aggregation function is simple, the energy consumption for data aggregation will be negligible.

1.1 Data Aggregation

A number of mechanisms called aggregation algorithms are suggested in order to omit the redundant data. Aggregation algorithms, after receiving data from several sensors, process data and omit the redundancy and send the result of aggregation to the sink (Liang & Liu, 2006). Due to the reduction in data volume, these algorithms decrease the energy consumption (Lee & Wong a, 2005).

Therefore the networks which perform aggregation have more life time (Eskandari et al a, 2008; Lee & Wong a, 2005) and draw more attention (Eskandari et al a, 2008; Lee & Wong a, 2005; Lee & Wong b, 2005). In addition to mentioned improvements, aggregation decreases collision and retransmission delay (Zhu et al, 2006).

Data aggregation is performed during routing in wireless sensor networks. Finding the route from several nodes to the sink in a way that maximizes the shared path and redundancy removing is one of the main objectives in these protocols (Liang & Liu, 2006).

In aggregation algorithms, we must construct aggregation spanning tree (Lee & Wong a, 2005). The spanning tree is a tree which contains all network nodes and doesn't have any loop.

Aggregation mechanism works as follow: each node senses data from the environment and receives other node's data, then aggregates these data, based on the aggregation function and transmits the aggregation result to the sink.

2. Aggregation Tree Construction

As a result of energy saving of data aggregation, different aggregation algorithms have been presented. In this section, we review them briefly and compare their efficiency, and then we introduce a new algorithm, describe it and evaluate its efficiency. Finally, we consider a new challenge, i.e. tree construction cost.

2.1 Recent Works

In (Krishnamachari et al, 2002), the authors investigate the computational complexity of optimal data aggregation in sensor networks and show that it is generally NP-hard; they present some suboptimal data aggregation tree generation heuristics, Center at Nearest Source (CNS), Shortest Paths Tree (SPT) and Greedy Incremental Tree (GIT) and show the existence of polynomial special cases.

As presented in (Zhang & Cao, 2004), DCTC algorithm dynamically constructs the aggregation tree for mobile target tracking. In the presented algorithm depending on the target location, a subset of nodes participates in tree construction.

In (Upadhyayula et al, 2003), the sink saves the entire network state and then by considering link cost, in centralized form, constructs the tree with minimum cost. In cluster algorithm (Younis & Fahmy, 2004), after partitioning the network into clusters, cluster's members construct aggregation tree and transmit data to cluster head. After aggregation, cluster heads transmit aggregated data to the sink in one hop or multihop manner (Chen et al, 2005).

Espan (Lee & Wong a, 2005) is an energy-aware spanning tree algorithm that constructs the aggregation tree to aggregate the data. In Espan, the source node which has the highest residual energy is chosen as the root and other nodes choose their corresponding parent node among their neighbors based on distance to the root and residual energy. Each node selects the closest neighbors to root as its parent. If there are multiple neighbors with equal distance, the node which has the most remaining energy is selected as parent.

As Espan protocol considers distance as main parameter and remaining energy as second, one of the most important problems of Espan is that the nodes with the least distance to root maybe selected as parent by many nodes. So these nodes consume their energy quickly and then they will fail sooner than other network nodes, so the network cannot cover region completely.

In LPT (Lee & Wong b, 2005) after selecting the node with most energy as root, each node selects neighbors with the most energy as parent and its parent forwards its data to the sink. In the mentioned algorithm, when a node in the tree fails, the tree will be reconstructed. LPT aims to prolong the lifetime of the sources which transmit data reports periodically. But in LPT, the parents may have higher distance to root and this cause more energy consumption. LPT does not consider the distance parameter in parent selection.

We have presented an energy efficient algorithm, which constructs the aggregation tree in (Eskandari et al a, 2008). To prevent failing of nodes and to increase the network lifetime, the algorithm considers both the remaining energy and the distance parameters. Each node selects a node which has the most energy within neighbors as its parent. Furthermore, the distance from this parent to the root must be reasonable. To balance the energy and distance parameters, the algorithm uses path's energy and length parameters.

2.2 An Efficient Aggregation Tree Construction Algorithm

In this section, we present an Energy Efficient Spanning tree (EEspan) algorithm which is a new energy efficient algorithm for wireless sensor networks. The current work is a modified version of our former published papers (Eskandari et al a, 2008; Eskandari et al b, 2008). Unlike the algorithms given in (Lee & Wong a, 2005; Lee & Wong b, 2005) which use only one of the distance and energy parameters as the main parameter, to decrease the number of failed nodes and to increase the network lifetime, this algorithm considers both remaining energy and distance parameters.

To control the energy and distance parameters, the algorithm uses path's energy and path's length parameters. Using this strategy, a node with low remaining energy can be alive more than that of Espan protocol. This increases the lifetime of the network and supports better coverage. Also, unlike the LPT algorithm, the presented algorithm prevents selecting a parent with high remaining energy, and far distance to the root.

In fact, the presented algorithm might select a node with higher energy but farther from root as its parent. If the selected neighbor with highest energy is in a distance farther than a threshold, the presented algorithm selects the less energy path. In addition, to provide fairness in energy consumption, the algorithm considers a third parameter which is the maximum number of children. In the presented algorithm, the nodes have a predetermined maximum number of children. Based on (Upadhyayula & Gupta, 2006), if the nodes have the same number of children, we can conclude that the nodes will be prepared to transmit data at the same time and their parent will have to be awake for a shorter duration to collect data from all its children.

An example which helps us to understand the details of the presented algorithm is given in Figure 1.

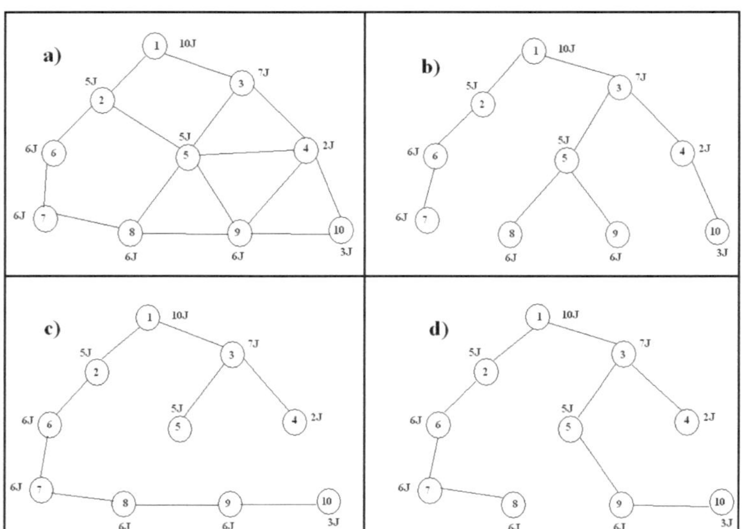

Fig. 1. The spanning trees of different algorithms a) connectivity graph, b) Espan's tree, c) LPT's tree, d) EEspan's tree

In this example, a connectivity graph with 10 different sensor nodes is used. The Espan, LPT and EEspan spanning trees are shown in figure 1. The remaining energy of nodes 1, 2, 3, 4, 5,

6, 7, 8, 9 and 10 are equal to 10J, 5J, 7J, 2J, 5J, 6J, 6J, 6J, 6J and 3J, respectively. Suppose that node 10 wants to select its parent.

Using Espan algorithm, node 4 which has the minimum distance to the root will be selected, while in the presented algorithm, node 9 which has more average path's energy is selected as the parent of node 10. The selected parent by Espan algorithm has low energy and fails quickly. As shown in Figure 1(c), LPT's tree has longer path length which causes more energy consumption.

The algorithm is a distributed algorithm which does not need to save global information about the entire network. This makes the presented algorithm more scalable. Furthermore, in the algorithm, routing is done in a multihop manner.

To verify the energy efficiency of the algorithm, here, we evaluate performance of the algorithm. Figure 2 shows the average path length of the three algorithms. At the beginning rounds, the Espan algorithm has lower energy consumption. This is because in this algorithm, nodes transmit data via shortest paths, but by ruining the low power nodes in these paths, data must be transmitted via other paths which may be longer. Since LPT algorithm selects paths by considering only the energy parameter, nodes transmit their data via longer paths which make higher energy consumption.

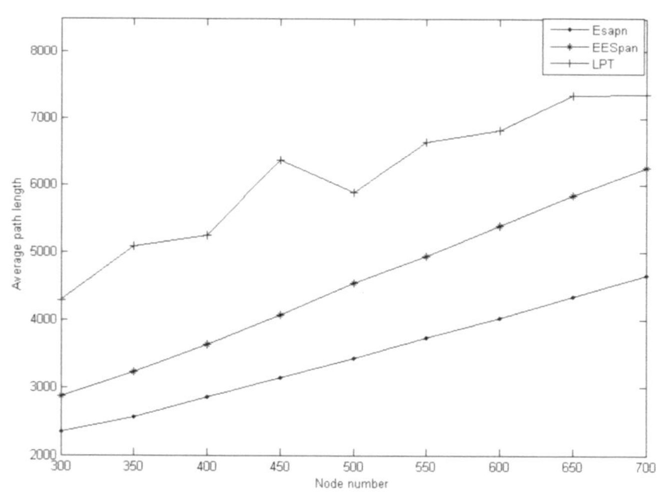

Fig. 2. The average path length of three algorithms

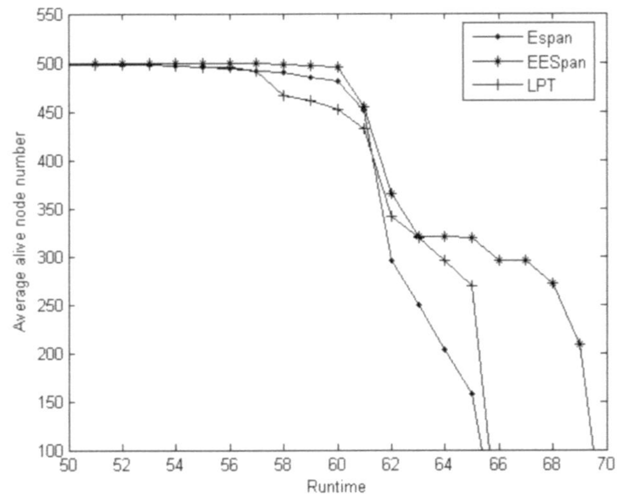

Fig. 3. Number of alive nodes at N=500

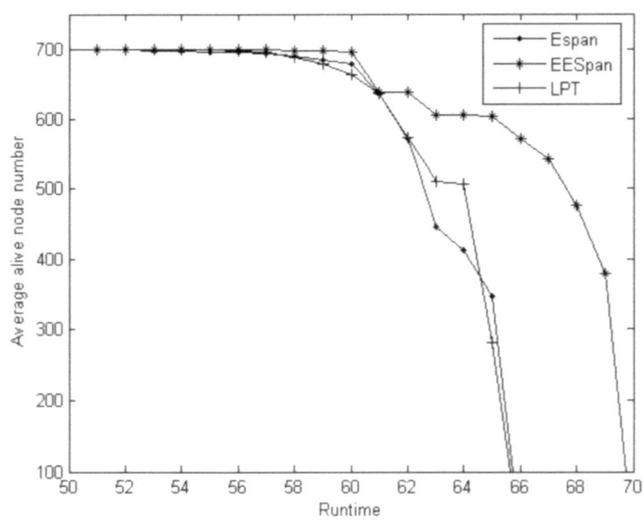

Fig. 4. Number of alive nodes at N=700

In Figures 3, 4, for different values of N, N = 500, and 700 nodes, the average number of alive nodes is plotted versus runtime. As the EEspan selects the nodes with high remaining energy, the nodes with low energy remain longer time in the network. Therefore the number of alive nodes is more than that of Espan algorithm. Furthermore, the LPT algorithm transmits data via the longer paths that leads to consume more energy and the failure of more nodes. More alive nodes can sense environment better, that means the network nodes have better coverage.

In Figure 5, for the three algorithms, the average lifetime is plotted versus the number of nodes. The main objective of all the algorithms is to achieve high energy efficiency. In addition to reducing the energy consumption, balancing energy consumption in nodes is important, too. In Espan algorithm, nodes transmit data via the smallest paths, but this leads the low power nodes in these paths to fail quickly and so the network's lifetime is decreased. To balance energy consumption in the network, the EEspan algorithm operates in an energy aware and transmit data via paths with more energy. Note that EEspan algorithm considers the path length to find the best tree.

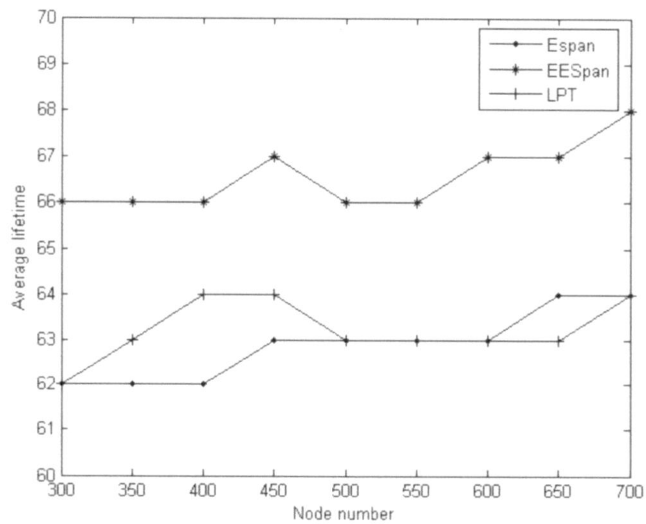

Fig. 5. Average lifetime comparison

2.3 Aggregation Tree Construction Cost

Since the status of the network is dynamic, like routing algorithms, aggregation algorithms should also be aware of the network topology and based on these information and queries which are propagated by root, network nodes select aggregation function and aggregate the data, and then forward the aggregated data to sink. And, also they should construct the aggregation tree periodically. To construct an aggregation tree, at the beginning of each period, routing packets are flooded into the entire network to inform all nodes. After this step, each node selects the best path towards the sink node and transmits data via the selected path until the next period. When a timer is expired or some nodes fail in the network, the new aggregation tree must be constructed (Lee & Wong a, 2005; Lee & Wong b, 2005). Since the node's energy is limited, transmitting and receiving this volume of routing information is not a good solution to construct an aggregation tree. This overhead causes a lot of energy consumption. So, some nodes run out of energy quickly and fail. This causes the network to be disconnected.

3. Reconfiguration

To solve the mentioned problems, in this section we introduce reconfiguration property; if a node in the aggregation tree fails, and a part of the tree is disconnected, only this part of tree starts to reconstruct locally, so it is not necessary to flood routing packets into the entire network. To do this, each node uses the environment feedbacks, and updates its information on its neighbors. In this section we introduce an automata-based algorithm to reconstruct spanning tree, the current work is published in [Eskandari et al b, 2008; Eskandari et al c, 2009).

3.1 Recent Works

Cluster based algorithms (Younis & Fahmy, 2004) needs only local information to construct the aggregation tree; therefore they transmit fewer packets to construct the aggregation tree.
In (Radivojac et al, 2003), the presented algorithm uses machine learning to transmit the sensed data to the sink. Learning algorithm is executed in the sink and its result is propagated throughout the network. In (Beyens et al, 2005) Q-leaner is used to construct aggregation tree to maximize aggregation ratio.
In (Esnaashari & Meybodi, 2007), an algorithm to construct the automata-based aggregation tree, is presented. In this algorithm, in which each node is equipped with an automaton, the automaton selects a path for transmitting data via the path whose aggregation ratio is maximized. In (Ankit et al, 2006), the algorithm considers an automaton for each node, which selects a path to transmit data to the sink in accordance with network conditions.

3.2 An automata Based Aggregation Tree Reconstruction Algorithm

Learning automata is an abstract model which has a finite set of actions as its input. Each member of the input set has a selection probability parameter. The automata select an input with highest selection probability as their output. Then the environment evaluates the selected action and responses to the automata. Automata use the response for learning process.
Learning process is as follows: if the environment response is unfavorable based on network parameter, the automata penalize the selected input by decreasing its selection probability and increasing selection probability of the other members of the input set. But if the environment response is favorable, the automata reward the selected input by increasing its selection probability and decreasing selection probability of the other members of the input set. The rewarding process increases selection probability of the awarded input for the next step. As shown in figure 6, an automaton is learned based on the feedback of the environment.

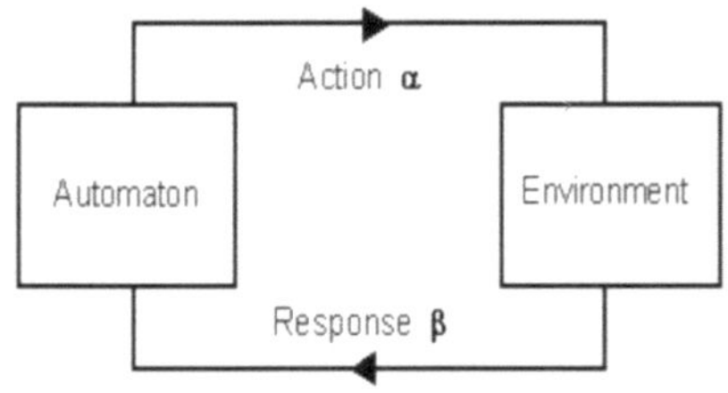

Fig. 6. learning automata

In automata-based algorithms (Ankit et al, 2006; Esnaashari & Meybodi, 2007), at the beginning, routing packets are flooded into the entire network. Each node considers each neighbor as entry in its routing table and then calculates the selection probability of each entry based on the algorithm's parameters, energy or distance and etc., and then each node selects the neighbor with highest selection probability as its parent and sends its data via this parent to the root.

In (Esnaashari & Meybodi, 2007) after receiving data, the root sends acknowledgment to the sender node; this acknowledgment has some information for automata. Based on acknowledgment information, automata penalize or reward the path's nodes, on the way that if the selected path was optimal based on the network parameters, the selection probability is increased for the next step, but if the selected path was not optimal, the selection probability is decreased for the next step. This process is called automata learning.

In the next steps, each node selects a new parent based on the updated selection probability of the nodes in the network and this process is repeated till the end of the network's lifetime. By using this learning property of automata, the algorithm prevents flooding the routing packets periodically, at the same time, by using ack information, nodes become aware of changes in network topology and paths are updated.

The presented algorithm in this section works as follows: at the beginning, routing packets are flooded into the network. Each neighbor, after receiving these packets, considers the sender as a new entry in its routing table.

This sending/receiving is performed in the entire network, so each node maintains neighbors information in its routing table. Then the routing table entries are considered as input set of automata and the automata calculate the selection probability of each entry as follow:

$$Sel - prob = C_i * \frac{energy_j}{distance_j} \tag{3}$$

In equation 3, Ci is a constant which is calculated by node and is dependent on the sum of energy and distances to the root of entries in routing table of node i.

Each node selects neighbor with highest selection probability as its parent, nodes in the network sense data and aggregate them with collected data from their child, then send the result of aggregation to their parents. Their parents forward data to the sink by repeating this process.

In order to update the automata, each node must collect some information from the network. By using this information, an automaton becomes aware of the network changing. In (Ankit et al, 2006) to be aware of the network state, each node after receiving data sends feedback or acknowledgment message to the sender of the data and as mentioned before, this message has some information. By using these feedbacks, automata penalize or reward the selected parent, but sending these acknowledgments have a lot of overhead. In (Esnaashari & Meybodi, 2007) to decrease this overhead, acknowledgment is sent after some data transmissions.

But, transmitting these additional data leads to waste of energy because parent's energy becomes less than other nodes in the neighborhood after some rounds. So, we can improve algorithm performance by working as follows: if a node in the aggregation tree fails or the node's energy is lower than a pre determined threshold, then the node's children select a

new parent from the nodes in their neighborhoods. Then, it is not necessary to reconstruct the aggregation tree globally and periodically.
By using this strategy the tree is reconstructed when it is needed, and reconstruction packet broadcasts locally. This leads to reduction in data transmission in the network and power saving.
Reconstruction property is an important section in the tree construction algorithm that is noted rarely. In this work, we try to achieve two main goals:

- Construct an energy efficient tree by considering both energy and distance parameters.
- Add the reconstruction property, to prevent from flooding packets globally.

In this section, to evaluate the performance of the presented algorithm, we compare it with other algorithms (Lee & Wong a, 2005; Lee & Wong b, 2005; Eskandari et al a, 2008).
At the first simulation trial, to evaluate the energy efficiency of the presented algorithm, the automata-based Energy Efficient Spanning tree (AEEspan), we measure remained energy of the network nodes. In figure 7, sum of the remaining energy of all nodes in network is plotted versus the number of nodes for four algorithms.
Since LPT algorithm selects paths by considering only energy parameter, nodes transmit their data via longer paths which cause higher energy consumption. In Espan algorithm, nodes transmit data via shortest paths, but by failing low power nodes in these paths, data must be transmitted via other paths which may be longer. While in EEspan (Eskandari et al a, 2008) and AEEspan, nodes consume less energy, because in these algorithms, the tree is constructed by applying a reasonable relation between energy and distance parameters.

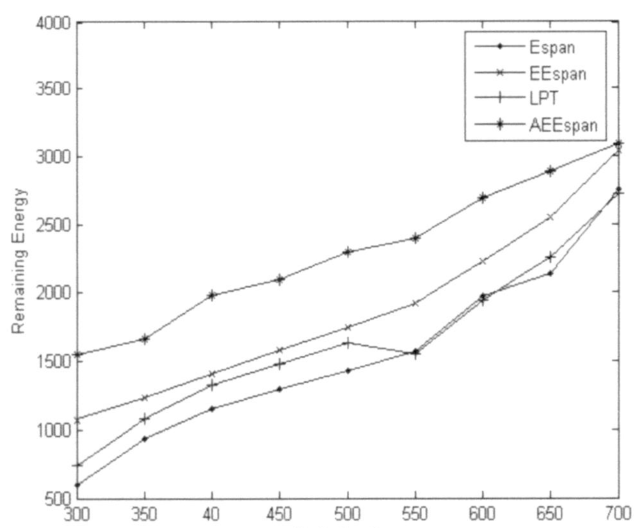

Fig. 7. The remaining energy of algorithms without considering tree reconstruction cost

In figure 8, the average path length is plotted versus the number of nodes. As in AEEspan, automata select their parents with the highest selection probability, and this value has converse relation to distance parameter, so the node with less distance has higher priority to be selected as parent that causes the parent with higher energy and less distance is selected.

As shown above, LPT tree has longer branches, because of not regarding distance parameter at all, while in Espan which regard distance as main parameter, the tree has shorter branches. While in this work, branches are between these two bounds.
As described earlier, the algorithm with automata learning property consumes less energy as a result of preventing from flooding routing packet. By considering learning property, transmission volume is decreased, that leads to more power saving. To show this, the remaining energy of the network nodes is measured. In figure 9, the sum of the remained energy of all nodes in the network is plotted versus the number of nodes.

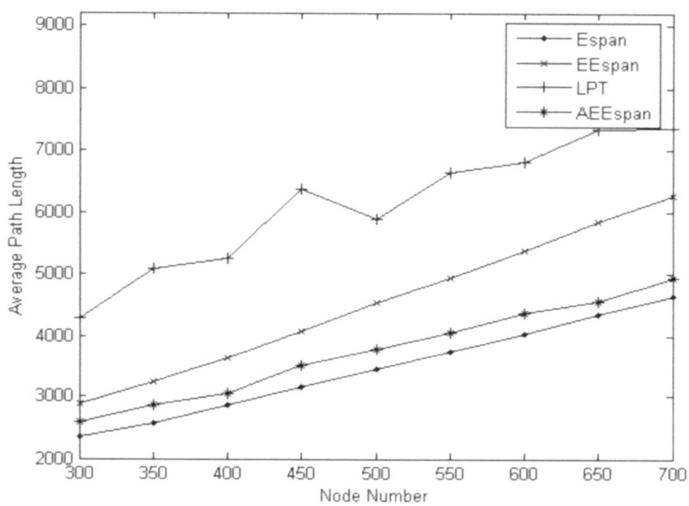

Fig. 8. Average hop count to root

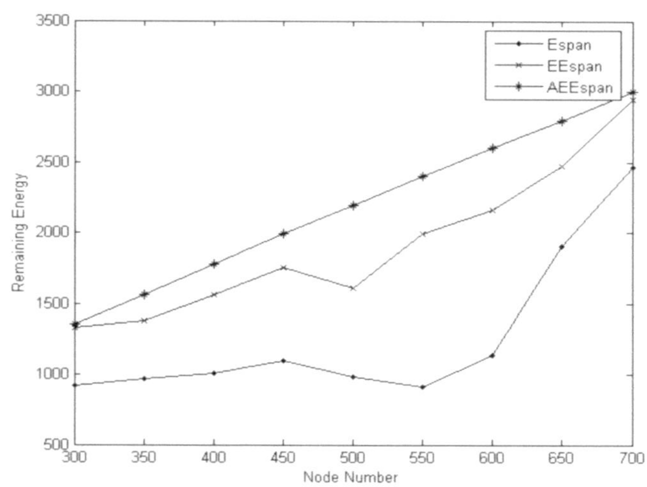

Fig. 9. The remaining energy of distributed algorithms with considering tree reconstruction cost

We measure the number of alive nodes after each simulation round in figures 10 and 11 when N = 300, and 500 nodes, respectively. As in AEEspan, the automata select a parent with the highest selection probability which has direct relation to energy parameter, so the nodes with low energy remain a longer time in the network rather than the other algorithms.

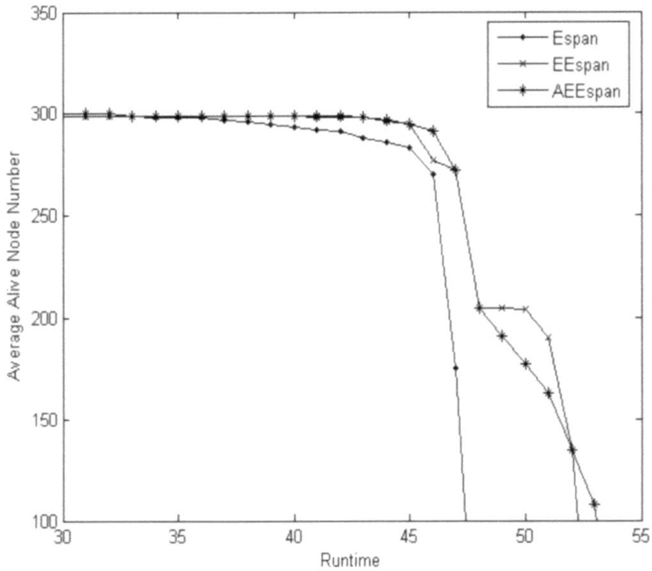

Fig. 10. Number of alive nodes at N=300

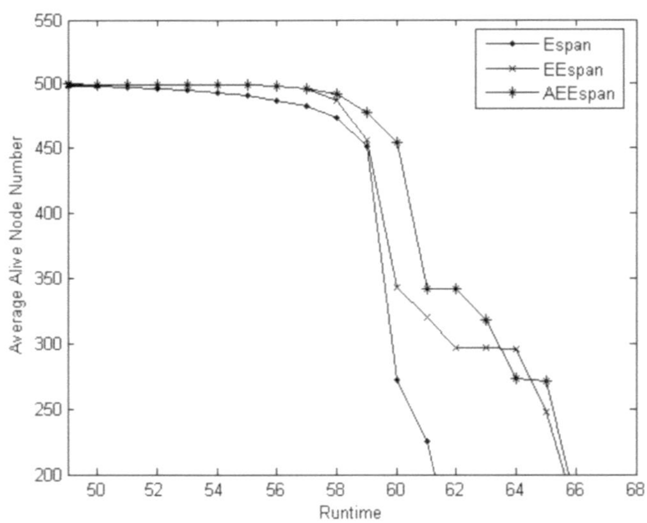

Fig. 11. Number of alive nodes at N=500

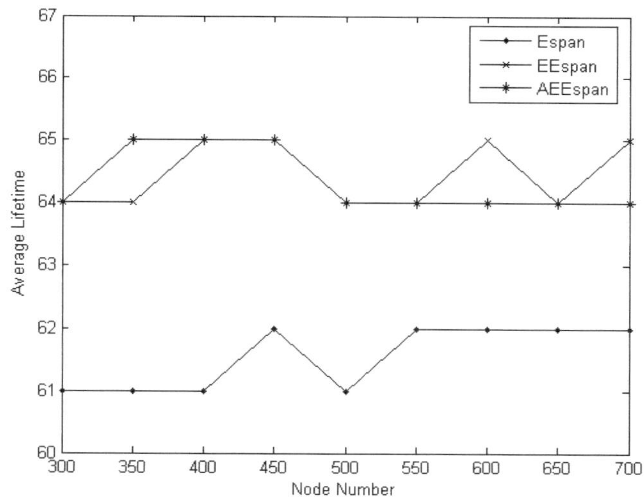

Fig. 12. Average lifetime comparison

As mentioned before, energy efficiency is a main goal of algorithms in wireless sensor networks. By decreasing energy consumption that leads to prevent from failing network nodes, network's coverage whether spatial or temporal is supported better and the network's lifetime increases. AEEspan algorithm by decreasing transmission volume, can meet this goal.

In figure 12, for these algorithms, the average lifetime is plotted versus the number of nodes. The results are obtained after 20 different simulation trials. As shown in figure 8, the presented algorithm has higher lifetime than the other algorithms. Based on the lifetime definition, lifetime has direct relation to alive node numbers.

4. Conclusion

One of the most important constraints in wireless sensor networks is the energy consumption. Aggregation algorithms have a considerable role in decreasing the energy consumption due to the reduction of the transmitted data volume. Data aggregation has been put forward as an essential paradigm for wireless routing in sensor networks. The idea is to combine the data coming from different sources, eliminating redundancy, minimizing the number of transmissions and thus saving energy. In this work, an energy efficient algorithm to construct the aggregation tree is presented. The algorithm considers both energy and distance to construct the aggregation tree. Simulation results show that the algorithm has better performance than the existing algorithms and also, the algorithm decreases the number of failed nodes and provides higher network lifetime and better coverage. To construct the aggregation tree, routing packets are flooded into the network periodically that leads to waste of energy. To omit this overhead, we introduce automata-based reconfiguration property. An automaton is an able-to-learn structure which tries to choose the best path to send the data to the root by getting feedback from the environment.

Also, by preventing from flooding the routing packet into the entire network, the presented algorithm consumes less energy.

5. References

Akyildiz, F.; Su, W.; Sankarasubramaniam, Y. & Cayirci, E. (2002), Wireless Sensor Networks: A Survey, *Computer Networks Journal*

Ankit, M.; Arpit, M.; Deepak, T.; Venkateswarlu, R. & Janakiram, D. (2006), TinyLAP: A Scalable Learning Automata- Based Energy Aware Routing Protocol for Sensor Networks, IEEE

Beyens, P.; Peeters, M.; Steenhaut, K. & Nowe, A. (2005), Routing with compression in wireless sensor networks: a Q-learning Approah, AAMAS

Cantoni, V.; Lombardi, L. & Lombardi, P. (2006), Challenges for Data Mining in Distributed Sensor Networks, IEEE Computer Society

Chen, Y.; Liestman, A. & Liu, J. (2005), Energy-Efficient Data Aggregation Hierarchy for Wireless Sensor Networks, the 2nd Int'l Conf. on Quality of Service in Heterogeneous Wired/Wireless Networks

Chlamtac, I. & Kutten, S. (1987), Tree-based broadcasting in multi-hop radio networks, *IEEE Transactions on Computers*

Chlamtac, I. & Weinstein, O. (1991), The wave expansion approach to broadcasting in multi-hop radio networks, *IEEE Transactions on Communications*

Eskandari, Z.; Yaghmaee, M.H.; Mohajerzade, A.M. (2008), Energy Efficient Spanning Tree for Data Aggregation in Wireless Sensor Networks, IEEE ICCCN

Eskandari, Z.; Yaghmaee, M.H & Mohajerzade, A.M. (2008), Automata based Energy Efficient Spanning Tree for Data Aggregation in Wireless Sensor Networks, IEEE ICCS

Eskandari, Z.; Yaghmaee, M.H. (2009), AEESPAN: Automata based Energy Efficient Spanning Tree for Data Aggregation in Wireless Sensor Networks, *International Wireless Sensor Network (WSN) journal*

Esnaashari, M. & Meybodi, M. (2007), a learning automata based data aggregation method doe sensor networks, CSICC

Heinzelman, W.R; Chandrakasan, A. & Balakrishnan, H. (2000), Energy-efficient communication protocol for wireless micro sensor networks, International Conference on System Science

Intanagonwiwat, C.; Estrin, D.; Govindan, R. & Heidemann, J. (2002), Impact of network density on data aggregation in wireless sensor networks, International Conference on Distributed Computing Systems

Kamimura, J.; Wakamiya, N. & Murata, M. (2004), Energy-Efficient Clustering Method for Data Gathering in Sensor Networks, BROADNETS

Krishnamachari, B.; Estrin, D. & Wicker, S. (2002), The Impact of Data Aggregation in Wireless Sensor Networks, International Workshop on Distributed Event-Based Systems

Lee, M. & Wong, V. (2005), An Energy-Aware Spanning Tree Algorithm for data aggregation in wireless sensor networks, IEEE

Lee, M. & Wong, V. (2005), LPT for Data Aggregation in Wireless Sensor Networks, IEEE GLOBECOM

Lee, W.M. & Wong, V. (2006), E-Span and LPT for data aggregation in wireless sensor networks, *Elsevier*

Liang, B. & Liu, Q. (2006), A Data Fusion Approach for Power Saving in Wireless Sensor Networks, IEEE

Min, R. & Chandrakasan, A. (2001), Energy-efficient communication for ad-hoc wireless sensor networks, the Thirty-Fifth Asilomar Conference on Signals Systems and Computers

Radivojac, P.; Korad, U.; Sivalingam, K.M. & Obradovic, Z. (2003), learning from class-imbalanced data in wireless sensor networks, IEEE VTC

Upadhyayula, S.; Annamalai, V. & Gupta, S. (2003), A low-latency and energy-efficient algorithm for convergecast in wireless sensor networks, IEEE Global Communications Conference

Upadhyayula, S. & Gupta, S.K.S. (2006), Spanning tree based algorithms for low latency and energy efficient data aggregation enhanced convergecast (DAC) in wireless sensor networks, *Elsevier*

Younis, O. & Fahmy, S. (2004), HEED: A hybrid, energy- efficient, distributed clustering approach for ad hoc sensor networks, IEEE

Zhang, W. & Cao, G. (2004), DCTC: Dynamic convoy tree-based collaboration for target tracking in sensor networks, IEEE

Zhu, J.; Papavassiliou, S.; Kafetzoglou, S. & Yang, J. (2006), An Efficient QoS-Constrained Data Aggregation and Processing Approach in Distributed Wireless Sensor Networks, IEEE

8

Distributed Localization Algorithms for Wireless Sensor Networks: From Design Methodology to Experimental Validation

Stefano Tennina
Department of Electrical and Information Engineering, Center of Excellence in Research DEWS, University of L'Aquila, 67040 L'Aquila
Corresponding author
Italy

Marco Di Renzo
Institute for Digital Communications (IDCOM), School of Engineering, College of Science and Engineering, The University of Edinburgh, Alexander Graham Bell Building, The King's Buildings, Mayfield Road, Edinburgh EH9 3JL, Scotland
United Kingdom (UK)

Fabio Graziosi and Fortunato Santucci
Department of Electrical and Information Engineering, Center of Excellence in Research DEWS, University of L'Aquila, 67040 L'Aquila
Italy

Abstract

Recent advances in the technology of wireless electronic devices have made possible to build ad–hoc Wireless Sensor Networks (WSNs) using inexpensive nodes, consisting of low–power processors, a modest amount of memory, and simple wireless transceivers. Over the last years, many novel applications have been envisaged for distributed WSNs in the area of monitoring, communication, and control. Sensing and controlling the environment by using many embedded devices forming a WSN often require the measured physical parameters to be associated with the position of the sensing device. As a consequence, one of the key enabling and indispensable services in WSNs is localization (i.e., positioning).

Moreover, the design of various components of the protocol stack (e.g., routing and Medium Access Control, MAC, algorithms) might take advantage of nodes' location, thus resulting in WSNs with improved performance. However, typical protocol design methodologies have shown significant limitations when applied to the field of embedded systems, like WSNs. As a matter of fact, the layered nature of typical design approaches limits their practical usefulness for the design of WSNs, where any *vertical* information (like, e.g., the actual node's position) should be efficiently shared in such resource constrained devices. Among the proposed solutions to address this problem, we believe that the Platform–Based Design (PBD) approach Sangiovanni-Vincentelli (2002), which is a relatively new methodology for the design of embedded systems, is a very promising paradigm for the efficient design of WSNs.

In particular, the PBD methodology allows to define a standard set of services and interface primitives (called Sensor Network Services Platform or SNSP) that can be made available to an application programmer independently from implementation issues on any (wireless) sensor network platform.
In the depicted context, the present contribution reports our recent research advances along two main directions. Firstly, we exploit the PBD methodology for the efficient design of ad–hoc WSNs with localization capabilities. In particular, the PBD paradigm is used to derive a fully distributed positioning algorithm, and a general protocol architecture for WSNs. Secondly, we validate the suitability of a practical implementation of the proposed solutions onto commercially available WSN platforms, and analyze their achievable performance in realistic propagation environments.
More specifically, the contributions of the present research work are as follows: 1) we will define a PBD–inspired Location Service (LS) along with its parameters and service primitives, which collects and provides network–wide information about the nodes' spatial position, 2) we will introduce a novel iterative positioning algorithm, which is called Enhanced Steepest Descent – ESD Tennina et al. (n.d.), and will show, by using computer–based simulations, that it can outperform other well–known distributed localization algorithms in terms of estimation accuracy and numerical complexity, 3) we will analyze the implementation issues related on mapping the ESD algorithm onto the CrossBow's MICAz sensor node platform, and investigate, via experimental activities, the effect of network topology and ranging errors on the performance of the proposed distributed localization algorithm, and 4) we will test the performance of the ESD algorithm during an extensive campaign of measurements conducted by using the Texas Instruments (TI)/Chipcon CC2431's hardware location–finder engine in a realistic and dynamic indoor propagation environment. We will show that the ESD algorithm can be efficiently used to improve the localization accuracy provided by the CC2431's location–finder engine. Moreover, as a byproduct of this latter experimental activity, we will show that the need of site–specific parameters for the correct operation of the CC2431's location–finder engine may severely reduce the localization accuracy of the system in dynamic environments, as well as propose and validate a simple solution to counteract this problem.

Keywords: Platform Based Design (PBD), positioning, ad–hoc wireless sensor networks.

1. Introduction

1.1 Overview

Wireless Sensor Networks (WSNs) are distributed networked embedded systems where each node combines sensing, computing, communication, and storage capabilities Goldsmith & Wicker (2002). Due to their unprecedented design challenges and potentially large revenues, in recent years WSNs have witnessed a tremendous upsurge in interest and activities in both academia and industry Dohler (2008). In particular, they have become increasingly popular in military and civilian sectors, and have been proposed for a wide range of application domains, e.g., control and automation, logistics and transportation, environmental monitoring, healthcare and surveillance.
In general, WSNs are required to possess self–organizing capabilities, so that little or no human intervention for network deployment and setup is required. A fundamental component

of self–organization is the ability of sensor nodes to "sense" their location in space, i.e., determining where a given node is physically located in a network Bachrach & Taylor (2005); Wang & Xiao (2007). In particular, node localization is a key enabling capability to support a rich set of geographically aware protocols for distributed and self–organizing WSNs Mauve & Widmer (2001), and for achieving context–awareness.
It is well–known Hofmann-Wellenhof et al. (1997), that the Global Positioning System (GPS) can greatly facilitate the task of location estimation by potentially allowing every GPS–equipped receiver to accurately localize itself in any point located on or above the Earth surface. However, GPS–based localization solutions are often considered a non–completely viable and well–suited solution for position estimation in WSNs, as sensor nodes are supposed to operate at low–complexity and low–power consumptions Bulusu et al. (2000). Moreover, GPS–based solutions have the undesirable side–effect that they cannot provide reliable location estimates in indoor environments, and in the presence of dense vegetation Perkins et al. (2006); Savvides et al. (2001). As a consequence of the above, much research has been done in the WSNs community to develop new techniques for localization in those environments where GPS–aided positioning is either unfeasible or does not meet the design requirements and paradigms of networked embedded systems, i.e., the so–called GPS–denied (or GPS–less) environments. The result of this intensive research work has been the proposal of many new solutions (alternative to GPS) to address the problem of distributed network location discovery (see, e.g, Santucci et al. (2006) and references therein). However, in Langendoen & Reijers (2003); Wang & Xiao (2007) the authors have clearly shown that among the existing algorithms none seems to perform better than the others, and claim that the definition of location algorithms with accurate positioning capabilities and low communication and computation costs for GPS–denied environments is still an ongoing area of research at both theoretical and experimental levels.
Furthermore, existing solutions for location estimation have often been obtained without considering the fundamental interactions of positioning algorithms with other entities of the protocol stack: in other words, current solutions do not adopt a methodological view of the whole protocol stack for system optimization. As a matter of fact, the traditional design approach is based upon the ISO–OSI layered model, i.e., the whole system is decomposed in a layered fashion, and the design of each layer follows the *isolation* principle. In general, lower layers are abstracted by means of a set of service primitives, while the higher layers in term of service requirements. This approach greatly simplifies the design task, but may lead to sub-optimal design solutions Kawadia & Kumar (2005). Moving from this consideration, novel design approaches are being developed by several researchers with the aim to design more efficient protocol solutions. Among the various alternatives, cross–layer design methodologies Srivastava & Motani (2005) are receiving a significant interest by the research community. In particular, the cross–layer approach advocates the benefits, in terms of costs and performance, of a joint design of the functionalities at different layers. In fact, it allows to reduce the duplication of functionalities, which may arise when designing each layer in isolation, and provides a joint optimization of system parameters. Nevertheless, cross–layer design is known to raise the design complexity, and to reduce the modularity and thus the re–use of system components Kawadia & Kumar (2005).

1.2 Aim and Motivation

In the light of the above overview, the main aim of the present manuscript is threefold: i) to propose the adoption of a novel methodology to design an efficient Location Service for

WSNs, thus overcoming the limitations of current design methodologies based on ISO–OSI and cross–layer paradigms, ii) to introduce a novel atomic localization algorithm with improved performance with respect to current solutions, and iii) to offer a solid proof of concept of the proposed methodologies and algorithms by means of computer simulations and experimental activities conducted with some WSNs testbeds.

1.2.1 The Need for A Novel Methodological Approach

In general, the WSN domain presents several challenging problems: it is characterized by hard real-time constraints, it has to be fault tolerant and design–error free, and it has to react to a nondeterministic adversary environment. Although existing cross–layer design paradigms seem to solve the limitations shown by the ISO–OSI approach, we emphasize a methodology that favours re–use at all levels of abstraction to keep the design complexity at a moderate level. The goal is to design a sensor node which is able to reconfigure itself and to form a network without any need for expensive infrastructure.

To meet the above design goals and requirements of WSNs, we adopt a recently proposed design methodology for embedded wireless systems, which is called Platform Based Design (PBD) Sangiovanni-Vincentelli (2002). The basic tenets of this methodology are: i) an orthogonalization of concerns, i.e., the separation of the various aspects of design to allow more effective exploration of alternative solutions, and ii) a meet–in–the–middle process, where successive refinements of specifications meet with abstractions of potential implementations. Basically, orthogonalization of concerns pushes to identify parts of the system which are independent enough (*orthogonal*) to be designed in separate steps. This is the same approach pursued in the traditional ISO/OSI model, where orthogonal functionalities of a network node have been identified and grouped in the well–known 7 layers (application, presentation, session, transport, network, data link and physical layer). Moreover, the meet-in-the-middle process advocates a richer abstraction of a layer, where services are exposed together with a model of cost/performance. The expression *meet–in–the–middle* thus comes from the fact that the design of a layer is neither subject to the higher layer requirements, as in a top–down approach, nor to the lower layer features, as in a bottom–up approach. Instead, service requirements are defined with a notion of the potential capabilities, performance and related costs of the lower layers (called *platforms* in the methodology). In other words, the meet–in–the–middle view of the design process defines an approach that maximizes re–usability and verifiability, while mantaining constraints on performance, cost and power consumption.

Furthermore, in recent years, the adoption of PBD has been proposed for the design of communication protocols Sgroi et al. (2000) and communication infrastructures Pinto (2008); Pinto et al. (2008), with particular emphasis on the challenges of wireless communications da Silva et al. (2000). In Bonivento et al. (2005), the methodology is applied to wireless networked control systems, with the definition of a flow based on three layers of abstractions, which takes into account both the design of the control algorithm and of the distributed architecture, as well as the definition of the control application to be mapped over the network nodes. In Balluchi et al. (2004), the platform–based design approach has been applied to the design of wireless sensor networks, with the definition of a *Network Platform* as a collection of services. Motivated by the above considerations, the first aim of this research work is to show how the PBD tenets can be applied for the design of an efficient distributed Location Service for WSNs.

1.2.2 The Need for Improved Localization Algorithms

Although several optimization algorithms for location estimation have been proposed in the literature to date, in Wang & Xiao (2007) the authors have recently shown that each of them exhibits advantages and disadvantages in terms of computational cost, overall accuracy, and suitability to be deployed onto today's available WSNs' devices. Accordingly, one aim of this contribution is to introduce a novel and more efficient (in terms of computational cost and accuracy) optimization algorithm suitable for distributed WSNs localization.

Among the various solutions so far proposed in the literature, many authors agree that a promising approach for distributed sensor node localization is the so–called "recursive positioning methods", see e.g. Santucci et al. (2006); Savvides et al. (2001); Wang & Xiao (2007). Loosely speaking, recursive algorithms are often employed to overcome the limits related to the short–range communication capabilities of sensor nodes, by enabling the position estimation process to be composed by many subsequent steps/phases through which all the sensors in the network localize themselves in a distributed fashion Santucci et al. (2006); Savarese (2002). These techniques have several positive features, e.g., i) they appear to be a good solution for sensor nodes with limited range capabilities, ii) they may efficiently counteract the sparse anchor node problem, and iii) they are distributed by nature. However, they still present several critical design issues, e.g., i) in Savvides et al. (2001) authors have shown that in recursive approaches the positioning error may accumulate along the iterative process, thus severally corrupting the final estimates of sensor nodes located in remote areas, i.e., regions of the network where "startup anchors" (i.e., nodes that are aware of their exact location) are sparse, and ii) in Dulman et al. (2008); Savarese (2002) authors have verified that some bad network topologies may introduce significant errors even with accurate distance estimates. In particular, to cope with error accumulation, accurate optimization algorithms have to be used for position estimation, but typically with high computational costs and long time, which may represent a serious limitation for handling e.g. nodes' mobility.

Motivated by these considerations, the second aim of the present contribution is twofold: i) to propose a comparative study of various optimization algorithms Nocedal & Wright (2006) that can be used for position estimation, and ii) to propose an enhanced version of the classical Steepest Descent algorithm, which we call Enhanced Steepest Descent (ESD), for improving the efficiency of position estimation.

1.2.3 The Need for Experimental Analysis and Validation

Although most analysis about the performance of WSNs are often conducted via computer–based (numerical) simulations, such a kind of analysis typically show significant limitations to assess the actual improvement and implementation issues of the proposed solutions when the algorithms need to be implemented onto today's available sensor nodes platforms, and when the WSN needs to be deployed in a realistic propagation environment. A couple of examples of these issues may be as follows: i) most analysis conducted via numerical simulations do not take into account the actual and limited capabilities of commercially available sensor nodes, which often results in the development of novel solutions that, even providing improved performance, are not implementable onto sensor nodes platforms due to their high computational complexity and memory requirements, and ii) numerical simulations typically rely on important assumptions to reproduce, e.g., ranging (i.e., the distance estimation between pairs of nodes) error models and the wireless propagation conditions, which may not represent in a consistent way the actual technique used for ranging computation, as well as the actual characteristics of the wireless propagation channel (e.g., the presence of obstacles, non–line–of–sight propagation scenarios, and dynamic motion of objects or people around the area of interest), respectively. Actually, in the recent period the problem of understanding the real impact of the assumptions typically done for the analysis of ad–hoc networks via computer–based simulations is receiving a growing attention by the research community. In particular, recent papers, e.g., Newport et al. (2007), have claimed and verified via experiments that wrong or simplistic assumptions of how radios work may result in a completely different behavior and performance between simulation and experimentation. Accordingly, the authors suggest to either use real data as input to simulators or cross–validating simulated results with accurate experimental activities.

Motivated by the above considerations, the third aim of the present contribution is to validate the applicability and efficiency of the proposed PBD methodology by means of a WSN testbed deployed in a realistic propagation environment, as well as to analyze the performance improvement provided by the proposed ESD algorithm via experimental activities. In particular, we will describe two campaign of measurements aiming at analyzing the achievable performance (i.e., localization accuracy and reliability) of two WSNs testbed platforms implemented using commercially available sensor nodes. The measurement campaigns are performed in two typical GPS–denied environments represented by static and dynamic indoor scenarios. The WSNs testbed platforms are currently available at the Center of Excellence in Research DEWS (University of L'Aquila, Italy) – www.dews.ing.univaq.it/dews, and the Networked Control Systems Laboratory (NCSlab) (the Italian node of the European Embedded Control Institute (EECI) at the University of L'Aquila) – www.eeci-institute.eu, and are being extensively used for the analysis and design of WSNs for positioning applications.

1.3 Contribution

Motivated by the above considerations, the specific contributions of the present chapter are as follows: i) we will present a PBD–based Location Service for WSNs and define the set of primitives required for its implementation, ii) we will propose a novel distributed optimization algorithm for nodes' position estimation, which is an enhanced version of the classical Steepest Descent and is called ESD, iii) the proposed solution will be compared, via computer–based simulations, with other well–known optimization algorithms available in the open technical literature, and its improved performance in terms of error accuracy, computational complexity (i.e., time required to estimate the final position), algorithm initialization, and network topology will be investigated and discussed, iv) we will show that the ESD algorithm can be readily implemented onto the CrossBow's MICAz sensor node platform Cro (2008), and will substantiate and validate, via experimental activities, the results obtained via simulation when realistic ranging measurements are used at the input of the algorithm, and v) by means of off–line computer simulations performed on real captures acquired with the TI/Chipcon's CC2431[1] testbed Tex (2008) in a highly dynamic indoor environment, we will prove that the ESD algorithm can be effectively used to further refine the position estimated by the CC2431's location engine, thus yielding a non–negligible improvement in estimating the actual position of a sensor node with a modest increment in computational complexity.

[1] CC2431 sensor nodes developed by Texas Instruments (TI)/Chipcon are widely recognized as the first commercially available System–on–Chip (SoC) solution with a hardware RSS–based (Received Signal Strength) location–finder engine targeting ZigBee/IEEE 802.15.4 wireless sensor networking applications

1.4 Paper Organization

The reminder of this paper is organized as follows. In Section 2 the PBD–based Location Service is described; the functional decomposition into several platforms and their primitives are then provided. By focusing on the positioning algorithm, Section 3 will describe several optimization algorithms for WSNs position estimation, and will introduce the proposed ESD algorithm. In Section 4, simulation results will be presented and commented. In Section 5, the testbed platforms deployed using both Crossbow's MICAz and TI/Chipcon's CC2431 sensor nodes will be introduced, practical implementation issues of the ESD algorithm will be addressed, and experimental results will be discussed either in static and dynamic indoor environments. Finally, Section 6 will conclude the paper.

2. Location Service Design for WSNs: A PBD–based Approach

Moving from the basic tenets of the PBD approach described in Sgroi et al. (2003), we consider a node architecture as depicted in Fig. 1, where i) an Application Interface (API) exposes the set of relevant services and hides lower networking details; and ii) the Sensor Network Service Platform (SNSP) is a middleware layer of services, which implements the exposed functionalities by resorting to the underlying protocol stack entities. Among the SNSP Sgroi et al. (2003),

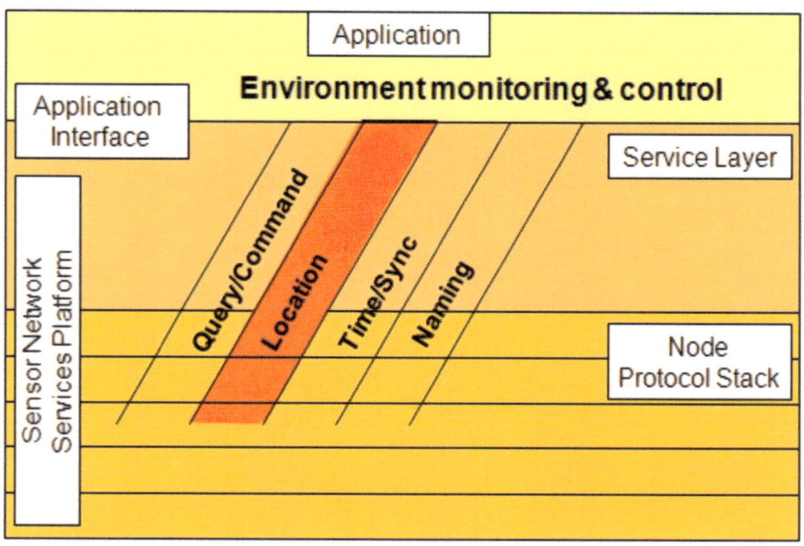

Fig. 1. Sensor Network Service Platform and Application Interface.

the Location Service (LS) collects and provides information about the spatial position of the nodes in the network. A *point location* is defined as a t–ple of values, which identify the position of the node within a reference system. Assuming, e.g., a common 3D cartesian reference system, a location (i.e., node's position) is a *struct* type collecting fields such as: i) the nodes' coordinates (x, y, z); ii) a scale factor, which defines the resolution, and iii) the accuracy level yielding the reliability indicator of an estimated position.

Fig. 2 shows the functional decomposition of our designed Location Service into several PBD platforms, each one characterized by the relevant set of primitives (i.e., services) exposed towards the upper layer, hiding lower level details. In other words, in this framework, the level of details increases when moving from the top to the bottom of the protocol stack. In what follows, we will briefly outline the set of primitives of each defined platform.

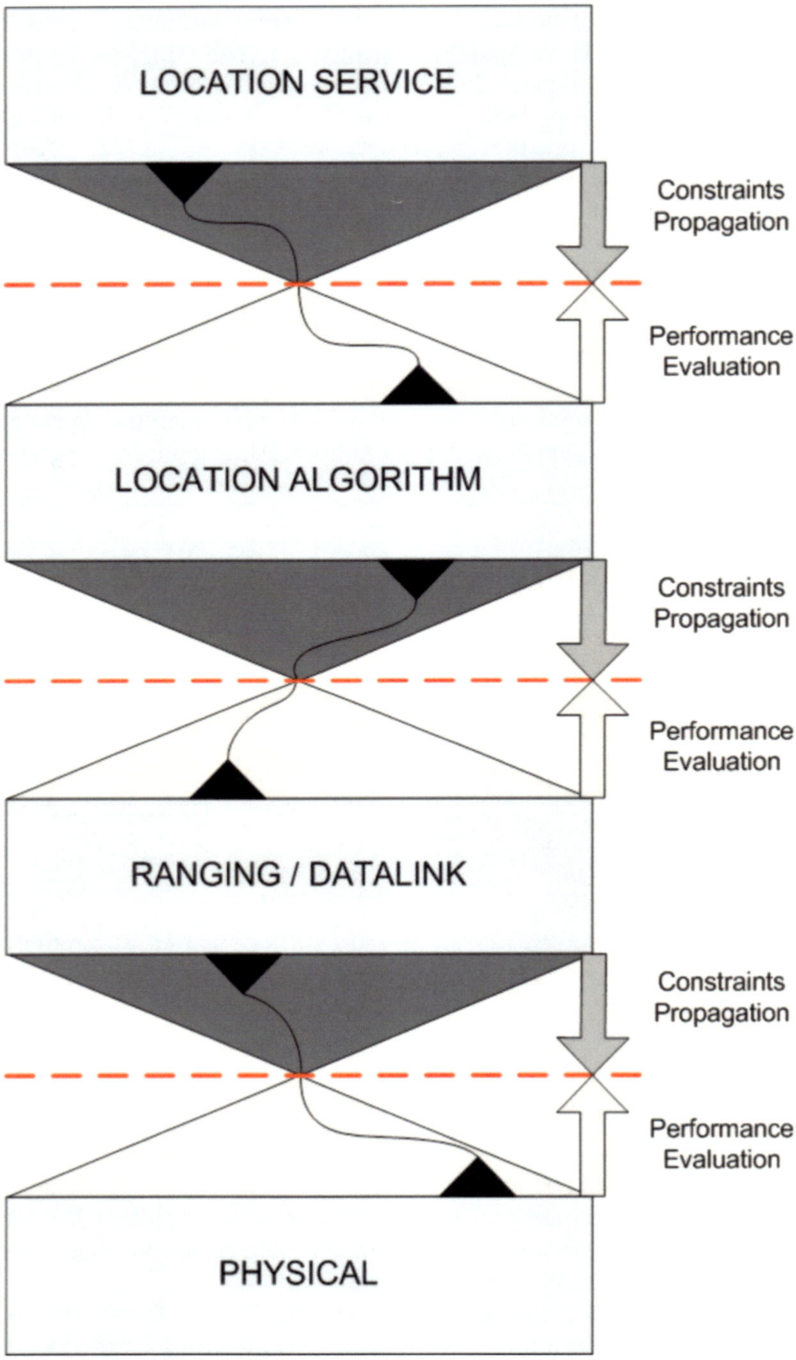

Fig. 2. Location Service Platform Stack.

2.1 Location Service Platform

According to the general setup introduced in Sgroi et al. (2003), the following set of LS primitives and related parameters are assumed at the application interface.

- *int LSSetup(struct resolution *r, struct accuracy *a, struct reference *rs, int Time Tmax)* sets the resolution and the accuracy of location data, the reference system and the maximum time interval for obtaining the location data. A call to this primitive also starts the LS service.
- *int LSUpdate(struct resolution *r, struct accuracy *a, struct reference *rs, int Time Tmax).* Similar to the *LSSetup()*, but at run–time.
- *struct location LSGetLocation(int NodeID)* returns the location of the node with $ID = NodeID$.

Accordingly, at the highest layer, we consider a set of primitives which simply consists in the attempt of each node to be aware of its position as soon as it starts operating into the network. Furthermore, this layer imposes a set of requirements which propagate deeply in the stack and which has to be met by the lower levels, i.e., by choosing the proper solutions. In this case, these requirements typically deal with i) the maximum allowable accuracy of the final position estimation of each node, ii) the maximum percentage of nodes allowed to remain unlocalized, and iii) the maximum time required to complete the position estimation algorithm.

2.2 Location Algorithm Platform

This is the core platform, where the mathematical model of the positioning algorithm is defined and performance is evaluated in order to meet the previous application's requirements. In this platform, we can consider the class of distributed and cooperative recursive positioning algorithms briefly described in Section 1.2.2. A set of primitives is listed and briefly described in what follows.

- *float distance LAGetRange(int NodeID)* operates a cooperative ranging[2] procedure between a node and the neighbor having $ID = NodeID$. NodeID denotes a node identifier, which is used by a node to identify its neighbors;
- *struct location LAInitialEstimation()* returns the initial position estimation according to a predefined criterion. Alternatives for initial estimation include the simple random guess, as well as a smarter, but more complicated, solution like in Savarese (2002);
- *struct location LAStep(struct location *arrNeighsLoc, struct refinementParameters par, int Time T_{up})* proceeds one step ahead with the positioning algorithm once new information about positions of neighbors is collected. It returns the updated position estimation of the present node. When a stop criterion is reached[3], node starts broadcasting its actual estimation;
- *void LABroadcast(struct location *loc)* locally broadcasts the present position and accuracy of the estimate as well;
- *LACoordination(struct location *loc)* is invoked when a node with insufficient connectivity cannot resolve an ambiguity in position estimation and requires cooperation of its neighbors.

---

[2] *Ranging* is the process of estimating the distance between a pair of nodes Tennina et al. (n.d.).

[3] A stop criterion may deal with the fact that the desired accuracy has been reached or that the timeout (T_{up}) is expired

2.2.1 Recursive Positioning Method

As discussed in Section 1, we will consider a recursive positioning method for network location discovery. In particular, the well–known recursive and hierarchical method proposed by Santucci et al. (2006); Savvides et al. (2001) is analyzed for the sake of illustration. The following notation is used: i) a *blind* node is a node not aware of its position, ii) a *startup anchor* is a node aware of its position since the beginning of the location discovery procedure, and iii) a *converted anchor* is a blind node that has estimated its position the location discovery process. The basic version of the algorithm involves the following steps:

- Phase 0: At the beginning, "startup anchors" broadcast their position.
- Phase 1: Blind nodes that are connected (i.e., they are in the neighborhood) to at least four[4] "startup anchors" compute their position.
- Phase 2: Once a blind node has estimated its position, it becomes a "converted anchor" and broadcasts its estimated position to other nearby blind nodes, thus enabling them to estimate their positions.
- Phase 3: This process is repeated until the positions of all the nodes that eventually can have either four "startup anchors" or "converted anchors" are estimated.
- Phase 4: In this phase, an attempt is performed to solve eventual ambiguities for those nodes that do not have sufficient connectivity, by assuming cooperative decisions within the set of neighboring nodes (*LACoordination()*).

As a consequence, depending on the current step of the algorithm, the four anchor nodes with known positions may be either "startup anchors" or "converted anchors". Of course, differently from "startup anchors", the position of the "converted anchors" is affected by a certain error. In what follows, we will denote with "reference nodes" both "startup" and "converted" anchors.

2.2.2 Position Computation

The recursive positioning method described in Section 2.2.1 requires a technique to compute the location of a blind node from the position of four "reference nodes", which may be in part "startup anchors" and in part "converted anchors". In general, the computation of the position of a blind node involves two basic steps: i) measuring the distances between pairs of sensors using the *LAGetRange()* function, and ii) estimating the node's position via the optimization of a given cost function obtained from the measured distances, using the *LAInitialEstimation()* function first and the *LAStep()* function after.

With regard to position computation from range estimates, in the literature two basic family of algorithms are often considered: i) *triangulation*, which foresees to estimate the position of the unknown node by finding the intersection of four spheres in a three–dimensional environment, and ii) *multilateration*, according to which the estimated position is obtained by reducing the difference between the actual measured distances and the estimated Euclidean distances between blind and reference nodes, i.e., via the minimization of an error cost function. According to Wang & Xiao (2007), the main difference between the two approaches is that multilateration algorithms are more robust to noisy range measurements. Both methods will be analyzed and compared in the present manuscript.

---

[4] In order to compute the position of a blind node we need, at least, ranging measurements from four anchor nodes in a three–dimensional space Tennina et al. (n.d.).

2.3 Ranging / DataLink Platform

With regard to distance estimation between pairs of nodes, several methods have been proposed in the literature (see, e.g., Patwari et al. (2005) for a survey). In our design stack, this role is managed by the Ranging/Datalink platform, which joins the design of the ranging capabilities of a node (i.e., which kind of algorithm is used to estimate the mutual distances) with the design of the communication protocol (i.e., which kind of medium access control is adopted and how the message passing is accomplished). A subset of primitives is listed and briefly described in the following.

- *float distance RDLGetRange(object source, float tol, float accuracy)* is the primitive by which a distance measurement (i.e., ranging) is obtained from the measurement of a physical parameter, which can be, e.g., the received signal strength (RSS) or the signal time of arrival (ToA). In addition, this primitive prepares the underlying physical entities for starting ranging operations, which typically require better accuracy if compared to the one adopted for communication purposes;
- *int RDLSetBOParameters(int BO, int α, int β)* is the primitive which sets the maximum value of the actual BackOff counter, as well as the parameters α and β, which represent the increasing and decreasing steps for the back off counter, respectively;
- *int RDLIncreaseBOValue(int BO, int α)* is the primitive that increases the BackOff counter based on a parameter α and some rules, as Binary Exponential Backoff or Multiplicative Increase Linear Decrease (MILD) Bhargahavan et al. (1994);
- *int RDLDecreaseBOValue(int BO, int β)*. Similar to the previous one, this primitive decreases the BackOff counter based on a quantity β and some rule;
- *int RDLSend(int NodeID, object Data, int Time T_{trans})* is a primitive which allows the node to send Data to the neighbor having $ID = NodeID$, subject to a timeout (T_{trans}) for the transaction;
- *int RDLReceive(object *Data)* is a primitive alerting the node about the arrival of Data from a neighbor.

As a matter of fact, at this level we have designed a subset of basic communication primitives, i.e., those derived from the class of Carrier Sense Multiple Access (CSMA) with Collision Avoidance (CA) MAC protocols, adopted e.g. in *IEEE Std 802.15.4: Wireless Medium Access Control (MAC) and Physical Layer (PHY) Specifications for Low-Rate Wireless Personal Area Networks (WPANs)* (2006), jointly with the definition of some primitives structured to support cooperative ranging in the contest of the mentioned localization application.

2.4 Physical Platform

This platform allows physical connectivity among nodes within the transmission range. For positioning purposes, we can define the following service primitives.

- *int PhySetup(int PTX, float T_{RES}, float δ)* sets values of physical parameters, such as transmission power level, maximum time resolution and delay for a synchronization process.
- *int PhyUpdate(int PTX, float T_{RES}, float δ)*. Similar to the *PhySetup()*, but at run–time.
- *object sourceVal PhyGetRange(object source, float T_{RES})* is the primitive which supports *RDLGetRange()*. It gets the value of a physical parameter, so that *RDLGetRange()* can convert this value in a distance estimate based on the type of physical parameter and

a mathematical model of conversion. For example, if ranging measure is done by RSS, the source would likely be an acquisition via ADC (Analog to Digital Converter) of the incoming signal strength.

At this interface several other internal parameters can be considered, due to the complexity of the transceiver design. Most of these parameters are also handled by Resource Management Service "which allows an Application or a Service to get or set the state of the physical elements of the hardware" Sgroi et al. (2003). Due to likely tight constraints on spatial resolution for ranging measurements, wideband or Ultra Wide–Band are interesting opportunities for signal design at the physical layer. If compared to other technologies for ranging, e.g., ultrasound *Calamari Project* (n.d.), UWB may provide highest resolution because it relies on very short impulses and large bandwidth, and ranging can be somehow embedded in a synchronization process with tuneable settings[5].
However, in the present contribution we consider the RSS measurements at the *PhyGetRange()* function (and in turn at the *RDLGetRange()* function), as it is nowadays a measure easily available on many commercial off–the–shelf sensor node platforms, such as the CrossBow's MICAz and the TI/Chipcon's CC2431 ones, which are used in our experimental activities and measurements. To be used in practice (see Section 5.1), RSS–based techniques need a calibration phase to estimate the path loss low, a relation between the received signal power and the actual distance between the nodes (by assuming the transmit power is known and fixed). These calibration issues will be analyzed in the present paper, as well as the impact of outdated measurements on the system performance.

3. ESD: A Novel Localization Algorithm for WSNs

3.1 Notation

The aim of this section is to introduce a novel localization algorithm for WSNs. To do so, let us first introduce some basic notations useful for analytical formulation. By assuming an area with N_A, $\{A_i\}_{i=1}^{N_A}$, "startup anchors" and N_U, $\left\{U_j\right\}_{j=1}^{N_U}$, blind nodes, the following notation will be used throughout this chapter: i) bold symbols will be used to denote vectors and matrices, ii) $(\cdot)^T$ will denote transpose operation, iii) $\nabla(\cdot)$ will be the gradient, iv) $\|\cdot\|$ will be the Euclidean distance and $|\cdot|$ the absolute value, v) $\angle(\cdot,\cdot)$ will be the phase angle between two vectors, vi) $(\cdot)^{-1}$ will denote matrix inversion, vii) $\hat{\mathbf{u}}_j = \left[\hat{u}_{j,x}, \hat{u}_{j,y}, \hat{u}_{j,z}\right]^T$ will denote the estimated position of the blind node $\left\{U_j\right\}_{j=1}^{N_U}$, viii) $\mathbf{u}_j = \left[u_{j,x}, u_{j,y}, u_{j,z}\right]^T$ will be the trial solution of the optimization algorithm, ix) $\bar{\mathbf{u}}_i = [x_i, y_i, z_i]^T$ will be the positions of the reference nodes $\{A_i\}_{i=1}^{N_A}$, and x) $\hat{d}_{j,i}$ will denote the estimated (via ranging measurements) distance between reference node $\{A_i\}_{i=1}^{N_A}$ and blind node $\left\{U_j\right\}_{j=1}^{N_U}$. Moreover, for analytical simplicity, but without loss of generality, we will present the optimization algorithms by assuming $N_U = 1$ and $N_A = 4$.

[5] At the receiver, synchronization can be done by using a correlation mechanism between the received signal and a local signal (template) Stiffler (1968) or a delayed version of the received signal itself (differential receiver Alesii, Antonini, Di Renzo, Graziosi & Santucci (2004); Alesii, Di Renzo, Graziosi & Santucci (2004))

Before going into the details of the novel ESD algorithm, let us also summarize some basic localization methods with the aim to highlight the main advantages and superiority of the proposed solution.

3.2 Triangulation Method

In this method, the position of node U_1 is obtained by inferring a geometric triangulation among estimated and actual distances. Accordingly, the unknown position is obtained by finding a solution that simultaneously solve the following set of equations:

$$\begin{cases} (x_1 - u_{1,x})^2 + (y_1 - u_{1,y})^2 + (z_1 - u_{1,z})^2 = \hat{d}_{1,1}^2 \\ (x_2 - u_{1,x})^2 + (y_2 - u_{1,y})^2 + (z_2 - u_{1,z})^2 = \hat{d}_{1,2}^2 \\ (x_3 - u_{1,x})^2 + (y_3 - u_{1,y})^2 + (z_3 - u_{1,z})^2 = \hat{d}_{1,3}^2 \\ (x_4 - u_{1,x})^2 + (y_4 - u_{1,y})^2 + (z_4 - u_{1,z})^2 = \hat{d}_{1,4}^2 \end{cases} \tag{1}$$

This system of equations can be solved using a Least Squares solution, which yields $\hat{\mathbf{u}}_1 = \left(\mathbf{A}^T\mathbf{A}\right)^{-1}\mathbf{A}^T\mathbf{b}$, where matrix $\mathbf{A}$ and vector $\mathbf{b}$ can be found in Savarese (2002). In general, triangulation methods may fail to find a solution for the system in (1) when range and reference position estimates are noisy. Multilateration methods are, in general, preferred in this case. The triangulation method will be denoted as the INV method throughout the paper.

3.3 Multilateration Method

In this method, the position of node U_1 is obtained by minimizing the error cost function $F(\cdot)$ defined as follows:

$$F(\mathbf{u}_1) = \sum_{i=1}^{N_A} \left(\hat{d}_{1,i} - \|\mathbf{u}_1 - \bar{\mathbf{u}}_i\|\right)^2 \tag{2}$$

such that $\hat{\mathbf{u}}_1 = \arg\min_{\mathbf{u}_1} \{F(\mathbf{u}_1)\}$. The minimization of (2) can be done using a variety of numerical optimization techniques, each one having its own advantages and disadvantages in terms of accuracy, robustness, convergence speed, complexity, and storage requirements Nocedal & Wright (2006). Note that as optimization methods are iterative by nature, we will denote with index k the k–th iteration of the algorithm and with $F(\mathbf{u}_1(k))$ and $\mathbf{u}_1(k)$ the error cost function and the estimated position at the k–th iteration, respectively. The final estimated position will be denoted by $\hat{\mathbf{u}}_1 = \mathbf{u}_1(\bar{k})$, where $\bar{k}$ is such that:

$$F(\mathbf{u}_1(\bar{k})) < \Phi \quad \text{or} \quad \bar{k} = \text{MAX}_{\text{iter}} \tag{3}$$

with Φ being the desired accuracy computed on the error function in (2) and MAX_{iter} being the maximum number of iterations allowed for the algorithm.
Basically, Equation (3) represents the stop criterion mentioned in Section 2.2; then both design parameters Φ and MAX_{iter} are application–dependent.

3.3.1 Classical Steepest Descent (SD)

The classical Steepest Descent (SD) is an iterative line search method which allows to find the (local) minimum of the cost function in (2) at step $k+1$ as follows (Nocedal & Wright, 2006, pp. 22, sec. 2.2):

$$\mathbf{u}_1(k+1) = \mathbf{u}_1(k) + \alpha_k \mathbf{p}(k) \tag{4}$$

where α_k is a step length factor, which can be chosen as described in (Nocedal & Wright, 2006, pp. 36, ch. 3) and $\mathbf{p}(k) = -\nabla F(\mathbf{u}_1(k))$ is the search direction of the algorithm.
In particular, when the optimization problem is linear, in the literature there exist some expressions to compute the optimal step length to improve the convergence speed of the algorithm. On the other hand, when the optimization problem is non–linear, as considered in this contribution, a fixed and small step value is in general preferred, in order to reduce the oscillatory effect when the algorithm approaches the solution. In such a case, we have $\alpha_k = 0.5\mu$, where μ is the learning speed Santucci et al. (2006).

3.3.2 Enhanced Steepest Descent (ESD)

The SD method provides, in general, a good accuracy in estimating the final solution. However, it may require a large number of iterations, which may result in a too slow convergence speed, especially for mobile ad–hoc wireless networks. In order to improve such convergence speed, we propose in this contribution an enhanced version of it, which we call Enhanced Steepest Descent (ESD).
The basic idea behind the ESD algorithm is to continuously adjust the step length value α_k as a function of the current and previous search directions $\mathbf{p}(k)$ and $\mathbf{p}(k-1)$, respectively. In particular, α_k is adjusted as follows:

$$\begin{cases} \alpha_k = \alpha_{k-1} + \gamma & \text{if} \quad \theta_k < \theta_{\text{min}} \\ \alpha_k = \alpha_{k-1}/\delta & \text{if} \quad \theta_k > \theta_{\text{max}} \\ \alpha_k = \alpha_{k-1} & \text{otherwise} \end{cases} \tag{5}$$

where $\theta_k = \angle(\mathbf{p}(k), \mathbf{p}(k-1))$, $0 < \gamma < 1$ is a linear increment factor, $\delta > 1$ is a multiplicative decrement factor, and θ_{min} and θ_{max} are two angular threshold values that control the step length update.
By using the four degrees of freedom γ, δ, θ_{min} and θ_{max}, we can simultaneously control the convergence rate of the algorithm and the oscillatory phenomenon when approaching the final solution in a simple way, and without appreciably increasing the complexity of the algorithm when compared to the classical SD method. Basically, the main advantage of the ESD algorithm is the adaptive optimization of the step length factor α_k at run time, which allows to dynamically either accelerate or decelerate the convergence speed of the algorithm as a function of the actual value of the function to be optimized. In the next sections we will show the performance improvement introduced by this algorithm.

4. Proof–of–Concept via Computer–based Simulations

In the frame of PBD approach, performance evaluation is a fundamental concern in the mapping process between *functional description* and *implementation* and it is intended to verify that a solution actually belongs to the design space defined by the *platform*, so that higher layer functional requirements can be met Sgroi et al. (2000). Due to the complexity of network scenario and the need of modeling various components, we have developed a flexible node model. We can test algorithms with a full view of the network while abstracting lower protocol layer (e.g. datalink) details. Furthermore, with the same framework, we can test specific node's behavior by restricting the attention to a reduced number of nodes.

4.1 Atomic Localization

In this section, we will describe some MATLAB simulation results with the aim to assess the performance of the proposed ESD algorithm in several operating conditions and compare its performance with other localization algorithms.

4.1.1 System Setup

The scenario depicted in Fig. 3, is used to have a common reference environment to analyze the improvement provided by the proposed ESD algorithm, and compare several optimization algorithms. For this setup, we assume that the anchor nodes are all "startup anchors", which allows to investigate the so–called atomic location discovery problem, i.e., only Phase 1 described in Section 2.2.1 is implicitly considered in this system setup.

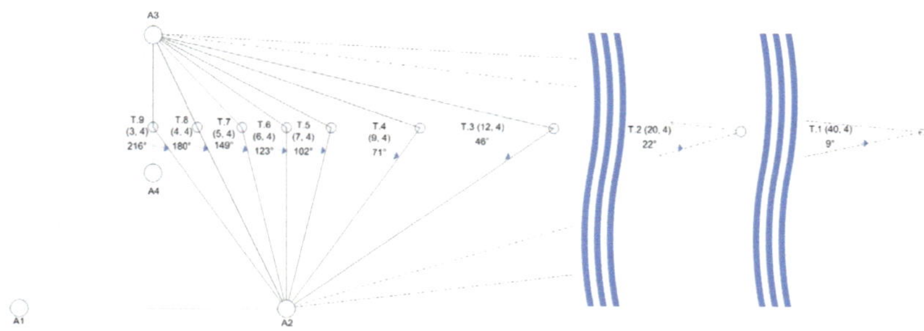

Fig. 3. Reference scenario and network topology (atomic localization step/phase).

In Fig. 3, we have three "startup" anchor nodes A_1, A_2, A_3, a non–complanar "startup" anchor node A_4, and a blind node U_1, which may be located in one of the positions T_h, with $h = 1, 2, \ldots, 9$. In order to analyze the impact of the network geometry/topology on the performance of the optimization algorithms, we have introduced a parameter similar to the so–called geometric dilution of precision factor Savvides et al. (2001). In particular, in every T_h position the unknown node sees the reference nodes with an increasing angle when moving from T_1 to T_9: this corresponds to moving from a scenario (T_1) with a bad geometry where ambiguities may arise during position estimation, towards a scenario (T_9) where the unknown node is surrounded by reference nodes, thus giving an ideally optimal network topology for position estimation, regardless of the specific algorithm Wang & Xiao (2007).

The main parameters used to obtain simulation results are as follows: i) $\bar{\mathbf{u}}_1 = [0,0,0]^T$m, $\bar{\mathbf{u}}_2 = [6,0,0]^T$ m, $\bar{\mathbf{u}}_3 = [3,6,0]^T$m, and $\bar{\mathbf{u}}_4 = [3,3,1]^T$m; ii) the blind node may occupy 9 positions, e.g., $\mathbf{u}_1 = [40,4,0]^T$ m in T_1 (9°) and $\mathbf{u}_1 = [3,4,0]^T$ m in T_9 (216°); iii) the ranging error will be modeled as a Gaussian random variable with mean value given by the actual distance between reference and blind nodes and a fixed standard deviation denoted by σ_R, which is supposed to be indipendent from the actual distance; iv) the position error statistics are obtained by averaging over 2500 realizations of the ranging error for every position of the blind node; v) in order to analyze the effect of both the initial guess and the network topology on the optimization algorithm, 36 starting points uniformly distributed on a circle on the plane $z = 0$ centered at $[0,0,0]^T$ and with radius 50m are considered; vi) the maximum number of iterations for each algorithm is $\text{MAX}_{\text{iter}} = 5000$; vii) the tolerance on the minimum of the error function is $\Phi = 0.05$; viii) the initial learning speed for SD and ESD is $\mu = 0.1$; and ix) the degrees of freedom for the ESD algorithm are: $\gamma = 0.1$, $\delta = 1.75$, $\theta_{\text{min}} = 5°$ and $\theta_{\text{max}} = 30°$.

Algorithm	Comp. Time (s)	Mean Error (m)	Std. Error (m)
CG_1	0.0253 (T_1) 0.0090 (T_5) 0.0060 (T_9)	7.47 (T_1) 1.93 (T_5) 1.21 (T_9)	6.28 (T_1) 1.17 (T_5) 0.56 (T_9)
CG_2	0.0255 (T_1) 0.0090 (T_5) 0.0058 (T_9)	7.44 (T_1) 1.93 (T_5) 1.21 (T_9)	6.23 (T_1) 1.18 (T_5) 0.56 (T_9)
SD	0.2206 (T_1) 0.0264 (T_5) 0.0115 (T_9)	6.65 (T_1) 1.93 (T_5) 1.26 (T_9)	4.14 (T_1) 1.07 (T_5) 0.61 (T_9)
ESD	0.0793 (T_1) 0.0096 (T_5) 0.0058 (T_9)	6.79 (T_1) 1.93 (T_5) 1.23 (T_9)	4.12 (T_1) 1.06 (T_5) 0.59 (T_9)
NLS	0.2615 (T_1) 0.0363 (T_5) 0.0202 (T_9)	6.72 (T_1) 1.92 (T_5) 1.23 (T_9)	4.12 (T_1) 1.03 (T_5) 0.58 (T_9)
INV	0.0001 (T_1) 0.0001 (T_5) 0.0001 (T_9)	15.67 (T_1) 3.50 (T_5) 2.26 (T_9)	9.96 (T_1) 2.19 (T_5) 1.36 (T_9)

Table 1. Comparison of optimization algorithms (CG_1 and CG_2 are the Fletcher–Reeves Polak–Ribière and Hestenes–Stiefel algorithms with secant method Tennina et al. (n.d.).

4.1.2 Numerical Results

In Table 1 we have reported a performance comparison of the optimization algorithms described in Section 3 in terms of computational time, mean and standard deviation of the positioning error. We observe that: i) the positioning error increases when moving the blind node from T_1 to T_9 due to network topology, as expected, ii) the triangulation algorithm (INV) provides the worst performance in terms of error accuracy, iii) the ESD algorithm provides the same accuracy as the SD and NLS[6] algorithms, but reaches the final solution faster (this is an important result for, e.g., mobile networks), iv) the ESD performs as well as the CG[7] algorithms in most scenarios, but outperforms them in those network topologies that are prone to ambiguities (e.g., when the blind node is located in T_1–T_4 positions).

Fig. 4 shows the performance of all simulated algorithms with respect to the Cramer–Rao Lower Bound (CRLB) as defined in Dulman et al. (2008). The results are related to a blind node located in position T_4 in Fig. 3, and the horizontal axis shows the starting position used to initialize every algorithm (i.e, initial guess point), which is an important parameter to be

[6] Non Linear Least Square Tennina et al. (n.d.). This is a sophisticated but quite complex solution, because matrix factorization and Hessian computation are required.

[7] Non–Linear Conjugate Gradient Tennina et al. (n.d.). These methods have been used extensively to solve non–linear optimization problems as they do not require matrix storage and need, in general, a smaller number of iterations than SD method.

investigated to analyze the robustness of every optimization algorithm. The results show that: i) the INV algorithm provides, on the average, the worst performance, which is also independent from the actual initialization point of the algorithm, ii) CG algorithms are very sensitive to the initial guess point, and in some scenarios the algorithm may fail to converge to the true position of the blind node (our experimental trials show that CG algorithms fail to converge when the initial guess is mirrored by 180° with respect to the true node's position), and iii) SD, ESD and NLS algorithms seem to perform globally better than the other ones, and have similar performance. Moreover, these latter algorithms provide results very close to the CRLB.

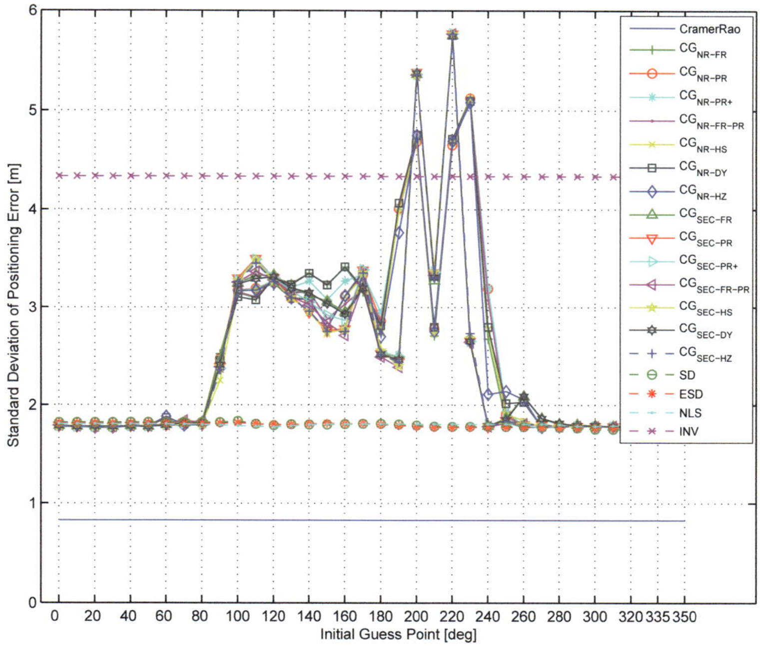

Fig. 4. Performance of the optimization algorithms with respect to the CRLB, and as a function of the initial guess point. The blind node is in position T_4 of Fig. 3.

4.2 Network–wide Localization

In this section we extend the results obtained at the atomic level to a network composed by several blind nodes to evaluate the performance of our proposed ESD algorithm, i.e. considering all the phases described in Section 2.2.1.

4.2.1 System Setup and Numerical Results

Accordingly, moving from the architectural view of the nodes already presented in Sgroi et al. (2003), we developed a node model as shown in Fig. 5, where at the application interface a set of services for implementing e.g. several kinds of control algorithms over WSNs are exposed. By focusing on the Network Platform, i.e. the blocks under such application interface, the introduction of a vertical module should be noted. The vertical nature of this data structure is specifically intended to let all layers may have access to the information stored within (e.g. distance, position estimation and residual energy of batteries for each neighbor). This structure is intended to be shared also in the simulation code, since various layers use a pointer for access. Performance evaluation at network level has been carried out by resorting to the Discrete Event Simulator OMNeT++ Varga (n.d.), in which the node model shown in Fig. 5 has been implemented.

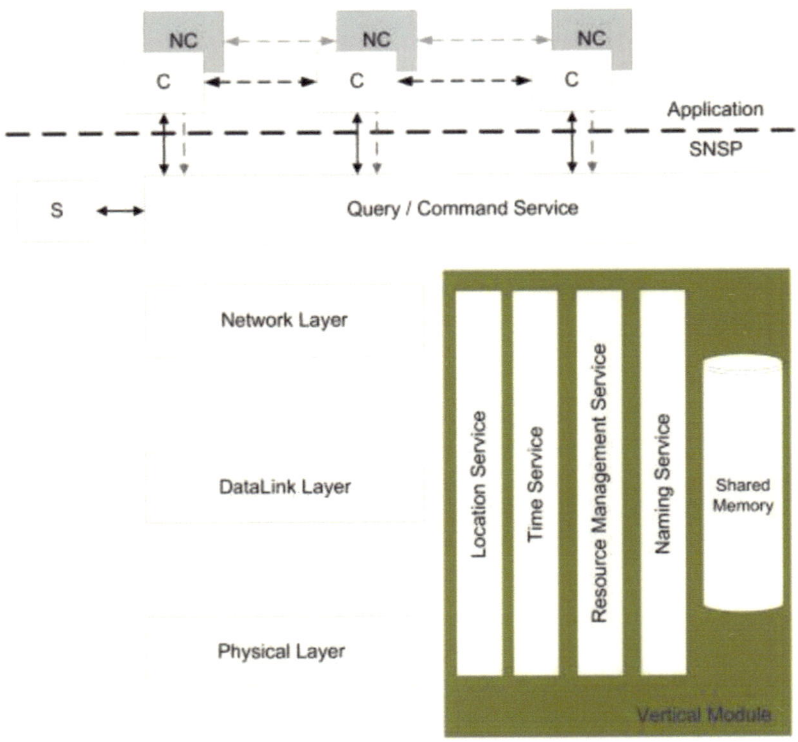

Fig. 5. Reference node architecture Santucci et al. (2006).

As an example, numerical results have been obtained in a network scenario with 100 nodes randomly (uniform distribution) deployed over a squared area with side length equals to $30m$. Five anchors are randomly placed along the perimeter of the network area and have a transmission range equal to $9m$, as large as those exhibited by normal sensor nodes. Moreover, the error on each distance measurement is modelled as a truncated (between -3σ and 3σ) zero–mean Gaussian random variable, with standard deviation $\sigma = 0.15m$. Nodes implement also the CSMA–CA algorithm whose primitives have been briefly depicted in Section 2.3.
While previous results showed that the proposed algorithm outperforms in many cases the solutions existent, in Fig. 6 we show that it allows effectively nodes to obtain good final position estimation. As a matter of fact, 83% of nodes has a final position estimation error less than transmission range and 99% of nodes estimate their position with an error less than twice of transmission range. Note that the density of nodes in this simulated scenario compensates for the low number of anchors in the network.

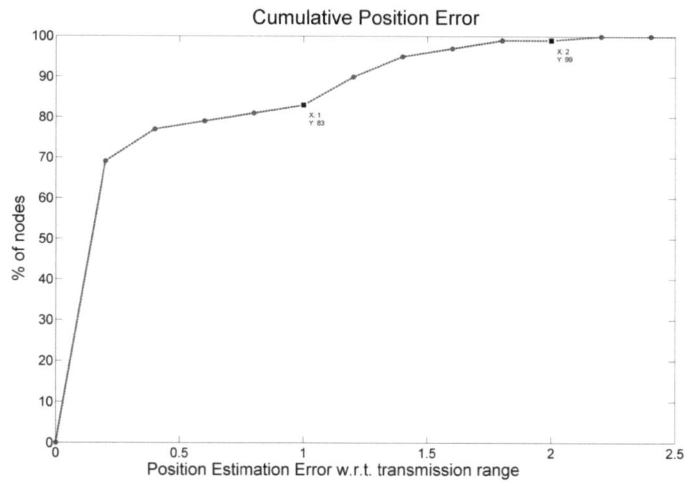

Fig. 6. Cumulative distribution of position error (x–axis scale is normalized to the nodes' radio range). 83% of nodes have a position error equal or less than transmission range, while 99% have a position error equal or less than twice of transmission range.

5. Proof–of–Concept via Experimental Tesbeds

In order to assess both implementation issues and performance of the proposed ESD algorithm via experiments besides computer simulations, we have implemented a testbed platform by using both CrossBow's MICAz (see Cro (2008)) and Texas Instruments/Chipcon CC2431 (see Tex (2007)) sensor nodes.

5.1 Ranging Model

Both sensor nodes platforms use a RSS–based ranging method, and requires a (known) RSS–to–distance calibration curve to estimate the distance between pairs of nodes from a RSS measurement Cro (2008), as follows:

$$d = 10^{\left[\frac{RSS-A}{10n}\right]} \tag{6}$$

where d denotes the transmitter–to–receiver distance, n is the propagation path–loss exponent, A represents the RSS value measured by a receiver that is located 1m away from the transmitter (i.e., reference distance), and RSS is the actual measured value.

In order to estimate this calibration curve, we use the standard procedure described in Aamodt (2008), which consists in deploying a grid of nodes in the area of interest and extracting the desired parameters by post–processing the gathered data. Accordingly, a 6m $\times$ 10m grid of sensor nodes has been deployed in the NCSlab, as shown in Fig. 7. The sensors located in the ground floor are receiver nodes, while transmitter nodes are deployed at the edge of the measurement area, thus yielding a minimum and maximum transmitter–to–receiver distance of 0.5m and 11.7m, respectively. Moreover, the transmitters can be located at different heights with respect to the ground floor (ranging from 5cm to 1.2m). To estimate the calibration curve, the transmitters broadcast packets in a time–scheduled fashion such that collisions are avoided, and the receivers collect RSS values for each received packet, and then send a report to the host PC.

Fig. 7. Deployed testbed using CrossBow's MICAz sensor nodes for ranging calibration.

The RSS–to–distance reference curve in Equation (6) is obtained via a least–squares best linear fitting from several collected RSS values (every receiver node measures RSS values during a 5 minutes acquisition window, resulting in approximately 2000 RSS values). The obtained result is shown in Fig. 8 along with real measurements. Note that, in Fig. 8: i) the RSS values are represented as absolute values in arbitrary units, as provided by the receiver nodes, ii) the distance d in the horizontal axis is normalized to the reference distance of $d_0 = 1$m, and iii) the computed fitting parameters are $A = 59.66$ and $n = 1.84$. Note that a path–loss exponent smaller than free space propagation is obtained (i.e., $n < 2$), which is probably due to the fact that the receiver nodes are located very close to ground floor, which provides a strong constructive reflected propagation path in addition to the direct one.

5.2 System Setup MICAz

In order to analyze implementation issues of the ESD algorithm, and validate simulative results of atomic localization with experimental activities, we have deployed CrossBow's MICAz sensor nodes with a similar setup as the one shown in Fig. 3. The testbed has been deployed in an empty conference room of our NCSlab.

The main parameters used in this testbed setup are as follows: i) the reference nodes' positions are $\bar{\mathbf{u}}_1 = [2,1,0]^T$m, $\bar{\mathbf{u}}_2 = [2,3,0]^T$m, $\bar{\mathbf{u}}_3 = [4,2,0]^T$ m, and $\bar{\mathbf{u}}_4 = [3,2,0.5]^T$m; ii) similar to Fig. 3, the blind node may occupy 16 positions, e.g., $\mathbf{u}_1 = [3,10,0]^T$m in T_1 and $\mathbf{u}_1 = [3,2.5,0]^T$m in T_{16}; iii) the statistics (e.g., mean value) of the positioning error are obtained by averaging over 40 independent runs (i.e., acquisitions) of the algorithm for each blind node; and iv) the maximum number of iterations for the ESD algorithm is 250. Finally, the ranging error is obtained from RSS measurements as described in Section 5.1. In order to compare experiments and simulations in a fair way, computer–based analysis having at the input the

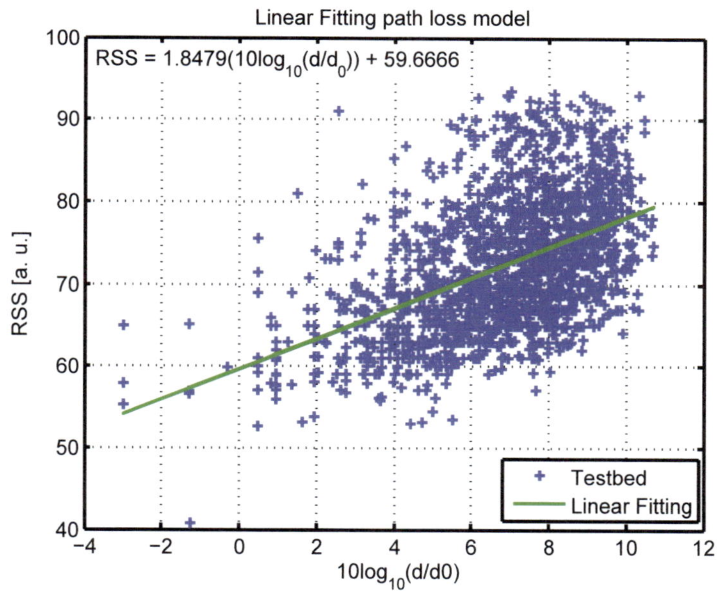

Fig. 8. RSS–to–distance ranging model.

ranging model derived in Section 5.1, and considering real RSS captures from each blind node have been simulated as well.

5.3 Results MICAz

In Fig. 9 we have reported the mean value of the positioning error with respect to the angle under which the unknown node sees the reference nodes (i.e., this curve is obtained by averaging over the 40 acquisitions), along with its standard deviation. Superimposed to the experimental results, we have also reported those obtained via computer–based simulations using the same experimental ranging model obtained in Section 5.1, and having at the input the real experimental captures taken with the testbed. The perfect overlap between the two curves substantiates the correct implementation of the ESD algorithm on the CrossBow's MICAz testbed platform using the NesC programming language Gay et al. (2003). This is an important result to use the testbed for further analysis aiming at quantifying, via experimental activities, other important performance indexes, such as power consumptions and complexity, as well as at judging the overall performance of the ESD algorithm.

5.4 System Setup CC2431

In order to try to overcome the issues related to the off–line RSS–to–distance ranging model calibration, we have deployed a second testbed in the NCSlab using TI/Chipcon's CC2431 sensor nodes. The goal of this study is to analyze the impact of an erroneous or outdated estimate of the propagation–dependent parameters, propose novel solutions to counteract this problem, and understand if the proposed ESD algorithm can be efficiently used to further refine the position estimation provided by the location–finder engine, available on TI/Chipcon's CC2431 sensor nodes, in a scenario with dynamic changes of the propagation conditions. To do so, and have a sound understanding of the performance of the ESD algorithm in a more

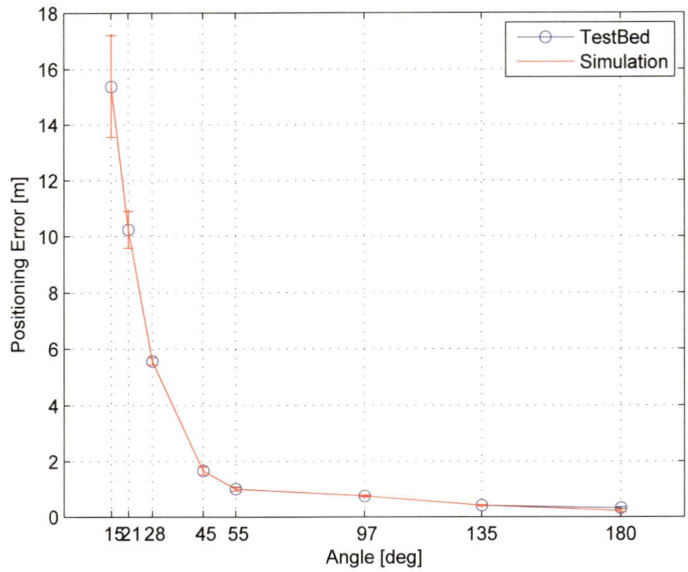

Fig. 9. Mean value and standard deviation of the positioning error: comparison between simulation and experimentation.

realistic scenario than the one analyzed in Section 5.2, we have conducted a campaign of measurements during the opening ceremony day of the NCSlab on March 27, 2008. The event was characterized by a half–day kick–off conference during which the past, present, and future activities of the laboratory were presented. The kick–off conference was attended by several people, and yielded a good occasion to test the performance of the deployed WSN, and, in particular, to test the achievable performance of the TI/Chipcon's CC2431 location engine in a realistic GPS–denied environment, where the propagation characteristics of the radio channel changed appreciably during the event due to the people's movement inside the room (i.e., dynamic indoor environment). The duration of the event was approximately three hours and forty minutes, thus providing enough statistical data to well support our findings and conclusions. The data collected during this measurement campaign have been used as an input to the ESD algorithm and its performance has been quantified via off–line computer–based simulations, while ongoing research activities concern with an efficient implementation of our ESD refinement algorithm onto the TI/Chipcon's CC2431 sensor node platform.

5.4.1 NCSlab Opening Ceremony

The opening ceremony of the NCSlab was characterized by four main phases, which well describe the dynamic nature of the event and, as a consequence, the dynamic nature of the propagation environment to be analyzed. In what follows there is a brief description of each phase:

1. The first phase, which took place before the starting of the ceremony, is characterized by a progressive increase of the number of people inside the room.

2. The second phase, which took place during the development of the ceremony, is characterized by several people (staying either seated or stand) inside the room, and some people coming in and going out the room.
3. The third phase, which took place at the end of the ceremony, is characterized by the vast majority of people staying stand and leaving the conference room.
4. The fourth phase corresponds to the scenario with no people in the room, thus giving a virtually static indoor scenario with almost fixed propagation characteristics.

The WSN's setup used during the event is characterized by the following main setting: i) nine anchor nodes distributed on the room's perimeter (i.e. in direct communication each other) broadcast their position every 800ms on a time division basis in order to avoid collisions, ii) a blind node fixed in the middle of the room estimates its position every 8s, averaging over 10 RSS acquisition per anchor, iii) the anchor nodes are located at 115cm above the ground floor on the top of wood supports, iv) the blind node is located 115cm above the ground floor during the first three phases, while it is 59cm above the ground floor during the last phase. Moreover, four case studies have been investigated and briefly described in the following.

5.4.2 Static Calibration with Measurement Grid – Conference Room Empty (1)

The first case study is related to a static estimation of the propagation parameters needed by the location engine. As described in Section 5.1, the parameters have been estimated in the conference room when it was empty, i.e., no chairs and desks were in the room, and with a grid of 44 "test" nodes deployed 115cm above the ground floor.
This off–line calibration leads to the definition of a curve similar to the one sown in Fig. 7, but whose fitting parameters for the present testbed platform are $A = 39.29$ and $n = 2.23$.

5.4.3 Static Calibration with Anchor Nodes – Conference Room with Furniture (2)

The second case study is still related to a static estimation of the propagation parameters needed by the location engine. However, with respect to the first case study, the propagation parameters are estimated in the conference room with furniture. Moreover, similar to the first case study, the propagation parameters are estimated just once, and are not updated during the progress of the opening ceremony.
However, the main difference with the previous case study is that A and n are not estimated by resorting to a grid of "test" nodes. In contrast to the usual method described by Aamodt (2008), we let anchor nodes performing an adaptive estimation of the propagation parameters A and n, by resorting to the knowledge of their positions, thus their mutual distances, and performing a least–squares best linear fitting of the couples $(RSS; d)$ of the Equation 6 Tennina et al. (n.d.).

5.4.4 Dynamic Calibration with Anchor Nodes – Continuous Training during the NCSlab Opening Ceremony (3)

In this third case study, we use the same approach as in Case 2 for the estimation of parameters A and n. However, these parameters are not estimated once, but are continuously updated on a regular basis during the whole development of the opening ceremony. In Fig. 10, the estimated propagation parameters are reported as a function of time. These parameters are those estimated by the blind node, and computed as the arithmetic average of those estimated by the anchor nodes. We can readily figure out that there is a significant fluctuation of these parameters during the progress of the conference. This figure qualitatively suggests that using an outdated estimate for the channel parameters may certainly yield less accurate estimates of the distances and thus of the final position estimation of the blind node.

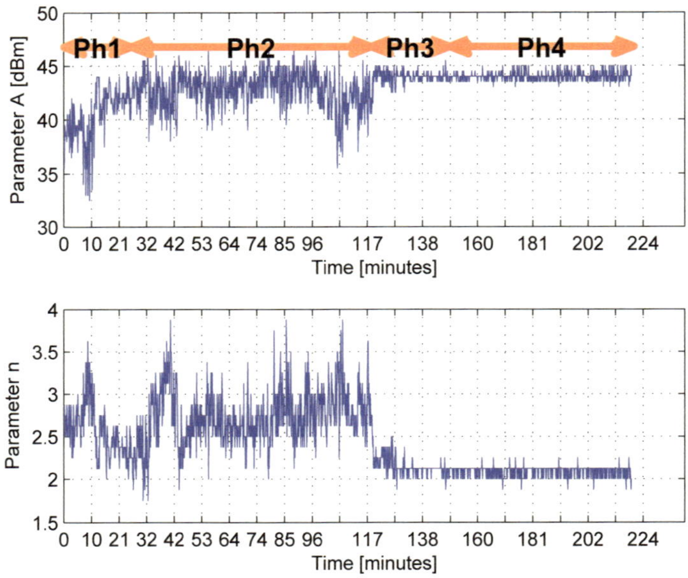

Fig. 10. Estimated propagation parameters during the NCSlab's opening ceremony.

5.4.5 Dynamic Calibration with Anchor Nodes – Off–Line Refinement using the ESD Algorithm (4)

The last case study foresees the same scenario and methods already described in Case 3. However, we introduce a refinement operation to improve the localization accuracy of the system. In particular, the position estimated by the location engine in Case 3 is not considered as the final estimated position of the blind node, but it represents the input for the ESD algorithm.

5.5 Results CC2431

In order to understand the improvement of dynamic updating the channel–dependent parameters, we can look at Table 2. The following conclusions can be drawn. i) For a fixed phase, the performance improves significantly when A and n are updated during the progress of the conference (third column). ii) The improvement is more remarkable during phase two, which is a very dynamic phase and where the dynamic adaptation is more important. iii) The continuous training is also beneficial in phases one and three, but the improvement is less evident due to the short duration of these two phases. iv) Apart from the case study described in Case 1 (first column), using the ESD algorithm to refine the estimated position is always beneficial to improve the accuracy. v) The reason why the ESD does not improve the performance in the first case study is due to the fact that the ESD needs the RSSI–to–distance curve to refine the position. Since this curve is not updated continuously in the first two case studies, the algorithm may diverge from the actual solution, as we have in column one. This conclusion is also confirmed by the fact that in an almost static scenario (phase four), the ESD improves the overall accuracy also without updating the channelŰ-dependent parameters. vi) The larger

Phase	Case 1	Case 2	Case 3	Case 4
1	2.69 (3.72)	2.85 (2.27)	2.20	2.00
2	3.04 (5.52)	3.19 (2.97)	1.22	1.03
3	2.77 (3.46)	2.94 (2.26)	2.72	1.93
4	3.04 (1.79)	3.04 (1.32)	2.11	1.28

Table 2. Average positioning error [meters] over the observation time. The value shown into the parentheses in the first two columns represents the improvement that can be obtained refining the search with the ESD algorithm (similar to Case 4).

error that can be observed in column two with respect to column one is probably due to the smaller number of points used to estimate the calibration curve (in both cases A and n are not updated during the progress of the conference). However, the difference is in the order of few tens of centimeters, and thus can be acceptable. vii) Finally, we note that, as described in Section 5.4, the results in phase four cannot be directly compared to the results in the other phases as the position of the blind node was different. However, also in this case the accuracy improves when moving from column one to column four.

6. Conclusions

In the present chapter, we report our recent research advances along two main directions. Firstly, we adopted the Platform Based Design methodology for the efficient design of ad-Űhoc WSNs with localization capabilities. In particular, the PBD paradigm has been used to derive a fully distributed positioning algorithm, that we call ESD, and a general protocol architecture for WSNs. The proposed solution has been compared with other well–known positioning algorithms available in the open technical literature, and the improvement provided by the proposed ESD algorithm has been clearly assessed by resorting to computer–based simulations in both network–wide and atomic scenarios. Secondly, we have validated the suitability of a practical implementation of the proposed solution onto commercially available WSN platforms, and analyzed their achievable performance in realistic propagation environments. Results have clearly shown that the ESD algorithm can be actually implemented in CrossBow's MICAz sensor node platforms with a modest computational complexity and with good localization performance even when using RSS–based ranging methods. The experiments conducted with the TI/Chipcon's CC2431 sensor node platform have confirmed this in typical and dynamic environments with any a priori knowledge of channel behavior. Although most of the results described in the present contribution are related to the performance of the localization algorithms in terms of accuracy, robustness, convergence speed, complexity, and storage requirements of static nodes, we are currently deploying a WSN testbed to focus our future analysis on i) the evaluation of the energy consumptions of the ESD algorithm, and its comparison with other positioning algorithms, and ii) the analysis of the performance when this solution is used to track people or objects moving in typical GPS–less environments.

Acknowledgements

The authors would like to express their gratitude to the research and technical staff of the Center of Excellence in Research DEWS and the Networked Control Systems Laboratory (NCSlab) for their assistance in conducting the experimental activity. Special thanks go to Alessia D'Alessandro (B.Sc., ECE) and Andrea Scarinci (B.Sc., ECE) for conducting part of the experimental measurements with the TI/Chipcon's CC2431 testbed platform.

7. References

Aamodt, K. (2008). CC2431 location engine, *Application note AN042*. Chipcon products for Texas Instruments, 20 pages.

Alesii, R., Antonini, F., Di Renzo, M., Graziosi, F. & Santucci, F. (2004). Performance of a Chip–Time Analog Differential Receiver for UWB Systems in a Log–Normal Frequency–Selective Fading Channel, *Wireless Personal Multimedia Communications WPMC04, Abiano Terme, Italy.*

Alesii, R., Di Renzo, M., Graziosi, F. & Santucci, F. (2004). A Low-Complexity Receiver for Ultra Wide Band Communications, *IEEE Euro Electromagnetics Congress, EuroEM04, Magdeburg, Germany.*

Bachrach, J. & Taylor, C. (2005). *Handbook of Sensor Networks: Algorithms and Architectures - Localization in Sensor Networks*, Wiley.

Balluchi, A., Di Benedetto, M., Ferrari, A., Girasole, G., Graziosi, F., Parasiliti, F., Pinto, A., Pesserone, R., Petrella, R., Sangiovanni-Vincentelli, A., Santucci, F., Sgroi, M., Tursini, M., Alesii, R. & Tennina, S. (2004). Platform Based Design: Applications and Flow, *Deliverable DPBD2, Version 1, Project IST-2001-38314 COLUMBUS.*
URL: *http://columbus.gr/documents/public/WPPBD/Columbus*$_D$*PBD2.pdf*

Bhargahavan, V., Demers, A., Shenker, S. & Zhang, L. (1994). MACAW: a media access protocol for wireless LANs, *ACM SIGCOMM.*

Bonivento, A., Carloni, L. & Sangiovanni-Vincentelli, A. (2005). Rialto: a Bridge between Description and Implementation of Control Algorithms for Wireless Sensor Networks, *ACM International Conference on Embedded Software, EMSOFT'05, Jersey City, NJ, USA.*

Bulusu, N., Heidemann, J. & Estrin, D. (2000). GPS–less low–cost outdoor localization for very small devices, *IEEE Wireless Communications* **7**(10): 28–34.

Calamari Project (n.d.). `http://www.cs.berkeley.edu/~kamin/calamari`.

Cro (2008). *CrossBow's Motes.*

da Silva, J., Sgroi, M., De Bernardinis, F., Li, S.-F., Sangiovanni-Vincentelli, A. & Rabaey, J. (2000). Wireless Protocols Design: Challenges and Opportunities, *8th IEEE International Workshop on Hardware/Software Codesign, S.Diego, CA, USA.*

Dohler, M. (2008). Wireless sensor networks: the biggest cross–community design exercise to–date, *Bentham Recent Patents on Computer Science*, Vol. 1, pp. 9–25.

Dulman, S., Baggio, A., Havinga, P. & Langendoen, K. (2008). A geometrical perspective on localization, *1st ACM International Workshop on Mobile Entity Localization and Tracking in GPS–less Environments*, pp. 85–90.

Gay, D., Levis, P., von Behren, R., Welsh, M., Brewer, E. & Culler, D. (2003). The NesC language: a holistic approach to networked embedded systems, *ACM SIGPLAN Conference on Programming Language Design and Implementation*, pp. 1–11. ISBN:1–58113–662–5.

Goldsmith, A. J. & Wicker, S. B. (2002). Design Challenges for Energy–Constrained Ad–Hoc Wireless Networks, *IEEE Communication Magazine* **9**: 8–27.

Hofmann-Wellenhof, B., Lichtenegger, H. & Collins, J. (1997). *Global positioning system: theory and practice.*

IEEE Std 802.15.4: Wireless Medium Access Control (MAC) and Physical Layer (PHY) Specifications for Low-Rate Wireless Personal Area Networks (WPANs) (2006). IEEE Computer Society.

Kawadia, V. & Kumar, P. (2005). A Cautionary Perspective on Cross Layer Design, *IEEE Communications Magazine* .

Langendoen, K. & Reijers, N. (2003). Distributed localization in wireless sensor networks: a quantitative comparison, *IEEE Computer Networks* **43**(4): 499–518.

Mauve, M. & Widmer, J. (2001). A survey on position–based routing in mobile ad hoc networks, *IEEE Networks* **15**(6): 30–39.

Newport, C., Kotz, D., Yuan, Y., Gray, R., Liu, J. & Elliott, C. (2007). Experimental evaluation of wireless simulation assumptions, *Simulation* **83**(9): 643–661.

Nocedal, J. & Wright, S. (2006). *Numerical Optimization*, 2nd edn, Springer.

Patwari, N., Ash, J. N., Kyperountas, S., Hero III, A. O., Moses, R. L. & Correal, N. S. (2005). Locating the Nodes, *IEEE Signal Processing Magazine* **22**(4): 54–69.

Perkins, D. D., Tumati, R., Wu, H. & Ajbar, I. (2006). Localization in Wireless Ad Hoc Networks, *Resource Management in Wireless Networking, Springer US*, Vol. 16, pp. 507–542.

Pinto, A. (2008). COSI: COmmunication Synthesis Infrastructure, *GSRC Annual Symposium.*

Pinto, A., D'Angelo, M., Fischione, C., Scholte, E. & Sangiovanni-Vincentelli, A. (2008). Synthesis of Embedded Networks for Building Automation and Control, *American Control Conference (ACC 08), Seattle, Washington.*

Sangiovanni-Vincentelli, A. (2002). Defining Platform–based Design, *EEDesign of EETimes.* (available for download at www.gigascale.org/pubs/141.html).

Santucci, F., Graziosi, F. & Tennina, S. (2006). Location Service Design and Simulation in Ad-Hoc Wireless Sensor Networks, *International Journal on Mobile Networks Design and Innovation* **1**(3/4): 208–214.

Savarese, C. (2002). *Robust positioning algorithms for distributed ad-hoc wireless sensor networks*, Degree of master of science, Department of Electrical Engineering and Computer Sciences, University of California at Berkeley.

Savvides, A., Han, C.-C. & Srivastava, M. (2001). Dynamic Fine-Grained Localization in Ad-Hoc Networks of Sensors, *International Conference on Mobile Computing and Networking, Rome, Italy*, pp. 166–179. ISBN: 1–58113–422–3.

Sgroi, M., da Silva, J., F., D. B., Burghardt, F., Sangiovanni-Vincentelli, A. & Rabaey, J. (2000). Designing Wireless Protocols: Methodology and Applications, *ICASSP*.

Sgroi, M., Wolisz, A., Sangiovanni-Vincentelli, A. & Rabaey, J. (2003). A service-based universal application interface for ad-hoc wireless sensor networks (draft).
URL: *http://www.gigascale.org/pubs/435.html*

Srivastava, V. & Motani, M. (2005). Cross-layer design: a survey and the road ahead, *IEEE Communications Magazine* .

Stiffler, J. (1968). Rapid Acquisition Sequences, *IEEE Transactions on Information Theory* **IT-14**(2).

Tennina, S., Di Renzo, M., Santucci, F. & Graziosi, F. (n.d.). ESD: A Novel Optimization Algorithm for Positioning Estimation of WSNs in GPS–denied Environments – From Simulation to Experimentation, *International Journal of Sensor Networks* . (in press).

Tex (2007). *CC2431 Data Sheet: System–on–chip for 2.4 GHz zigbee/IEEE 802.15.4 with location engine.* Chipcon products for Texas Instruments, 15 pages.

Tex (2008). *TI's website.*

Varga, A. (n.d.). The OMNeT++ Discrete Event Simulation System, *European Simulation Multiconference (ESM), Prague.*

Wang, C. & Xiao, L. (2007). Sensor Localization under Limited Measurement Capabilities, *IEEE Networks* **21**(3): 16–23.

9

Lightweight Event Detection Scheme using Distributed Hierarchical Graph Neuron in Wireless Sensor Networks

Asad I. Khan, Anang Hudaya Muhamad Amin
and Raja Azlina Raja Mahmood
Monash University
Australia

1. Introduction

Existing breakthrough in communication technologies have lead to the rapid growth of emerging networks in particular the wireless sensor networks (WSNs). These networks emerged from the confluence of wireless communication technology, extensive computational schemes, and advanced sensor technology. WSNs are created from a collection of self-organised wireless and battery-powered devices with sensing capabilities. The future of this kind of networks is promising, as been mentioned by (Stankovic, 2008), "The potential of these systems is nothing short of revolutionary. This technology will affect all aspects of our lives, bringing about substantial improvements in a broad spectrum of modern technologies ranging from healthcare to military surveillance".

The current scenario of WSN deployment is however is still far away from its fullest potential. To date, WSN has only been demonstrated for humble applications such as meter reading in buildings and basic form of ecological monitoring. In order to achieve its fullest potential, WSN requires an intelligent computational scheme which at present is still lacking.

Common approach implemented within existing WSN applications usually involve a number of processing steps including sensory data capture and conveyance of these data to a central entity known as the base station for further refinement and analysis. Consequently, this approach would lead to a system bottleneck, if it is scaled up for widespread use. Furthermore, processing delays would intermittently occur due to the high latency between data capture/aggregation and processing time. These limitations make WSN less suitable for real-time monitoring applications. We require a new approach for an improved data processing within WSN that has the abilities to process sensory data in situ within decentralised manner and to generate highly condensed and sophisticated outputs internally. These abilities will alleviate the bottleneck problem within WSN through on-site computations, and improve the detection performance by reducing the processing delays.

In this chapter, we will describe a lightweight and distributed event detection scheme within WSN infrastructure with one-shot learning pattern recognition capability. This

scheme is comprised of a distributed associative memory algorithm known as Distributed Hierarchical Graph Neuron (DHGN) (Khan and Muhamad Amin, 2007). It has the potential to recognise and classify multi-dimensional sensory input for identifying natural or man-made phenomena through clustered and hierarchical graph-based representation of input patterns for use within fully decentralized networks. In addition, DHGN divide and distribute recognition tasks throughout the network in a fine-grained manner for minimise the energy use. Thus, such scheme is highly-suited for resource-constrained networks such as WSN. In addition, DHGN is capable of integrating vast networks of sensors into intelligent macroscopes to observing our surroundings. These will bring unprecedented capabilities within the reach that transform the way we deal with phenomena occurring over large distances and inaccessible regions.
The outline for this chapter is as follows. Section 2 provides an overview of WSN technology and its current research trends. Section 3 explains the current event detection schemes within WSN and some significant issues related to it. Details of our proposed DHGN distributed pattern recognition scheme will be further described in section 4. Section 5 reports on our case study on forest fire detection using DHGN-WSN scheme. Section 6 entails further discussion on our proposed scheme and future direction of this research. Finally, section 7 concludes the chapter.

2. WSN Overview

The advancement of Wireless Sensor Network (WSN) technology has been driven by a massive development of wireless technology and an increasing miniaturisation of RF devices and microelectromechanical systems (MEMS) (Hafez et al., 2005). WSN is a collection of battery-powered and tiny electronic devices known as sensor nodes that are being used to capture sensory data from its surrounding environment. In addition, these sensor nodes are responsible for reporting the sensory readings to a centralised node, known as the base station. That possesses several orders of magnitude more processing capability than the other sensor nodes (Akyildiz et al., 2002). WSN has been used in a wide range of applications including event detection, environmental monitoring, smart home applications, and inventory management.

2.1 WSN Architecture

Every wireless sensor node is equipped with its own onboard processing, limited wireless communication, sensing module, as well as lightweight storage facilities. Each sensor node is built up from a number of electronic components including sensor, microcontroller unit (MCU) for signal controlling and processing, RF transceiver for signal transmissions (Rx and Tx), antenna, and power supply unit. Fig. 1 shows this generic wireless sensor node architecture. Currently, there is a number of commercially available wireless sensor nodes of different types for applications. These include Berkeley Mica Mote (http://www.xbow.com) and UCLA iBadge (Park et al., 2002). The specifications of the Berkeley Mica Mote sensor node that is used in many surveillance networks (Lewis, 2004, Levis and Culler, 2002) are also listed in Table 1.

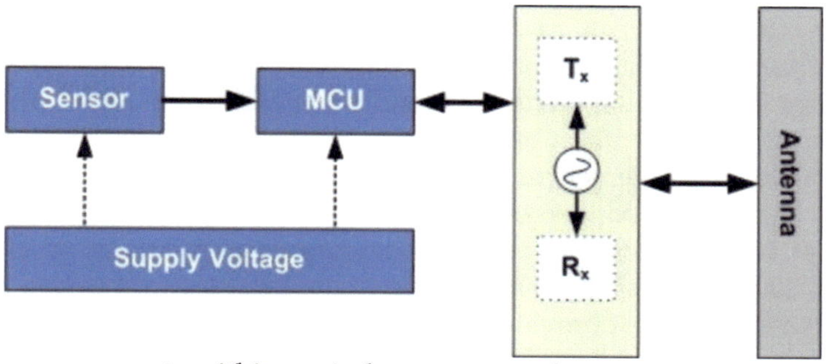

Fig. 1. Common components within a wireless sensor node

CPU:	8-bit 4 MHz
Memory :	128KB Flash and 4KB RAM
Communication :	916 MHz 40 Kbps Radio
Power :	2 AA Batteries

Table 1. Berkeley Mica Mote sensor node specifications

On a macro level, WSN is made up of of wireless sensor nodes that are linked together through a common entity known as the base station (also commonly known as sink). Due to limited power and processing capabilities, communications between sensor nodes and base station usually involve a series of data aggregation techniques to reduce the volume of traffic enroute to the base station. Data aggregation is a process which allows vast amounts of data to be communicated in an efficient manner through the use of data centric routing protocols and effective middleware. Protocols such as SPIN, LEACH, PEGASIS and SMECN are some of the known data aggregation solutions for WSNs. Fig. 2 shows a common infrastructure for WSN network.

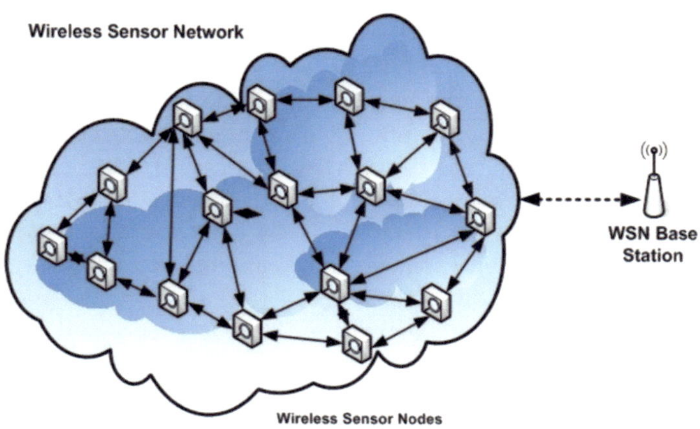

Fig. 2. Common infrastructure for WSN network

2.2 Advances in WSN technology

The advancement of WSN technology has also created a new network architecture known as wireless heterogeneous sensor network (WHSN) (Shih et al., 2006). WHSN is a new form of WSN network in which each sensor node may have a number of different sensors. An example for this kind of sensor node is the Crossbow's Mica2 mote (http://www.xbow.com) which commonly equipped with various sensors able to capture sensory readings such as temperature, humidity and light exposure. With WHSN, the capability of the sensor network to provide various range of sensor readings are extended. Consequently, this makes the network capable of detecting any occurrences of significant event by observing multiple parameters in comparison to a single parameter observation.

Another form of WSN network has recently being introduced, known as the mobile WSN (Benyuan et al., 2005, Wang et al., 2008). Mobile WSN derives its name from the presence of mobile sink or sensor nodes. It provides greater deployment flexibility in comparison to the static WSN architecture. Mobile WSNs have also been found to demonstrate enhanced performance over static WSNs (Munir et al., 2007).

2.3 WSN Deployment Issues

Issues with WSN deployment across wide area of applications encircled towards its resource-constrained characteristics, which include limited communication bandwidth, power, processing capability and memory capacity (Culler et al., 2004). In addition, any algorithm that entails computations, communications, or storage resources within a sensor node would lead to quick exhaustion of the limited battery power available per node. The limited energy and computational resources of sensors imply that the data processing and transmission must be kept to a minimum level in order to conserve energy.

In solving these issues, systems designers must be able to produce a well-managed design for WSN deployment in which it will provide long-term reliability to the network. An effective design should include some important principles such as data-centric mechanism, localised algorithms, and lightweight middleware. In this chapter, we propose a new design for WSN deployment for event detection which incorporates these principles for highly-reliable sensor networks.

3. WSN Event Detection

One of the primary purposes of the existence of WSN is to provide capabilities for monitoring, detecting, and reporting various significant occurrences of events in the sensory domain. An event can be defined as a behavioural change over time on a certain dynamic phenomenon (Guralnik and Srivastava, 1999). An example of event is the change in rainfall amount, ranging from light to heavy to extreme. The behavioural change mentioned here could be either a change involving single environmental parameter value or changes involving composite parameters. In explaining this, (Li et al., 2004) proposed an event hierarchy terminology that differentiates between atomic events and compound events. Atomic event can be determined based on an observation of a sensor, while compound event cannot be determined from a single observation. Rather, compound event is a collection of observations on different types of sensors. For instance, forest fire is a compound event in which observations could be made on four different parameters including temperature, relative humidity, wind speed, and rainfall.

In relation to event detection using WSN, the direction of research in this area is more towards developing energy-efficient, scalable, and reliable scheme to be used within this resource-constrained network.

Research in the area of event detection using WSN could commonly be classified into two groups: performance-specific research and application-specific research. The performance-specific research concerns with the efficiency of the event detection scheme. The main research goal is to develop an event detection scheme with minimum energy consumption and extended lifetime of the WSN network. Alternatively, application-specific research focuses on the development of event detection mechanism that provides accurate and reliable detection for predefined applications such as intrusion detection or phenomenon detection. The common goal of this research area is to obtain efficient mechanism for event detection that deploys specific data processing algorithm that is able to provide accurate and reliable detection using WSN network. This section will further describe the two common research areas of event detection.

3.1 Performance-specific Event Detection Schemes

Most of the recent works on performance-specific event detection schemes are focusing on efficient localisation method and routing mechanism that could be deployed within a WSN network. Localisation and routing are the two important factors in determining the optimum coverage and performance of WSN network. Furthermore, these works have also considered multiple event detection scenarios.

A collaborative event detection and tracking in wireless heterogeneous sensor networks (WHSN) has been proposed by (Shih et al., 2008). In their work, emphasis has been put into tracking procedure and localisation of sensors' attribute region for event detection. Event detection scheme known as CollECT (Collaborative Event Detection and Tracking) has been introduced. A collaboration of different types of sensor nodes is used for event detection and tracking. The three main procedures involved are vicinity triangulation, event determination, and border sensor node selection. The scheme allows event detection and tracking to be conducted simultaneously. However, the scheme requires significant distinction of sensor nodes and their attributes according to its sensing capability. Furthermore, it also requires extensive collaboration of sensor nodes to derive towards maximum accuracy in the event detection within WHSN.

(Banerjee et al., 2008) introduces multiple-event detection scheme with fault tolerant within WSN. They propose the use of polynomial-based scheme that addresses the problems of Event Region Detection (PERD). There are two important components involved, which are the event recognition and event report with boundary detection. For event recognition, (Banerjee et al., 2008) adopts min-max classification scheme which classifies event according to the sensor reading values. These values would then be transformed into polynomial coefficients and passed through a data aggregation scheme. The proposed event detection scheme has enabled a 33% savings in the communication overhead experienced by the network.

Another important contribution in this event detection with performance-specific research is on the works conducted by (Ai et al., 2009) in Authentic Delay Bounded Event Detection System (ADBEDS) for WHSN. ADBEDS implements iterative event detection scheme using event detection tree. This system does simultaneous event detection and packet routing. ADBEDS support singular and composite event monitoring. Important aspects within ADBEDS implementation include energy efficiency and authenticity within WSN deployment for event detection. ADBEDS implements user-specified bounded delay for event detection. Energy efficiency is achieved through sleep-awake alternation between sensor nodes.

3.2 Application-specific Event Detection Schemes

Application-specific schemes for event detection refer to the area of research involving development of application middleware for WSN. This middleware provides enhanced capability and accuracy for event detection using sensor networks. Several machine learning algorithms have been applied including Fuzzy-ART neural network, multi-layer perceptrons (MLPs), and Self-Organising Maps (SOMs).

The use of Adaptive Resonance Theory (ART) neural network for event tracking was introduced by (Kulakov and Davcev, 2005b). Further classification scheme for event detection within WSN has also been introduced in (Kulakov and Davcev, 2005a). In these works, the use of artificial neural networks (ANNs) in the form of ART network has been used as pattern classifier for event detection and classification. The scheme offers reduction in communication overhead with only cluster labels being sent to the sink, instead of the overall sensory data. However, the implementation of ART neural network incurs excessive iterative cycle to achieve optimum cluster matches.

The works by (Kulakov and Davcev, 2005b) on ART neural network for event tracking has also been further researched by (Li and Parker, 2008) in their works on intruder detection using a WSN with fuzzy-ART neural networks.

Self-organisation for event detection has also been a major focus in application specific research within WSN networks. (Elaine et al., 2003) propose a concept of distributed event classification through the use of Kohonen self-organising map (SOM) approach (Kohonen, 2001). The occurrence of events, which are signified by changes in sensor parameter values, could be mapped into clusters representation. The proposed scheme however, imposes significant iterative learning procedure and the classification process is carried out on each input unit, rather than collective input units.

3.3 Summary

Existing event detection schemes within wireless sensor network commonly involve centralised processing at the sink or base station. Efforts to minimise the tendency for this centralised or singular processing have been shown by the works of both performance and application-specific research works. However, a complete decentralisation processing for event detection has yet to be achieved. There are several factors related to this issue. These include complex learning algorithms for event detection and tightly-coupled schemes being deployed for event detection.

We can see by the works of (Kulakov and Davcev, 2005b) and (Elaine et al., 2003) that extensive learning procedures are required in order to derive clusters of events. Consequently, the inputs from the sensors would need to be processed separately and thus incur additional communication overhead for inter-nodes communication. In addition, the proposed schemes do not take into account the variable data processing latency for each sensor node, that is some nodes may require longer processing time than the others.

The works conducted by (Shih et al., 2008) and (Banerjee et al., 2008) offer significant contribution in the efficiency of communication schemes for event detection using WSN. However, the tendency for centralised processing is somewhat undeniable. Furthermore, approaches for distinguishing different roles of specific nodes within WSN are still within a scope of further discussion, due to the nature of WSN network which consists of uniformly-equivalent resource-constrained sensor nodes.

In this chapter, we propose a holistic solution for event detection using WSN. It incorporates a distributed pattern recognition scheme within WSN network and provides on-site and localised computation. We implement a single-cycle learning distributed pattern recognition algorithm known as Distributed Hierarchical Graph Neuron (DHGN) (Khan and Muhamad Amin, 2007). Within this scheme, a dimensionality reduction approach has been employed for minimising the need for complex computation, as well as the incurrence of communication overhead within the network. Our proposed scheme is also capable of providing scalable detection in which we are able to cater for the outgrowth of event classes. Furthermore, integration with computational grid for complex event analysis is viable through this scheme is. Finally, our proposed lightweight event detection scheme also equipped with a detailed workflow of the event detection process. The following sections will provide further descriptions of our proposed solution.

4. Distributed Hierarchical Graph Neuron (DHGN)

DHGN is a novel distributed associative memory (AM) algorithm for pattern recognition. The main idea behind this algorithm is that common pattern recognition approaches for various kinds of patterns would be able to be conducted within a body of a network. DHGN shifts the recognition algorithm paradigm from employing CPU-centric processing towards network-centric processing approach. It also adopts single-cycle learning with in-network processing capability for fast and accurate recognition scheme.

The basic foundation of DHGN algorithm is based upon the functionalities and capabilities of two other associative memory algorithms known as Graph Neuron (GN) (Khan, 2002) and Hierarchical Graph Neuron (HGN) (Nasution and Khan, 2008). It eliminates the crosstalk issue in GN implementation, as well as reduces the complexity of HGN algorithm by reducing the number of processors required for its execution. DHGN is also a lightweight pattern recogniser that supports adaptive granularity of the computational network, ranging from fine-grained networks such as WSN to coarse-grained networks including computational grid.

DHGN network consists of a collection of DHGN processing clusters (PCs) that are interconnected through an important processing entity known as Collective Recognition Unit (CRU), which is responsible for collection of recognition results from each DHGN PC. Fig. 3 shows the basic architecture for DHGN network.

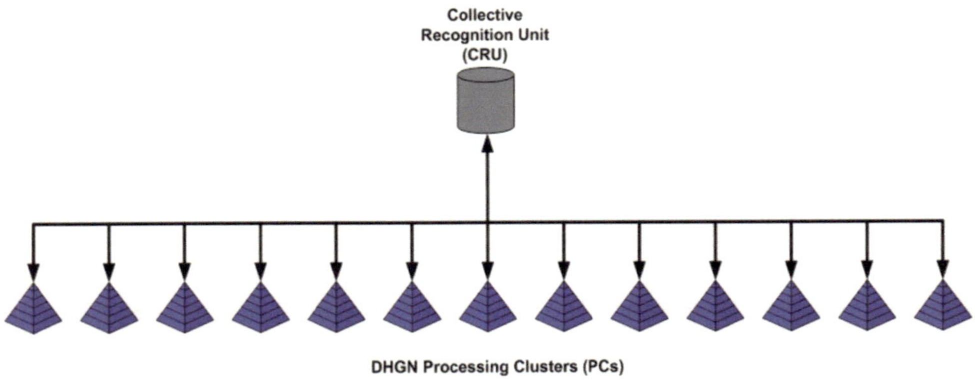

Fig. 3. DHGN network architecture

DHGN processing cluster (PC) is a structural formation of recognition entities called processing elements (PEs) as shown in Fig. 4. The formation is a pyramid-like composition where the base of the structure represents the input patterns. Pattern representation within DHGN network is in the form of *[value, position]* format. Fig. 5 shows how character pattern "AABCABC" is represented in DHGN algorithm.

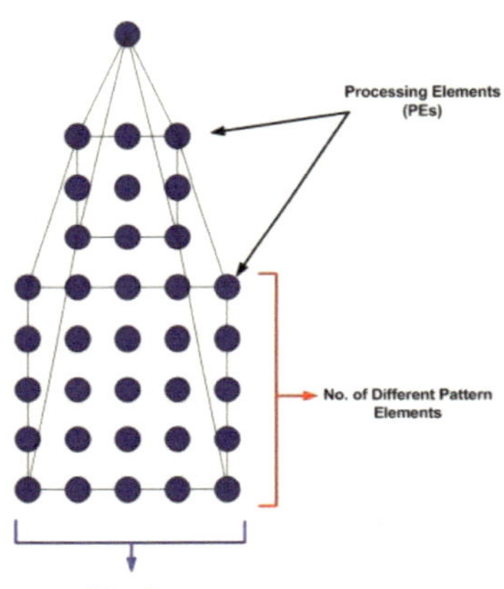

Fig. 4. DHGN processing cluster (PC) formation consists of a number of processing elements (PEs)

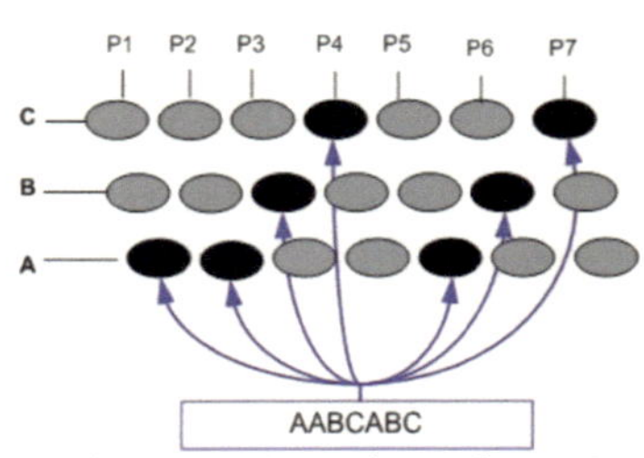

Fig. 5. Pattern representation within DHGN algorithm. Each element within a pattern is represented with *[value, position]* format,

Each row in this representation forms the pattern's possible values v, while each column represents the position of each value within the pattern, p. Therefore, the number of columns within this formation is equivalent to the size of the pattern. In this manner, each location-assigned PE will hold a single value. The formation of the input representation at the base of DHGN processing cluster could be derived from the number of PEs, n_{PE} at the base level of the PC, as shown in Equation (1):

$$n_{PE} = pv \tag{1}$$

4.1 DHGN Recognition Process

Recognition process within DHGN involves a single-cycle learning of patterns on a distributed processing manner. Unlike other pattern recognition algorithms such as Hopfield Neural Network (HNN) (Hopfield and Tank, 1985) and Kohonen SOM (Kohonen, 2001), DHGN employs in-network processing feature within the recognition process. This processing capability allows the recognition process to be performed by a collection of lightweight processors (referred to PEs). PE is an abstract representation of processor that could be in the form of a specific memory location or a single processing node.

At macro level, DHGN pattern recognition algorithm works by applying a divide-and-distribute approach to the input patterns. It involves a process of dividing a pattern into a number of subpatterns and the distribution of these subpatterns within the DHGN network as shown in Fig. 6.

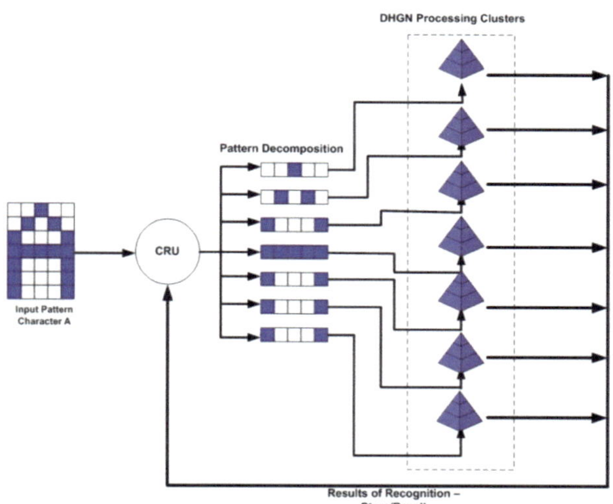

Fig. 6. Divide-and-distribute approach in DHGN distributed pattern recognition algorithm. Character Pattern 'A' is decomposed into similar size subpatterns

In this work, we have made an assumption that a pattern P is a series of data in the form of *[value, position]*, as shown in Equation (2):

$$P = \{v_1, v_2, \ldots, v_x\}, \quad x \in \mathbb{N}^+ \tag{2}$$

Where v represents element within a pattern and x represents the maximum length of the given pattern. For an equal distribution of subpatterns into DHGN network, the Collective Recognition Unit (CRU) firstly needs to determine the capacity of each processing cluster. The following equation shows the derivation of the size of subpattern for each processing cluster from the pattern size x and the number of processing clusters s_n available, assuming that each processing cluster has equal processing capacity:

$$s_{size} = \frac{x}{s_n} \tag{3}$$

Each DHGN processing cluster holds a number of processing elements (PEs). The number of PEs required, PE_n is directly related to the size of the subpattern, s_{size} and the number of possible values, v:

$$PE_n = v\left(\frac{s_{size}+1}{2}\right)^2 \tag{4}$$

Within each DHGN processing cluster, PEs could be categorised into three categories as shown in Table 2.

Type	Description
Base-Layer PE	Responsible for pattern initialisation. Pattern is introduced to DHGN PC at the base layer. Each PE holds a respective element value on specific location within the pattern structure.
Middle-Layer PE	Core processing PE. Responsible to keep track on any changes on the activated PEs at the base-layer and/or its lower middle-layer.
Top-Layer PE	Pre-decision making PE. Responsible for producing final index for a given pattern.

Table 2. Processing element (PE) categories

At micro level, DHGN adopts an adjacency comparison approach in its recognition procedures. This approach involves comparison of values between each processing elements (PEs). Each PE contains a memory-like structure known as bias array, which holds the information from its adjacent PE within the processing cluster. The information kept in this array is known as bias entry with the format *[index, value, position]*. Fig. 7 shows the representation of PE with bias array structure.

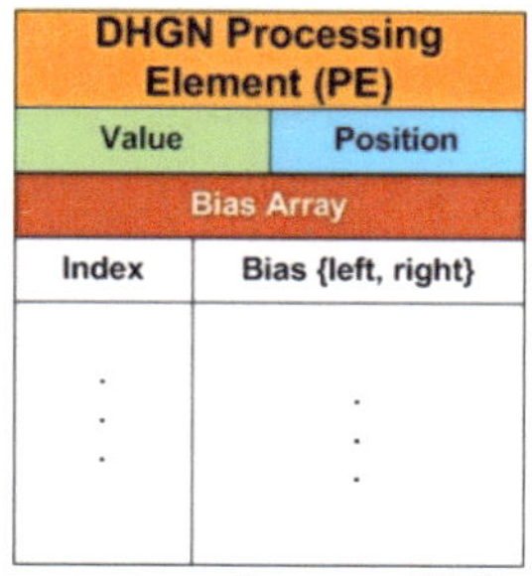

Fig. 7. Data structure for DHGN processing element (PE)

Fig. 8 shows inter-PE communication within a single DHGN processing cluster. The activation of base-layer PE involves matching process between PE's and the pattern element's *[value, position]*. Each activated PE will then initiate communication between its adjacent PEs and conducting bias array update. Consequently, each activated PE will send its recalled/stored index to the PE at the layer above it, with similar position, with exception of the PEs at the edges of the map.

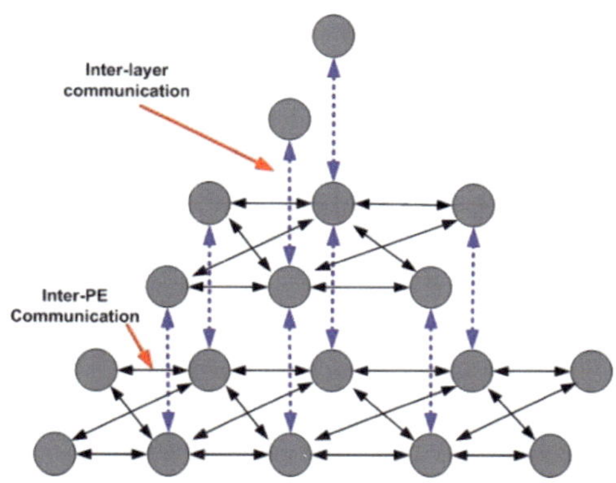

Fig. 8. Communications in DHGN processing cluster (PC)

Unlike other associative memory algorithms, DHGN learning mechanism does not involve iterative modification or adjustment of weight in determining the outcome of the recognition process. Therefore, fast recognition procedure could be obtained without affecting the accuracy of the scheme. Further literature on this adjacency comparison approach could be found in (Khan and Muhamad Amin, 2007, Muhamad Amin and Khan, 2008a, Muhamad Amin et al., 2008, Raja Mahmood et al., 2008).

4.2 Data Pre-processing using Dimensionality Reduction Technique

Event detection usually involves recognition of significant changes or abnormalities in sensory readings. In WHSN, specifically, sensory readings could be of different types and values, e.g. temperature, light intensity, and wind speed. In DHGN implementation, these data need to be pre-processed and transformed into an acceptable format, while maintaining the original values of the readings.

In order to achieve a standardised format for pattern input from various sensory readings, we propose the use of adaptive threshold binary signature scheme for dimensionality reduction and standardisation technique for multiple sensory data. This scheme has originally been developed by (Nascimento and Chitkara, 2002) in their studies on content-based image retrieval (CBIR). Binary signature is a compact representation form that capable of representing different types of data with different values using binary format.

Given a set of n sensory readings $S = (s_1, s_2, \ldots, s_n)$, each reading s_i would have its own set of k threshold values $P_{s_i} = (p_1, p_2, \ldots, p_k)$, representing different levels of acceptance. These values could also be in the form of acceptable range for the input. The following procedures show how the adaptive threshold binary signature scheme is being conducted:

a. For each sensor reading s_i, is discretised into j binary bins $\left(B^i = b_1^i\, b_2^i \ldots b_j^i\right)$ of equal or varying capacities. The number of bins used for each data is equivalent to the number of threshold values P_{s_i}. This bin is used to signify the presence of data which is equivalent to the threshold value or within a range of the specified p_i values using binary representation.

b. Each bin would correspond to each of the threshold values. Consider a simple data as shown in Table 3. If the temperature reading is between the range 20 – 25 degrees Celsius, the third bin would be activated. Thus, a signature for this reading is "01000".

c. The final format of the binary signature for all sensor readings would be a list of binary values that correspond to specific data, in the form of $S_{bin} = b_1^1 b_2^1 b_1^2 b_2^2 \ldots b_j^n$, where b_j^k represent the binary bin for k th sensor reading and j th threshold value.

Temperature Threshold Range (°C)	Binary Signature
0 – 20	10000
21 - 40	01000
41 – 60	00100
61 – 80	00010
81 - 100	00001

Table 3. Simple dataset with its respective binary signature

4.3 DHGN Integration for WSN

With distributed and lightweight features of DHGN, an event detection scheme for WSN network can be carried out at the sensor node level. It could act as a front-end middleware that could be deployed within each sensor nodes in the network, forming a network of event detectors. Hence, our proposed scheme minimises the processing load at the base station and provides near real-time detection capability. Preliminary work on DHGN integration for WSN has been conducted by (Muhamad Amin and Khan, 2008b). They have proposed two distinctive configurations for DHGN deployment within WSN.

In integrating DHGN within WSN for event detection, we have considered mapping each DHGN processing cluster into each sensor node. Our proposed scheme is composed of a collection of wireless sensor nodes and a sink. We consider a deployment of WSN in two-dimensional plane with w sensors, represented by a set $W = (w_1, w_2, \ldots, w_n)$, where w_i is the ith sensor. The placement for each of these sensors is uniformly located in a grid-like area, $A = (x \times y)$, where x represents the x-axis coordinate of the grid area and y represents the y-axis coordinate of the grid area. Each sensor node will be assigned to a specific grid area as shown in Fig. 9. The location of each sensor node is represented by the coordinates of its grid area (x_i, y_i).

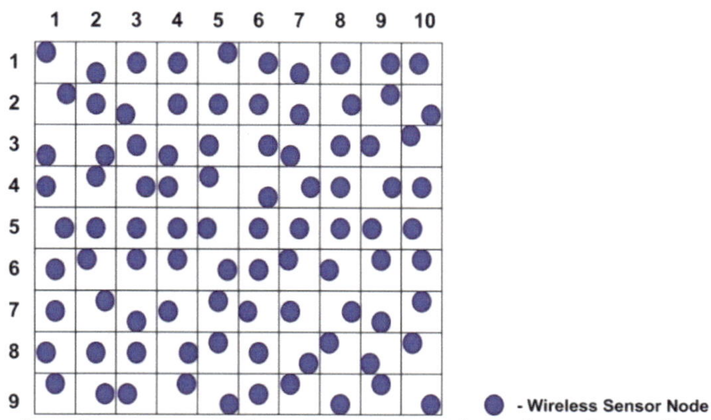

Fig. 9. Sensor nodes placement within a Cartesian grid. Each node is allocated to a specific grid area

For its communication model, we adopt a single-hop mechanism for data transmission from sensor node to the sink. We suggest the use of "autosend" approach as originally proposed by (Saha and Bajcsy, 2003) to minimise error due to the lost of packets during data transmission. Our proposed scheme does not involve massive transmission of sensor readings from sensor nodes to the sink, due to the ability for the front-end processing. Therefore, we believe that a single-hop mechanism is the most suitable approach for DHGN deployment. On the other hand, the communication between the sink and sensor nodes is done using broadcast method.

4.4 Event Classification using DHGN

DHGN distributed event detection scheme involves a bottom-up classification technique, in which the classification of events is determined from the sensory readings obtained through WSN. As been discussed before, our approach implements an adaptive threshold binary signature scheme for pattern pre-processing. These patterns would then be distributed to all the available DHGN processing clusters for recognition and classification purposes.

The recognition process involves finding dissimilarities of the input patterns from the previously stored patterns. Any dissimilar patterns will create a respond for further analysis, while similar patterns will be recalled. We conduct a supervised single-cycle learning approach within DHGN that employs recognition based on the stored patterns. The stored patterns in our proposed scheme include a set of ordinary events that could be translated into normal surrounding/environmental conditions. These patterns are derived from the results of an analysis conducted at the base station, based upon the continuous feedback from the sensor nodes. Fig. 10 shows our proposed workflow for event detection.

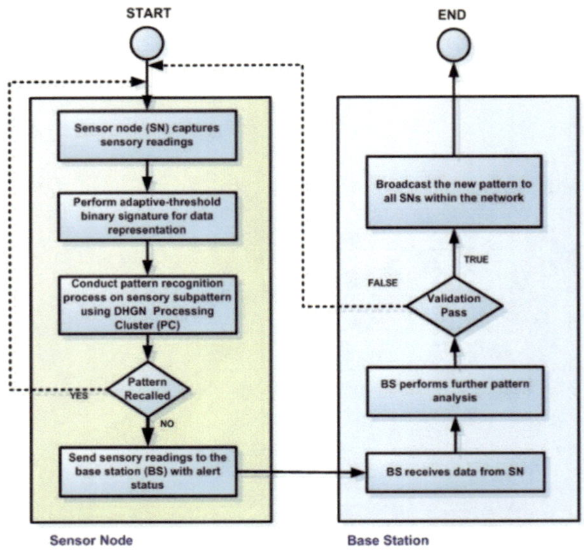

Fig. 10. DHGN distributed pattern recognition process workflow

Our proposed event detection scheme incorporates two-level recognition: front-end recognition and back-end recognition. Front-end recognition involves the process of determining whether the sensor readings obtained by the sensor nodes could be classified as extraordinary event or simply a normal surrounding condition. On the other hand, the spatial occurrence detection is conducted by the back-end recognition. In this approach, we consider the use of signals sent by sensor nodes as possible patterns for detecting event occurrences at specific area or location. In this chapter, we will explain in more details on our front-end recognition scheme.

4.5 Performance Metrics

DHGN pattern recognition scheme is a lightweight, robust, distributed algorithm that could be deployed in resource-constrained networks including WSN and Mobile Ad Hoc Network (MANET). In this type of networks, memory utilisation and computational complexity of the proposed scheme are two factors need to be highly considered. The performance of the scheme largely relies on these major factors.

A. Memory utilisation

Memory utilisation estimation for DHGN algorithm involves the analysis of bias array capacity for all the PEs within the distributed architecture, as well as the storage capacity of the Collective Recognition Unit (CRU). In analysing the capacity of the bias array, we observe the size of the bias array, as different patterns are being stored. The number of possible pattern combinations increases exponentially with an increase in the pattern size. The impact of the pattern size on the bias array storage is an important factor in bias array complexity analysis. In this regard, the analysis is conducted by segregating the bias arrays according to the layers within a particular DHGN processing cluster.

The following equations show the bias array size estimation for binary patterns. This bias array size is determined using the number of bias entries recorded for each processing element (PE). In this analysis, we have considered a DHGN implementation for one-dimensional binary patterns; wherein a two dimensional pattern is represented as a string of bits.

Base Layer. For each non-edge PE, the maximum size of the bias array:

$$bs_{ne}^{l_0} = n_r^2 \tag{5}$$

Where n_r represents the number of rows (different elements) within the pattern. For each PE at the edge of the layer:

$$bs_{e}^{l_0} = n_r \tag{6}$$

The cumulative maximum size of bias arrays at the base layer in each DHGN processing cluster could be derived as shown in Equation (7):

$$bs_{total}^{l_0} = n_r(bs_{ne}^{l_0}(s_{size} - 2) + 2bs_{e}^{l_0}) \tag{7}$$

The maximum size of bias array, i.e. the total number of bias entries at the base layer is mostly determined by the number of possible combinations of values within a pattern.

Middle Layers. The maximum size of the bias array at a middle layer depends on the maximum size of the bias array at the layer below it. For non-edge PE in a middle layer, the maximum size of its bias array may be derived as follows:

$$bs_{ne}^{l_i} = bs_{ne}^{l_i - 1} * n_r^2 \tag{8}$$

For each PE at the edge, the maximum size of its bias array could be derived as the following:

$$bs_{e}^{l_i} = bs_{e}^{l_i - 1} * n_r \tag{9}$$

Therefore, the cumulative maximum size of bias arrays in a middle layer (of a processing cluster) could be estimated using the following equation:

$$bs_{total}^{l_i} = \sum_{i=1}^{l_{top}-1} n_r \left(bs_{ne}^{l_i} \left(s_{size} - (2i + 2) \right) + 2bs_{e}^{l_i} \right) \tag{10}$$

Top Layer. At the top layer, the maximum size of the bias array could be derived from the preceding level non-edge PE's maximum bias array size. Hence, the maximum size of the bias array of PE at the top level is:

$$bs_{all}^{l_{top}} = n_r * bs_{ne}^{l_{top}-1} \tag{11}$$

From these equations, the total maximum size of all the bias arrays within a single DHGN processing cluster could be deduced as shown in Equation (12):

$$bs_{total}^{DHGN} = bs_{total}^{l_0} + \sum_{i=1}^{l_{top}-1} bs_{total}^{l_i} + bs_{all}^{l_{top}} \tag{12}$$

From these equations, one could derive the fact that DHGN offers efficient memory utilisation due to its efficient storage/recall mechanism. Furthermore, it only uses small memory space to store the newly-discovered patterns, rather than storing all pattern inputs. Fig. 11 shows the comparison between the estimated memory capacities for DHGN processing cluster with increasing subpattern size against the maximum memory size for a typical physical sensor node (referring to Table 1).

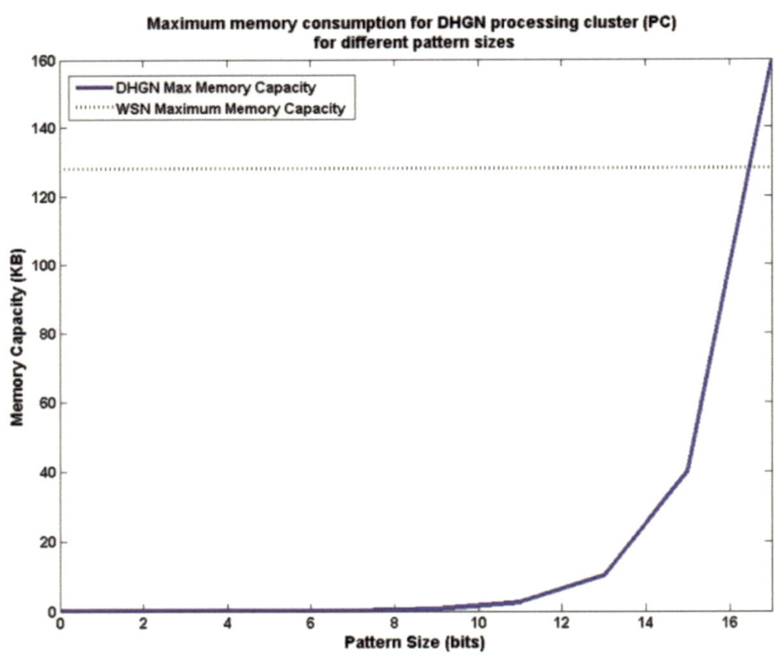

Fig. 11. Maximum memory consumption for each DHGN processing cluster (PC) for different pattern sizes. DHGN uses minimum memory space with small pattern size

As the size of subpattern increases, the requirement for memory space is considerably increases. It is best noted that small subpattern sizes only consume less than 1% of the total memory space available. Therefore, DHGN implementation is best to be deployed with small subpattern size.

B. Computational complexity

Computational complexity of DHGN distributed pattern recognition algorithm could be observed from its single-cycle learning approach. A comparison on computational complexity between DHGN and Kohonen's self-organising map (SOM) has been prescribed by (Raja Mahmood et al., 2008).

Within each DHGN processing cluster, the learning process consists of the following two steps: (i) submission of input vector x in orderly manner to the network array, and (ii) comparison between the subpattern with the bias index of the affected PE, and respond accordingly. There are two main processes in DHGN algorithm: (i) network initialisation, and (ii) recognition/classification. In the network initialisation stage, we are interested to find the number of created processors (PE) and the number of PEs that are initialised. In DHGN, the number of generated PEs is directly related to the input pattern's size. However, only the processors at the base layer of the hierarchy are initialised. Equation (13) shows the number of PEs in DHGN PE_{DHGN}, given the size of the pattern P_{size}, the size of each DHGN processing cluster N_{DHGN}, and the number of different elements within the pattern e:

$$PE_{DHGN} = e.\left(\frac{\frac{S}{N_{DHGN}} + 1}{2} \right)^2 \tag{13}$$

The computational complexity for the network initialisation stage, I_{DHGN} for n number of iterations, could be written as in Equation (14):

$$f\left(I_{DHGN}\right) = O(n) \tag{14}$$

This equation proves that DHGN's initialization stage is a low-computational process. Fig. 12 shows the estimated time for this process. Similar speed assumption of 1 microsecond (µs) per instruction is applied in this analysis. It can be seen that the time taken in the initialization process of DHGN takes approximately only 0.2 seconds to initialize 20,000 nodes.

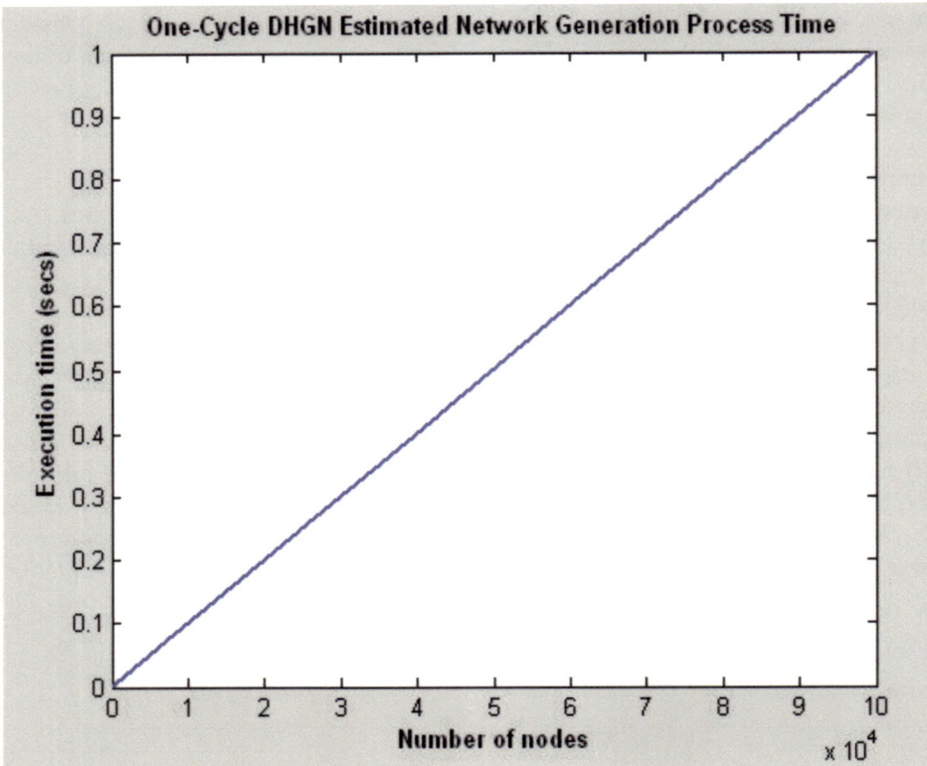

Fig. 12. Complexity performance of DHGN's network generation process (Adopted from (Raja Mahmood et al., 2008))

In the classification process, only few comparisons are made for each subpattern, i.e. comparing the input subpattern with the subpatterns of the respective bias index. The computational complexity for the classification process is somewhat similar to the network generation process, except an additional loop is required for the comparison purposes. The pseudo code of this process is as follows:

```
for each PE in the cluster
{
       recognition()
      {
             for each bias entry
             {
                  check whether input index is similar to stored index
              }
       }
       classification()
}
```

From this pseudo code, the complexity of the classification process C_{DHGN} for n number of iterations could be written as the following equation:

$$f\left(C_{DHGN}\right) = O\left(n^2\right) \tag{15}$$

It can be seen from Equation (15) that DHGN's classification process requires low computational complexity. The time taken for classification by DHGN in a network of 50,000 nodes is less than 3 seconds, as shown in Fig. 13. This exponential effect is still low in comparison to other classification algorithms, including SOM (Raja Mahmood et al., 2008).

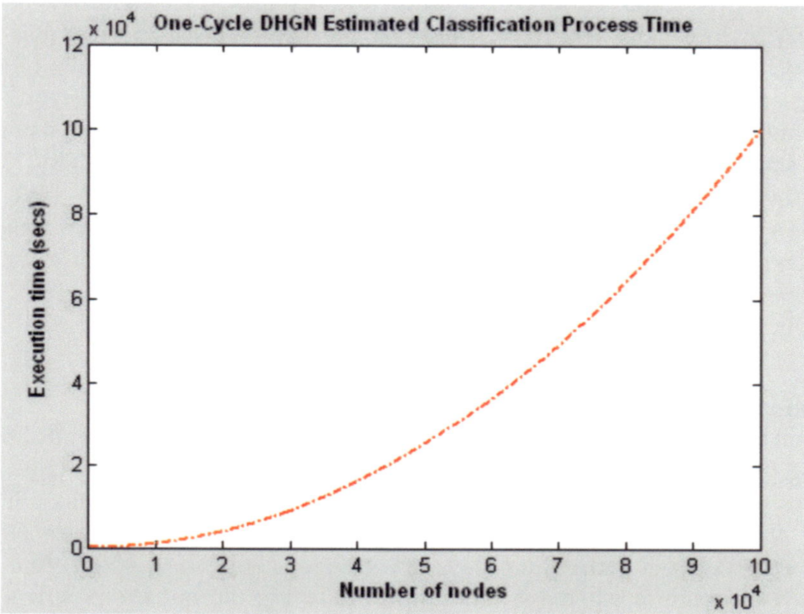

Fig. 13. Complexity performance of DHGN classification process (adopted from (Raja Mahmood et al., 2008))

In summary, we have shown in this chapter that our proposed scheme follows the requirements for effective classification scheme to be deployed over lightweight networks such as WSN. DHGN adopts a single-cycle learning approach with non-iterative procedures. Furthermore, our scheme implements an adjacency comparison approach, rather than iterative weight adjustment approach using Hebbian learning that has been adopted by numerous neural network classification schemes. In addition, DHGN performs recognition and classification processes with minimum memory utilisation, based upon the store/recall approach in pattern recognition.

5. Case Study: Forest Fire Detection using DHGN-WSN

In recent years, forest fire has become a phenomenon that largely affects both human and the environment. The damages incurred by this event cost millions of dollars in recovery. Current preventive measures seem to be limited, in terms of its capability and thus require active detection mechanism to provide early warnings for the occurrence of forest fire. In this chapter, we present a preliminary study on the adoption of DHGN distributed pattern recognition scheme for forest fire detection using WSN.

5.1 Existing Approaches

There are a number of distinct approaches that have been used in forest fire detection. These include the use of lookout towers using special devices such as Osborne fire finder (Fleming and Robertson, 2003) and video surveillance systems such as in the works of (Breejen et al., 1998).

There are also a few works on forest fire detection using WSN, including the works of (Hefeeda and Bagheri, 2007) in the implementation of forest fire detection using Fire Weather Index (FWI) and Fine Fuel Moisture Code (FFMC). In addition, the works of (Sahin, 2007) in forest fire detection suggested the use of animals as mobile biological sensors, which are equipped with wireless sensor nodes. On the other hand, (Zhang et al., 2008) proposed the use of ZigBee technique in WSN for forest fire detection.

Our main interest is in the works conducted by (Hefeeda and Bagheri, 2007) using FWI and FFMC for standard measurement for forest fire detection. FWI and FFMC have been introduced by Canadian Forest Service (CFS) and (De Groot et al., 2005). FWI is used to describe the spread and intensity of fires, while FFMC is used as a primary indicator for a potential forest fire. At this stage, our interest is mainly focuses on early detection for potential forest fire. Hence, our works basically concentrate on the use of FFMC values for fire detection.

5.2 Dimensionality Reduction on FFMC Values

The detection scheme proposed by (Hefeeda and Bagheri, 2007) involves centralised process of obtaining the FFMC values from the sensory readings. The readings obtained from the sensor nodes would be transmitted to the sink for FFMC value determination. FFMC value is derived from an extensive calculation involving environmental parameter values including temperature, relative humidity, precipitation, and wind speed. Our approach using DHGN recognition scheme is focusing on reducing the burden experienced by the back-end processing within the sink, by providing a front-end detection scheme that enables only valid (event-detected) readings that will be sent for further processing.

Table 4 shows the FFMC value versus ignition potential level. This FFMC value provides an indication of relative ease of ignition and flammability of fine fuels due to exposure to extreme heat. In general, fires usually begin to ignite at FFMC values around 70, and the highest probable value to be reached is 96 (De Groot et al., 2005).

Our DHGN implementation performs dimensionality reduction on the FFMC values, by combining the existing five ignition potential levels for ignition potential into two stages: High Risk and Low Risk, as shown in Table 5. Using this approach, we could determine the possibility of forest fire occurring, given certain values of sensory readings.

Ignition Potential	FFMC Value
Low	0 - 76
Moderate	77 - 84
High	85 - 88
Very High	89 - 91
Extreme	92+

Table 4. Ignition potential versus FFMC value

Ignition Potential	VVMC Value
Low Risk	0-84
High Risk	85+

Table 5. Modified FFMC classification for DHGN event detection scheme

5.3 Methodology

The implementation of DHGN for forest fire detection involves a series of steps that reduces the expensive computation of FFMC values at the base station. We have proposed a distributed detection scheme that enables each sensor node to perform a simple recognition process using DHGN to detect any abnormal readings obtained from its surroundings.

The first processing step in our recognition scheme is to reduce the sensory data dimension using adaptive threshold binary signature approach. In this approach, we assume that each sensor node is composed of multiple sensors including temperature, relative humidity, precipitation, and wind speed. The readings would be converted into binary string representation, using the conversion methods as discussed in Section 4.2.

The second step is the actual recognition process, in which the binary signature is treated as subpattern and being introduced into specific DHGN processing cluster within each of the sensor nodes. We assume that DHGN processing cluster in this context is taken place as a block of memory space that could be used for simple DHGN recognition process. In addition, we assume that each node is handling a subpattern (sensory readings) which collectively could become an overall pattern for the whole sensor nodes within the network. The recognition process is conducted by using reference patterns which consist of normal event subpattern/readings.

Once the sensor node detected an abnormal occurrence of subpattern (subpattern is not being recalled), it will send a signal to the base station for further analysis. This signal consists of all the sensory readings and event flag. The base station then compute the FFMC value of the readings. Continuous signals being sent to the base station could be interpreted as a high potential risk of forest fire and vice versa. Therefore, early process of prevention could be executed at a specific location within the area of the sensor nodes.

We have conducted a preliminary test on the accuracy of our scheme and a comparison with Kohonen's self-organizing map (SOM). We have taken a forest fire data from (Cortez and Morais, 2007) and performed our DHGN simulation on computational grid environment for this dataset with 517 items. We have taken three distinctive readings from the dataset, which include temperature, relative humidity and wind speed. We have ignored the precipitation (rainfall) values for this dataset as it has shown minimal effect to the FFMC values. Table 6 shows the bits allocation for each of the readings. This bits allocation eventually will be represented as a binary signature. The results of this test are presented in the following subsection.

Data	Bit Allocation
Temperature	2 bits
Relative Humidity	3 bits
Wind Speed	2 bits

Table 6. Sensory data with allocated binary signature bits

5.4 Classification/Recognition Results

In this study, we have performed a supervised classification test using DHGN event detection scheme. We then compared our results with Kohonen's self-organizing map (SOM) classifier. We have used the SOM toolbox for Matlab that has been developed by (Vesanto et al., 2000). The results from this test have shown that our approach non only produces equivalent recall accuracy in comparison to the SOM classifier but also requires minimum training and training data.

The training data used in the experiments only signifies the normal event data (FFMC values lower than 84). With similar number of training data used, DHGN produces higher classification accuracy as compared to Kohonen SOM. Table 7 shows the training data that have been used in this classification test. The classification accuracy obtained using DHGN reaches up to 88.78%, while SOM is only able to achieve accuracy percentage of 5.61%. Fig. 14 shows the results of this classification test.

Data	1	2	3
FFMC Value	≤ 84	≤ 84	≤ 84
Temperature (°C)	0-40 (10)	0-40 (10)	0-40 (10)
Relative Humidity (%)	> 70 (001)	≤ 40 (100)	> 70 (001)
Wind Speed (km/h)	≤ 3 (10)	≤ 3 (10)	> 3 (01)
Binary Signature	1000110	1010010	1000101

Table 7. Training data set in the form of specific threshold ranges used in classification test. Binary digits in brackets represent signature for the respective data range

We then extend our test on Kohonen SOM to observe the effect of the increased training data on its classification accuracy. We added abnormal event data (FFMC values higher than 84) in the training set. Fig. 15 presents the results of this extended test. Note that an increase in the number of training data improves the accuracy of SOM classifier.

The test reveals that DHGN offers higher accuracy with minimum training data, in comparison to SOM. Furthermore, our distributed approach requires no training iteration, as it adopts a single-cycle learning mechanism. Comparatively, SOM requires high training iteration in order to achieve high classification accuracy. Fig. 16 shows the number of iterations incurred for different number of training data being used in our test.

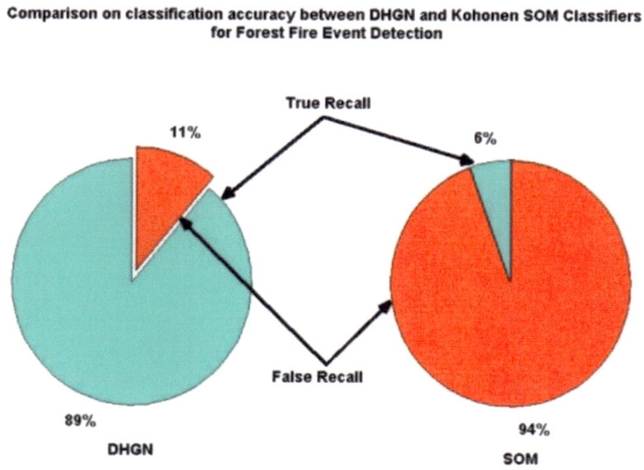

Fig. 14. Classification accuracy comparison between DHGN and Kohonen SOM classifiers for forest fire detection. We have used small number of training data (3 patterns) for each algorithm

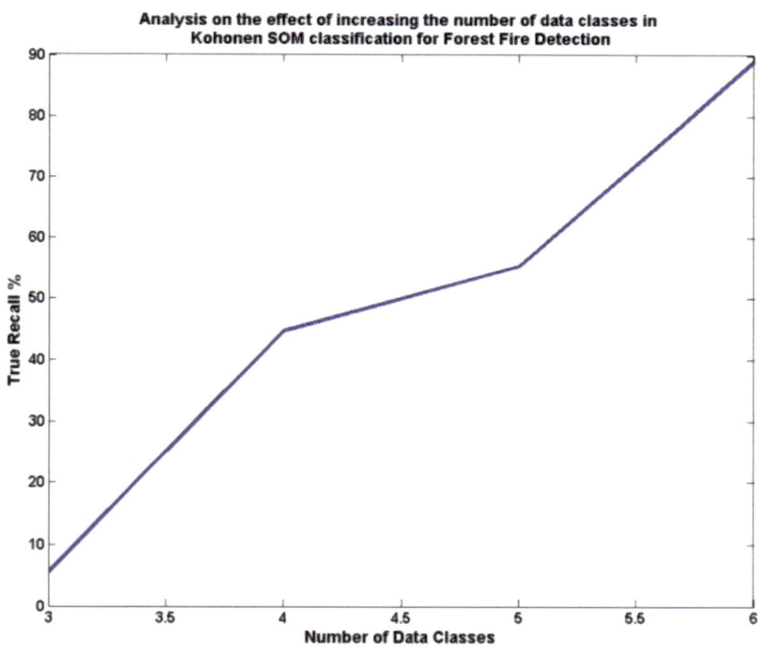

Fig. 15. Analyses of the effect of increasing number of training data set in Kohonen SOM for forest fire data Classification

The results of this test have also convinced us that our proposed scheme is capable of providing high classification accuracy while requiring minimal training effort. Thus, makes it highly suitable to be deployed over resource-constrained networks such as WSN.

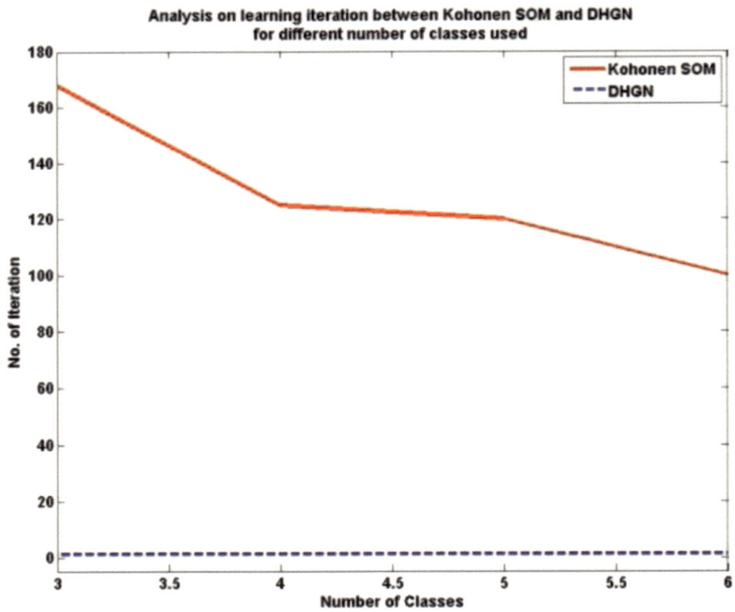

Fig. 16. Analysis of learning iteration between Kohonen SOM and DHGN for different number of classes used in training

5.5 Summary

We have presented our preliminary study on forest fire detection using DHGN distributed pattern recognition algorithm within WSN network. Our proposed implementation involves minimum modification towards existing WSN infrastructure. Furthermore, based on the classification test results, DHGN has shown to perform well with minimum training data and within a single-cycle learning mechanism. This makes our proposed approach more viable for WSN deployment in forest fire detection.

6. Discussion and Future Research

There are several benefits and advantages in our DHGN implementation for event detection within WSN network. Our approach offers low memory consumption for event data storage using simple bias array representation. Furthermore, this scheme only stores subpatterns/patterns that are related to normal event, rather than keeping the records of all occurring events. We have also shown that our approach is most effective for small subpattern size, since it uses only a small portion of the memory space in a typical physical sensor node in WSN network.

In addition to this efficient memory usage, DHGN also eliminates the need for complex computations for event classification technique. With the adoption of single-cycle learning and adjacency comparison approaches, DHGN implements a non-iterative and lightweight computational mechanism for event recognition and classification. The results of our performance analysis have also shown that DHGN recognition time increases linearly with an increase in the number of processing elements (PEs) within the network. This simply reveals that DHGN's computational complexity is also scalable with an increase in the size of the subpatterns.

DHGN is a distributed pattern recognition algorithm. By having this distributed characteristic, DHGN would be readily-deployable over a distributed network. With such feature, DHGN has the ability to perform as a front-end detection scheme for event detection within WSN. Through divide-and-distribute approach, complex events could be perceived as a composition of events occurring at specific time and location. Our approach eventually would be able to be used in event tracking. However, the discussion on event tracking is not within the scope of this chapter. Nevertheless, our proposed scheme has been demonstrated to perform efficiently within an event detection scheme such as forest fire detection using WSN.

Despite all its benefits, DHGN has its own limitations. Firstly, DHGN simple data representation would requires significant advanced pre-processing at the front-end of the system. This might not be viable for strictly-resource constrained sensor nodes, where processing capability is very limited. In addition, DHGN single-hop communication for event detection scheme is not viable for large area monitoring, due to high possibility of communication error due to data packet loss during transmission. Our existing DHGN implementation has also been focusing on supervised classification. However, there is a need for unsupervised classification technique to be deployed for rapid event detection scheme.

Overcoming the DHGN distributed event detection scheme limitations would be the path of our future research direction. We intend to look into event tracking scheme using DHGN distributed detection mechanism, as well as providing unsupervised classification capability for rapid and robust event detection scheme. Furthermore, we are looking forward into implementation of this scheme in large-area monitoring using multi-hop communication strategy.

7. Conclusion

The development of event detection scheme within WSN has been made viable with the advancement in communication, computational, and sensor technologies. However, existing detection/recognition algorithms fail to achieve optimum performance in a distributed environment, due to its tightly-coupled and computationally intensive nature. In this chapter, we have presented our readily-distributable event detection scheme for WSN network which is known as Distributed Hierarchical Graph Neuron (DHGN). Throughout our studies, we discover that DHGN is able to perform recognition and classification processes with limited training data and within a one-shot learning. These DHGN features have given added-value for implementing this scheme within a lightweight distributed network such as WSN. In addition, our proposed adaptive threshold binary signature scheme has the ability to provide generalisation and simplification of datasets to be used in DHGN distributed pattern recognition scheme.

Current implementation of DHGN in event detection using WSN has been focusing on the front-end processing, in which detection could be carried out earlier using the available wireless sensor nodes. Our approach differs from other existing event detection schemes in which major processing steps are conducted at the base station. By having a front-end

detection, our proposed scheme is able to alleviate the computational costs experienced by the centralised-processing undertaken by the base station.
In this chapter, we have also discussed the advantages and limitations of our proposed scheme. The future direction of this research lies in the development of a complete event detection scheme that incorporates front-end detection and back-end complex event analysis. We foresee our DHGN distributed pattern recognition scheme as a complete event detection and analysis tool that is deployable over different types of event detection schemes on WSN networks.

8. References

Ai, C., Hou, H., Li, Y. and Beyah, R. (2009) *Ad Hoc Networks,* **7,** 599-613.

Akyildiz, I. F., Su, W., Sankarasubramaniam, Y. and Cayirci, E. (2002) In *IEEE Communications,* Vol. 40 IEEE, pp. 102-114.

Banerjee, T., Xie, B. and Agrawal, D. P. (2008) *Journal of Parallel and Distributed Computing,* **68,** 1222-1234.

Benyuan, L., Peter, B., Olivier, D., Philippe, N. and Don, T. (2005) In *Proceedings of the 6th ACM international symposium on Mobile ad hoc networking and computing*ACM, Urbana-Champaign, IL, USA.

Breejen, E. d., Breuers, M., Cremer, F., Kemp, R., Roos, M., Schutte, K. and Vries, J. S. d. (1998) In *The Third International Conference on Forest Fire Research,* Vol. 2 Luso, Portugal, pp. 2003-2012.

Cortez, P. and Morais, A. (2007) In *Proceedings of the 13th EPIA 2007 - Portuguese Conference on Artificial Intelligence*(Eds, Neves, J., Santos, M. F. and Machado, J.) APPIA, Guimarães, Portugal, pp. 512-523.

Culler, D., Estrin, D. and Srivastava, M. (2004) In *Computer,* Vol. 37, pp. 41-49.

De Groot, W. J., Wardati and Wang, Y. (2005) *International Journal of Wildland Fire,* **14,** 161–168.

Elaine, C., Kristof Van, L. and Martin, S. (2003) In *Proceedings of the eighth international conference on Artificial life*MIT Press.

Fleming, J. and Robertson, R. G. (2003) In *Fire Management Tech Tips* SDTDC USDA Forest Service, San Dimas CA, USA.

Guralnik, V. and Srivastava, J. (1999) In *The Fifth ACM SIGKDD International Conference on Knowledge Discovery & Data Mining (KDD-99)*ACM, San Diego, CA, USA, pp. 33-42.

Hafez, R., Haroun, I. and Lambaridis, I. (2005) In *Systems & Subsystems,* Vol. 2008 Penton Media, Inc.

Hefeeda, M. and Bagheri, M. (2007) In *IEEE Internatonal Conference on Mobile Adhoc and Sensor Systems, 2007 (MASS 2007)* IEEE Press, Pisa, Italy, pp. 1-6.

Hopfield, J. J. and Tank, D. W. (1985) *Biological Cybernetics,* **52,** 141-152.

Khan, A. I. (2002) In *The Proceedings of The Thirteenth Australasian Conference on Information Systems*Melbourne, Australia.

Khan, A. I. and Muhamad Amin, A. H. (2007) In *AI 2007: Advances in Artificial Intelligence, 20th Australian Joint Conference on Artificial Intelligence, Gold Coast, Australia, December 2-6, 2007, Proceedings,* Vol. 4830 (Eds, Orgun, M. A. and Thornton, J.) Springer, pp. 705-709.

Kohonen, T. (2001) *Self-Organizing Maps,* Springer.

Kulakov, A. and Davcev, D. (2005a) In *Proceedings of the 10th IEEE Symposium on Computers and Communications (ISCC 2005)*IEEE Computer Society, Cartagena, Spain.

Kulakov, A. and Davcev, D. (2005b) In *Proceedings of International Conference on Information Technology: Coding and Computing, ITCC 2005*Las Vegas, NV, USA, pp. 534-539.

Levis, P. and Culler, D. (2002) In *Proceedings of the 10th International Conference on Architectural Support for Programming Languages and Operating Systems (ASPLOS X)*ACM, San Jose, CA, USA.

Lewis, F. (2004) In *Smart Environments: Technology, Protocols and Applications*(Eds, Cook, D. and Das, S.) Wiley.

Li, S., Lin, Y., Son, S. H., Stankovic, J. A. and Wei, Y. (2004) *Telecommunication Systems,* **26,** 351-368.

Li, Y. and Parker, L. E. (2008) In *IEEE/RSJ International Conference on Intelligent Robots and Systems, 2008 (IROS 2008),* vol., no., pp.3292-3298, 22-26 Sept. 2008 URL: http://ieeexplore.ieee.org/stamp/stamp.jsp?arnumber=4651031&isnumber=4650570, pp. 3292-3298

Muhamad Amin, A. H. and Khan, A. I. (2008a) In *Sixth Australasian Symposium on Grid Computing and e-Research (AusGrid 2008),* Vol. 82 (Eds, Kelly, W. and Roe, P.) ACS, Wollongong, NSW, Australia, pp. 27-34.

Muhamad Amin, A. H. and Khan, A. I. (2008b) In *Proceedings of the 2008 Ninth International Conference on Parallel and Distributed Computing, Applications and Technologies 2008 (PDCAT'08)*IEEE Computer Society, Dunedin, New Zealand, pp. 305-308.

Muhamad Amin, A. H., Raja Mahmood, R. A. and Khan, A. I. (2008) In *Computer and Information Technology Workshops, 2008. CIT Workshops 2008. IEEE 8th International Conference on,* pp. 153-158.

Munir, S. A., Ren, B., Jiao, W., Wang, B., Xie, D. and Ma, J. (2007) In *21st International Conference on Advanced Information Networking and Applications Workshops (AINAW'07)*IEEE Computer Society.

Nascimento, M. A. and Chitkara, V. (2002) In *Proceedings of the 2002 ACM Symposium on Applied Computing (SAC'02)*ACM, Madrid, Spain, pp. 687-692.

Nasution, B. B. and Khan, A. I. (2008) *IEEE Transactions on Neural Networks,* **19,** 212-229.

Park, S., Locher, I. and Srivastava, M. (2002) In *6th International Symposium on Wearable Computers (ISWC2002)*Seattle, WA, USA.

Raja Mahmood, R. A., Muhamad Amin, A. H. and Khan, A. I. (2008) In *Intelligent Networks and Intelligent Systems, 2008. ICINIS '08. First International Workshop on*Wuhan, China, pp. 675-680.

Saha, S. and Bajcsy, P. (2003) In *The IASTED International Conference on Communications, Internet and Information Technology 2003 (CIIT 2003)*(Ed, Hamza, M. H.) ACTA Press, Scottsdale, AZ, USA.

Sahin, Y. G. (2007) *Sensors,* **7,** 3084-3099.

Shih, K.-P., Wang, S.-S., Chen, H.-C. and Yang, P.-H. (2008) *Computer Communications,* **31,** 3124-3136.

Shih, K.-P., Wang, S.-S., Yang, P.-H. and Chang, C.-C. (2006) In *IEEE Symposium on Computers and Communications (ISCC),* pp. 935-940.

Stankovic, J. A. (2008) In *Computer,* Vol. 41 IEEE Computer Society, New York, NY, pp. 92-95.

Vesanto, J., Himberg, J., Alhoniemi, E. and Parhankangas, J. (2000) In *Proceedings of the Matlab DSP Conference,* pp. 35-40.

Wang, W., Srinivasan, V. and Chua, K.-C. (2008) *IEEE/ACM Transactions on Networking,* **16,** 1108-1120.

Zhang, J., Li, W., Han, N. and Kan, J. (2008) *Frontiers of Forestry in China,* **3,** 369-374.

10

Dynamic Hierarchical Communication Paradigm for Improved Lifespan in Wireless Sensor Networks

Suraiya Tarannum

Department of Telecommunication Engineering AMC Engineering College, Bangalore - 560 083, India ssuraiya@gmail.com

Abstract

Effective utilization of limited power resources by the sensors is pre-eminent to the Wireless Sensor Networks. Organizing the network into balanced clusters based on assigning equal number of sensors to each cluster may have the consequence of unbalanced load on the cluster heads. By-product of this is unbalanced consumption of the energy by the nodes which leads to minimization of network lifetime. We put forth a Sink administered Load balanced Dynamic Hierarchical Protocol (SLDHP) to balance the load on the principal nodes. Hierarchical layout of the sensors endows the network with increased lifespan. Outcome of this protocol also includes substantial saving of the energy consumed by the nodes. Simulation results indicate significant improvement of performance over Base station Controlled Dynamic Clustering Protocol (BCDCP).

Wireless Sensor Network, Sink, Principal Node, Superior Node, Network Lifetime.

1. Introduction

Wireless Sensor Network (WSN) is an ad-hoc wireless telecommunications network which embodies number of tiny, low-powered sensor nodes densely deployed either inside a phenomenon or close to it [1]. The multi-functioning sensor nodes operate in an unattended environment with limited sensing and computational capabilities. The advent of wireless sensor networks has marked a remarkable change in the field of information sensing and detection. It is a conjunction of sensor, distributed information processing, embedded and communication techniques. WSNs may in the near future be equally prominent by providing information of the physical phenomena of interest and ultimately being able to detect and control them or enable us to construct more meticulous models of the physical world.

WSNs are easier, faster and cheaper to deploy than other forms of wireless networks as there are no predetermined positions for the sensors. They have higher degree of fault-tolerance than other wireless networks and are self-configuring or self-organizing [2]. Sensors are deployed randomly and are expected to perform their mission properly and efficiently. Another unique feature of sensor networks is the co-operative effort of sensor nodes. These unique features have popularized the WSN in the world of communications.

A WSN is envisioned to consist of a large number of sensors and many Base Stations (BS). The sensors are supplied with transceivers to gather information from their environment and pass it on to one of the base stations. A typical sensor node consists of four major components: a data processor unit; a sensor; a radio communication subsystem that consists of transmitter/receiver electronics, antennas, an amplifier; and a power supply unit [3]. The sensors are compact in size which make them extremely energy constrained. Replacing batteries in such a large scale in harsh terrains is practically not feasible. Hence, it is well accepted that the key challenge in unlocking the potential of such networks is maximizing their post-deployment *active* lifetime. The lifetime of the sensors can be prolonged by ensuring that all aspects of the system are energy-efficient. Since communication in wireless sensor networks consume significant energy, nodes must spend as minimum amount of energy as possible for receiving and transmitting the data.

A web of sensor nodes can be deployed to gather productive information from the sensor field. The benefits of using WSNs include extended range of sensing, fault-tolerance, improved accuracy and lower cost. The sensor networks are expected to find extensive use in a variety of applications, including remote climate monitoring, seismic, acoustic, medical and intelligence data-gathering [4,5]. As a result, they are suitable for a wide range of applications like military, health, education, commerce and so on. Military applications may range from tracking enemy movements in the battlefield to guiding targetting systems. Bio-sensors are used for monitoring patients blood sugar level. Commercial applications may range from tracking postal packages to monitoring product quality on an assembly line. Environmental applications include forest-fire detection, flood detection, tracking movements of birds etc. Sensors are also used to simulate home automation and build smart environments.

Efficient utilization of energy is crucial to the WSN. Wireless microsensor network protocols should therefore be self-configuring, to enable ease of deployment of the nodes, latency aware, qualitative, robust and to extend the system lifetime. The sensors being extremely energy bounded, the communication devices on these sensors are small and have limited power and sensing range. A routing protocol coordinates the activities of individual nodes in the network to achieve the goals and does it in an efficient manner. The simplest is the Direct Communication Routing Protocol, where each node transmits the sensed information directly to the base station. The nodes consume considerable amounts of energy, if the communication path is long resulting in early death of the distant nodes. To overcome this drawback, Minimum Transmission Energy Protocol uses a multihop routing scheme. Here, nodes close to the BS drain their energy rapidly as they are involved in the transmission of messages on behalf of others. Hierarchical routing groups all the sensors into clusters. It aims at reduction of energy consumption by localizing data communications within a cluster and aggregating data to decrease the transmissions to base station.

In Low Energy Adaptive Cluster Hierarchy (LEACH), the operation is framed in iterations, with each iteration consisting of a setup phase and a data transmission phase. During the setup phase, nodes organize themselves into clusters with a predetermined number of nodes serving as cluster heads. In the data transmission phase, the self-elected cluster heads aggregate data received from nodes in their cluster, before forwarding to the base station. The role of cluster heads is randomly rotated among all the nodes in the network. LEACH serves as a basic model for other hierarchical routing protocols. A centralized version of the adaptive approach comprises of a hierarchical structure in which the base station has control over the cluster formation. The base station uses the location and energy information sent by the nodes to select the predetermined number of cluster heads. Efficient clustering is achieved as the base station possess the global knowledge of the network and hence shows improvement over the adaptive approach.

In Power Efficient Gathering in Sensor Information Systems (PEGASIS), the nodes function co-operatively to optimize network lifetime. A greedy algorithm configures the network into chains. In each iteration, a randomly chosen leader node, directs the aggregated data to the base station. A centralized energy efficient routing protocol called Base Station Controlled Dynamic Clustering Protocol (BCDCP), was proposed which widened the area for research in hierarchical routing. Here, many of the functionalities like formation of clusters and routing paths are performed by the high energy base station which in turn lightens the load of sensor nodes. This protocol configures the network into balanced clusters where each cluster head serves an approximately equal number of member nodes. Cluster head-to-cluster head multihop routing is employed in this protocol to transfer the data to the base station. The topologies of hierarchical routing protocols is depicted in Figure 1.

Efficient energy management deserves much of the attention in the WSNs. Routing protocols designed for WSNs must effectively tackle this issue in order to enhance the lifetime of the network. Hierarchical routing techniques are preferable in this direction. The arrangement of nodes in the form of a load balanced hierarchy proves to be beneficial. In the present study, an energy-efficient hierarchical routing protocol, Sink Administered Load Balanced Dynamic Hierarchical Protocol (SLDHP) is proposed to increase the lifetime of WSNs. SLDHP achieves a load balanced hierarchical arrangement of nodes in the network, and which performs significantly better than other hierarchical routing protocols.

In this work, an energy-efficient hierarchical routing protocol, SLDHP is proposed to increase the lifetime of homogeneous and heterogeneous WSNs. SLDHP achieves a load balanced hierarchical arrangement of nodes in the network which performs better than other hierarchical routing protocols.

2. Literature Survey

Heinzelman et al. [6] have proposed an adaptive clustering protocol, LEACH which employs the technique of randomly changing the role of the cluster head among all the nodes. A centralized scheme is described in [7], where the base station determines the cluster heads. The results show improvement over [6]. A chain based protocol, PEGASIS is presented in [8], where each node communicates only with a close neighbour and takes turns to transmit to the base station. A greedy algorithm ensures that the nodes already on the chain are not revisited. A centralized clustering based routing protocol, BCDCP is discussed in [9]. This protocol configures the network into balanced clusters, i.e., the number of nodes in each cluster are same. Such equal clustering results in unequal load on the cluster head.

Huang et al. [10] have reviewed the energy efficiency of cluster based routing protocols, with extended complexity of data fusion and data compression. Geographic and energy aware routing algorithm developed by Yu et al. [11] propagates a query to the appropriate geographical region without using flooding. The protocol uses energy aware and geographically informed neighbour selection to route a packet towards the target region. The protocol exhibits noticeably longer network lifetime than non-energy aware geographic routing algorithms. A novel algorithm proposed by Depedri in [12] performs the three main functions of configuring the network into optimum number of clusters, decentralised cluster head selection and cluster formation. An adaptive strategy is used for cluster head selection and the cluster formation uses total path energy dissipation instead of energy lost in the path for the node to reach its cluster head.

A cost based comparision of homogeneous and heterogeneous clustered sensor networks is presented in [13]. Here the authors propose and analyze a multihop variant of the adaptive approach where communication radius for in-cluster communication and size of clusters are taken into consideration. An energy-efficient, distributed clustering approach for adhoc sensor networks is developed in paper [14]. Here the cluster heads are chosen randomly based on their residual energy and nodes participate in cluster operation such that communication cost is minimized. In [16], a cluster-based query protocol for wireless sensor networks functions using self-organized sensor clusters to register queries, process queries and disseminate data within the network is proposed. This protocol uses cluster heads as data storage and aggregation points. Energy efficiency is achieved by reducing the number of data transmissions over the network during the course of the data collection and query processing.

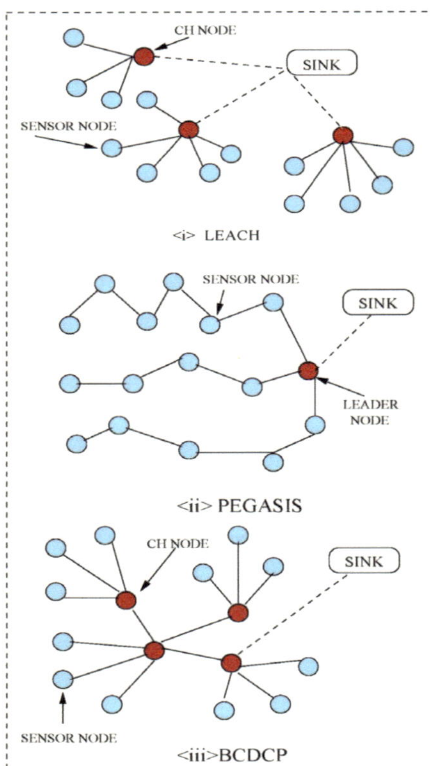

Fig. 1. Main Topologies of Hierarchical Routing Protocols

The stable election protocol described in [17] is a heterogeneous-energy-aware protocol. The weighted election probabilities, based on remaining energy of each node, is used to determine the formation of cluster head. The protocol does not consider the optimal assignment of nodes to the cluster heads. In [18], a balanced k-clustering algorithm for clustering sensor nodes into k clusters is proposed. Each cluster is balanced and the total distance between sensor nodes and the head nodes is minimized. The number of nodes is assumed to be a multiple of k at all times, which may not be feasible. A cluster based routing algorithm of [19] aims to extend the lifetime of the sensor network by maintaining uniform consumption of energy by the nodes. This protocol performs better than the adaptive approach. In [2], the authors focus on the design criteria for routing protocols and issues and challenges of cluster-based routing in WSNs.. Yunfeng et al. [20] have devised a protocol, the basic idea of which is that instead of source-initiated or destination-initiated route discovery, it is the base station that finds multipath to the source of the data and selects one of them. The problem of energy-aware routing in networks with renewable energy sources is adressed by Lin et al. [21]. The proposed static routing scheme utilizes the knowledge of traffic patterns and energy consumption, and does not demand the instantaneous information about the node energy.

3. Model

3.1 The Nomenclature

The terminology used in our study are:

Homogeneous Network Homogeneous network consists of sensors possessing uniform initial energy.

Heterogeneous Network The network in which the initial energy of the sensors is different.

$\mathbb{N}$ Set of all the sensor nodes deployed in the sensor field of the network. E_{avg} This is defined as the average energy of the wireless sensor network.

$$E_{avg} = \frac{1}{n}\sum_{k=1}^{n} E_k \quad (1)$$

where n is the number of the sensors and E_k is the energy of the kth sensor. $\mathbb{P}$ Set consisting of sensor nodes with energy equal to or greater than E_{avg}, and is a subset of set $\mathbb{N}$, which is a set of all the sensor nodes deployed in the network.

Principal Node This receives the sensed data from other nodes in its hierarchy, aggregates it to forward either to another principal node or to the *Superior Node.*

Superior Node Functions as the root of the hierarchy and sends the aggregated message to the sink.

3.2 Radio Power Model

A typical sensor node is depicted in Figure 2 and consists of four major components: a data processor unit; a micro-sensor; a radio communication subsystem that consists of transmitter/receiver electronics, antennas and an amplifier; and a power supply unit. Although energy is dissipated in all of the first three components of a sensor node, energy

dissipations associated with the radio component is considered since the core objective of this study is to develop an energy-efficient network layer protocol to improve the network lifetime. In addition to this, the energy dissipated during data aggregation is the cluster heads is also accounted.
The radio energy model [9] employed in our study is described in terms of the energy dissipated in transmitting k-bits of data between two nodes separated by a distance r meters and so also the energy spent for receiving at the destination sensor node and is given by,

$$E_T(k,r) = E_{Tx} * k + Eamp(r) * k \tag{2}$$
$$E_{amp}(r) = \varepsilon_{FS} * r^2 \tag{3}$$

The energy cost incurred in the receiver is given by,

$$E_{amp}(r) = \varepsilon_{FS} * r^2 \tag{4}$$

where E_{amp} denote energy dissipated in the transmitter of the source node is required to maintain an acceptable signal-to-noise ratio for reliable transfer of data messages. We use free space propagation model and hence the energy dissipation of the amplifier is given by:

$$E_{amp}(r) = \varepsilon_{FS} * r^2 \tag{5}$$

where ε_{FS} denotes the transmit amplifier parameter corresponding to free space.
The assumed values for the various parameters is as given below.

$$E_{Tx} = E_{Rx} = 50nJ/bit$$
$$\varepsilon_{FS} = 10pJ/bit/m^2$$

The energy spent for data aggregation is $E_{DA} = 5\text{nJ/bit/message}$.

4. Problem Definition

A sensor network is described by means of an edge-weighted graph, G_{WSN}($\mathbb{N}$, D, Sink), where $\mathbb{N} = \{n_1, n_2 .... n_n\}$ is a set of sensor nodes and $D = \{d_1, d_2 .... d_n\}$ is a set containing the inter-node distances existing between any two nodes.

4.1 Objectives

The objectives of our work are:
1. To design and develop an energy-efficient hierarchical routing algorithm which minimizes energy consumption of the wireless sensor network.
2. Maximizing the network lifetime.

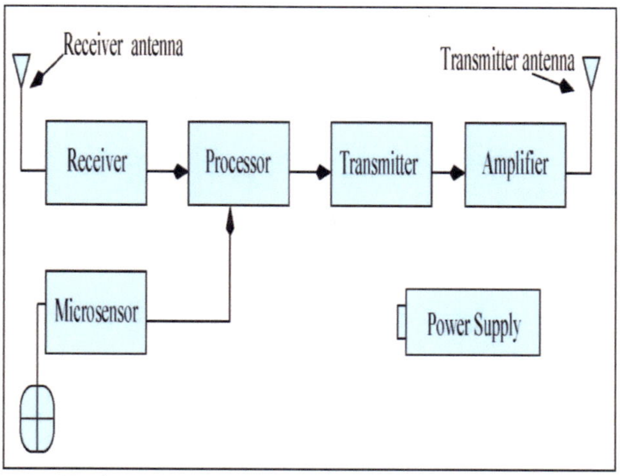

Fig. 2. A Typical Sensor Node

4.2 Assumptions

- A WSN consisting of a fixed sink with unlimited supply of energy and n wireless sensor nodes having limited power resources.
- The wireless sensor network can be either homogeneous or heterogeneous in nature.
- The sensor nodes are equipped with Global Positioning Systems (GPS).
- The nodes are equipped with power control capabilities to vary their transmitted power.
- Each node senses the environment at a fixed rate and always has data to send to the sink.

5. Sink Administered Load Balanced Dynamic Hierarchical Protocol (SLDHP)

This section focuses on the design details of our proposed protocol SLDHP, which is a hierarchical wireless sensor network routing protocol. Here the sink with unrestrained energy plays a vital role by performing energy intensive tasks thereby bringing out the energy efficiency of the sensors and rendering the network endurable. The pattern of the hierarchy varies dynamically as it is based on energy levels of the sensors in each iteration. SLDHP functions in two phases namely:

1. Network Configuring Phase
2. Communication Phase.

The algorithm steps are described in Table 1.

5.1 Network Configuring Phase

The goal of this phase is to establish optimal routing paths for all the sensors in the network. The key factors considered are balancing the load on the principal nodes and minimization of energy consumption for data communication. In this phase, the sink probes and beckons the sensors to send the status message that encapsulates information regarding their geographical position and current energy level. The sink upon receiving this, stores the information in its data structures to facilitate further computations. To construct the routing path, first the sink traces the node with minimum energy, n_{min} from the set $\mathbb{N}$. The minimum energy node n_{min} will be alloted to the principal node, which will be selected based on the following criteria:

- The sink reckons the set $\mathbb{P}$, that contains nodes with energy above E_{avg}, which is a subset of set $\mathbb{N}$.
- It then computes the *Euclidean Distance* between n_{min} and each of the nodes in $\mathbb{P}$. This distance between two nodes $u = (x_1, y_1)$ and $y = (x_2, y_2)$, is described by the equation,

$$dist(u - v) = ||u - v|| \tag{6}$$

This is in turn expanded as follows:

$$\sqrt{(|x_1 - x_2|)^2 + (|y_1 - y_2|)^2} \tag{7}$$

- The node in the set $\mathbb{P}$ which has minimum distance to n_{min} is selected as the principal node.

To aid further computations, the amount of energy spent by the principal node on receiving and aggregating message sent from n_{min} is virtually reduced. The minimum energy node is then removed from the set $\mathbb{N}$. This phase repeats until all the nodes in the network are assigned to principal nodes. The last node that remains in set $\mathbb{N}$ is the node with maximum energy, designated as the superior node and has the job of sending the aggregated message to the sink.

The protocol gives prime importance to achieve balancing of load on the principal nodes. The minimum energy nodes will be assigned to a principal node as long as this node has the capability to handle them. Once the energy of the principal node falls below E_{avg}, it will be treated as a normal node and hence will be assigned to another principal node. In this way, multihop minimal spanning tree is constructed without a need for running a separate minimal spanning tree algorithm. Figure 3 depicts the hierarchical setup of the proposed protocol.

SLDHP eliminates the necessity of knowing the optimum number of clusters in the network. The load is evenly balanced depending upon the capacity of the principal nodes. The protocol starts with a chaining setup and ends in a hierarchical model. In this way, multihop, load balanced network is achieved. The concluding task of this phase is to determine the TDMA slots for all the nodes within the hierarchy. Once all the computations are over, the sink sends messages to all the sensors indicating their principal nodes and the TDMA slots.

5.2 Communication Phase

The sensors send their sensed data to their respective principal nodes. Each principal node gathers data from the nodes down in its hierarchy, fuses it and forwards either to another principal node or to the sink. This phase inturn comprises of three activities.

Data gathering utilizes a time-division multiple access scheduling scheme to minimize collisions between sensor nodes trying to transmit data to the principal node.

Data fusion or aggregation Once data from all sensor nodes have been received, the principal node combines them into a target entity to greatly reduce the amount of redundant data sent to the sink.

Data routing Transfers the data along the principal node-to-principal node routing to the superior node, which transmits the fused data to the sink.

I *Network Configuring Phase*

(i) Initialize

- The *Sink* queries all the nodes regarding their status.
- Nodes reply by sending the status message.

(ii) Main Processing

- Sink computes the average energy of the network.
- begin
  - In addition to computing the average energy, the sink also does the following operations.
    * It traces n_{min}, which is the node with minimum energy. It then computes the set $\mathbb{P}$ which contains the node-ids of all the nodes with energy above or equal to E_{avg}.
    * Finds distance between n_{min} and elements of $\mathbb{P}$.
    * An element of $\mathbb{P}$, p_i having minimum distance to n_{min} is assigned as the principal node.
    * Checks whether p_i is still eligible to be in set $\mathbb{P}$. If not, it is eliminated from the set.
    * n_{min} is discarded from $\mathbb{N}$.
  - Repeat this until all the nodes have been assigned to principal nodes.
  - Schedules $TDMA$ slots for all the nodes.

(iii) Finalize

- Sink sends messages to all the nodes indicating their principal nodes and the $TDMA$ slots.

II *Communication Phase*

- Data Gathering
- Data fusion or aggregation
- Data routing

Table 1. SLDHP Algorithm

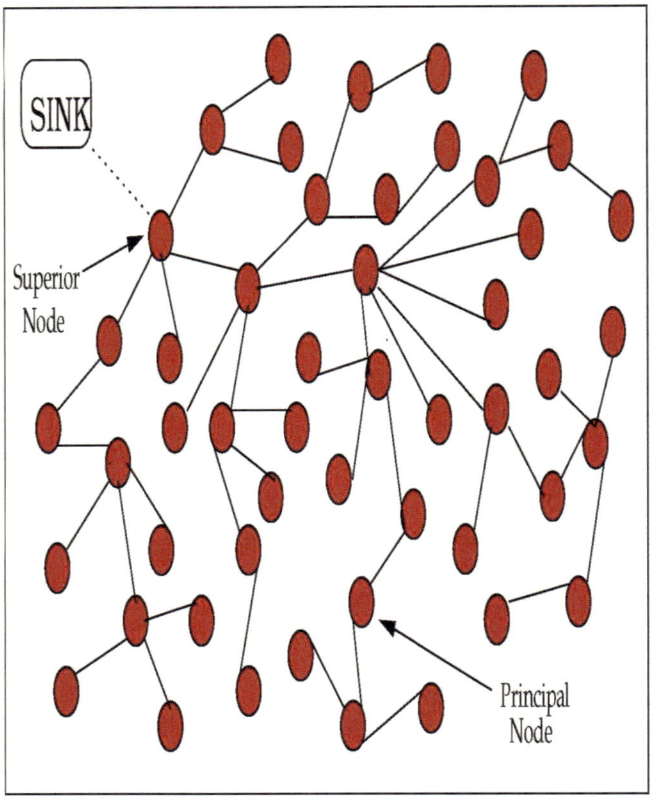

Fig. 3. Hierarchical Setup of SLDHP

6. Simulation and Numerical Results

6.1 The Test-Bed

A homogenous sensor network was set up with the simulation environment comprising 100 nodes, with all nodes possesing the same initial energy of 2J. The simulations were carried out using the OMNeT++ simulator. The sensor nodes were deployed randomly in a sensor field of a grid size of 500mx500m. The simulations were carried out several times, for different network configurations in order to obtain consistent results. The performance metrics considered are Average Energy Consumption by the nodes and Network Lifetime. The proposed protocol is compared with BCDCP.

6.2 Average Energy Consumption of the Sensor Network

Figure 4 shows the Average Energy Consumption of the sensor network, as a variation with reference to number of iterations of the network. The simulation environment is setup with the initial battery energy of all nodes being 2J and a message length of 4 kbits/packet. We observe that the protocol greatly reduces the energy consumed and hence outperforms others in terms of battery efficiency. This is due to the minimum-spanning tree hierarchical structure formed by SLDHP as compared to the cluster-based structure which consists of equal number of member nodes with unequal distribution of energy. BCDCP achieves balancing by assigning equal number of nodes to each of the clusters which results in overloading the already overloaded cluster-heads to drain out much of their energy on receiving, aggregating and transmitting the data at a much faster rate. In comparison, the proposed algorithm comprises of unequal member nodes within the hierarchy, but load balanced in terms of energy resources, which contributes significantly to the increased energy efficiency of the algorithm. Hence the packet transmission time in our algorithm is predominantly short as compared to others. From the plot, it is observed that initially when the number of iterations is less, energy consumption in both the schemes is found to be almost the same, with no conspicuous results. This is due to the fact that the hierarchical structure at this point of time seems almost the same. The real advantage comes to light when the number of iterations increases, with the hierarchical structure adapting itself dynamically to the changing scenario. The superior performance offered by SLDHP enables to achieve a reduction of energy consumption by about 21% as compared to the earlier algorithms.

6.3 Sensor Network Lifespan

The energy consumption rate can directly influence the lifespan of the sensor nodes as the depletion of battery resources will eventually cause failure of the nodes. Hence the wireless engineer is always entrusted with the task of prolonging the lifespan of the network by improving the longevity of the sensor nodes.

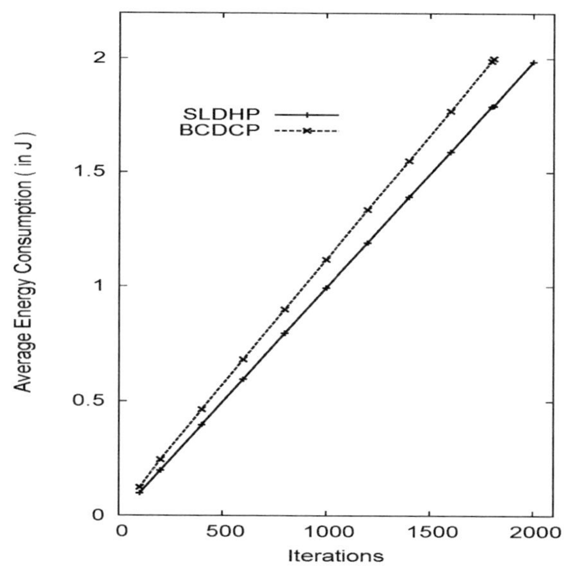

Fig. 4. Comparison of Average Energy Consumption

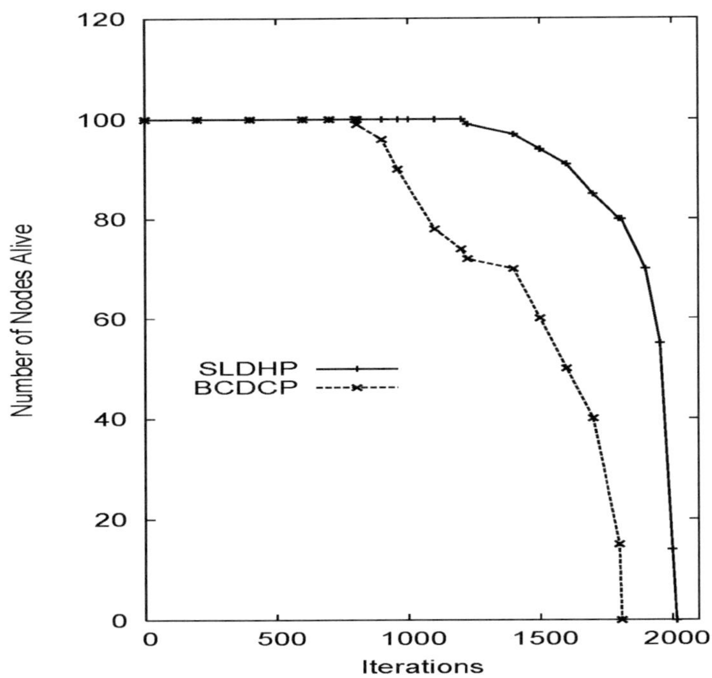

Fig. 5. Comparison of Lifespan

The simulation results of number of nodes alive over a period of time are presented in Figure 5. The simulation environment is the same, i.e., initial energy of nodes being 2J, message length being 4 kbits/packet and the initial node density being 100. Both the protocols are based on a hierarchical structure in which all the nodes rotate to take responsibility for being the cluster-head and hence no particular sensor is unfairly exploited in battery consumption. Due to the hierarchical structure, it is found that till the 806th iteration, the number of nodes that are alive is almost the same in both schemes and equals 100. This implies that the time duration between the first exhausted node and the last one is quite short or the difference in energy levels from node to node does not vary greatly for lower number of iterations. After this critical point, both the curves in the Figure drop indicating the fall in the number of alive nodes. It is evident from the plot that the number of alive nodes is significantly more in our protocol as compared to other and which agrees with the results obtained in the previous simulation. This algorihm can extend the lifespan of the network by about 34% as compared to the earlier algorithm. It is observed that the number of alive nodes in earlier algorithm is a maximum of 100, dropping at a steady rate till none of the nodes are found to be alive at the 1800th iteration. In comparison, the nodes of SLDHP are very much live and active even for a little beyond the 2000th iteration, once again indicating the superior performance of the algorithm. The reason for this is again the same, the difference in hierarchical structure, plus the added advantage of dynamically having a load balancing scheme.

6.4 Average Energy Consumption for varying message lengths

Figure 6 shows the average energy consumption of the network when SLDHP is run with the data communication phase transmitting data at varying message lengths of 4kbits/packet and 8kbits/packet respectively. From the plot, it is observed that when the message length is 4 kbits/packet, the behaviour is exactly similar to the one depicted in Figure 4 for SLDHP due to the similarities of the simulation environment set up. When the message length is doubled, the average energy consumption of the sensor network is much more as observed from the simulation results. This is quite obvious because of greater overhead involved in aggregating and transmitting a larger sized message. From the plot, it is seen that at the end of the 2000th iteration, the energy consumed for transmitting a smaller message is close to 2J while the same energy level is reached in the 1620th iteration itself, for a larger message transmission. A message length of 4 kbits/packet seems ideal as lesser length message may not be in a position to carry out the desired task and a larger length may unnecessarily contribute to additional overhead which can degrade the performance of the network.

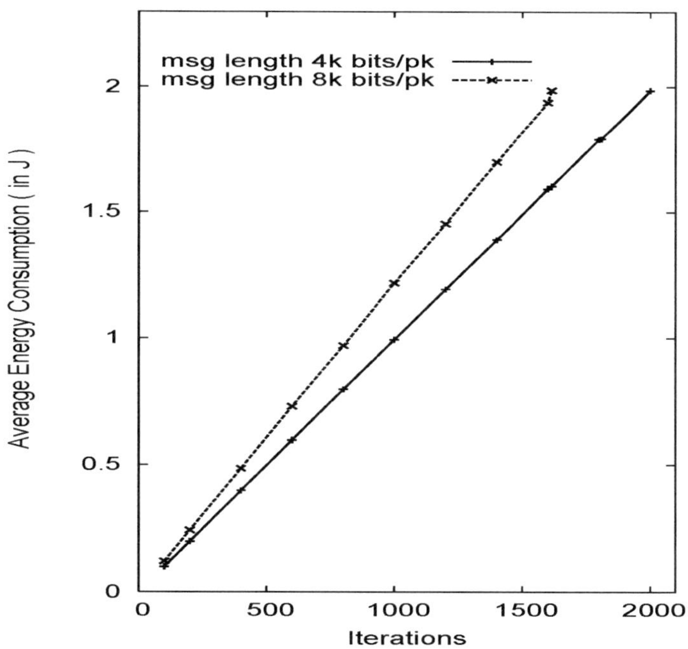

Fig. 6. Average Energy Consumption (SLDHP) with variable packet size

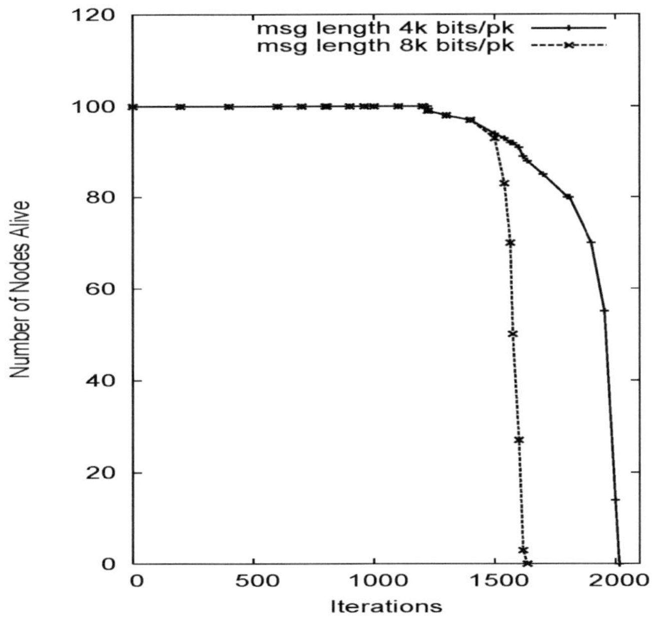

Fig. 7. Lifespan of the Wireless Sensor Network (SLDHP) with variable packet size

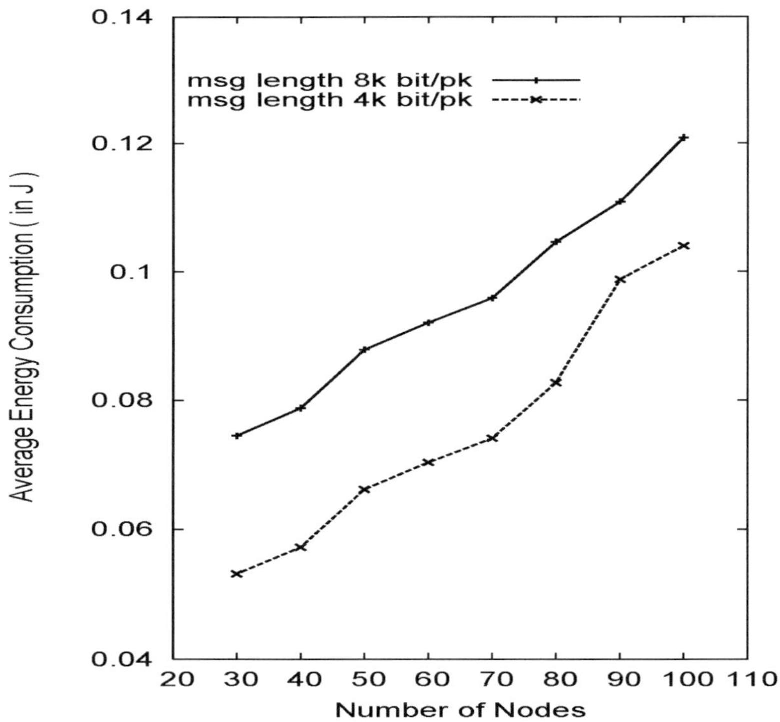

Fig. 8. Average Energy Consumption (SLDHP) for varying node density

6.5 Network Lifespan for varying message lengths

Figure 7 shows another performance run when communications in SLDHP, take place by transmitting varying length messages of 4 kbits/packet and 8 kbits/packet The simulations are carried out under similar conditions. As seen from the plot, when the message length is 4 kbits/packet, larger number of nodes are alive and the same is confirmed by the results obtained in Figure 5. When the message length is doubled, saturation of the network takes place at a faster rate due to increased overhead on the sensor nodes and the principal nodes in particular. This manifests in nodes consuming larger energy, resulting in a larger transmission cost, leading to a shorter lifespan of the network. The smaller the message length, greater is the lifespan of the network with the number of live nodes prolonging the network lifespan to as long as the 2000th iteration. Till the 1400th iteration, the number of alive nodes in both cases seems exactly the same, but drops abruptly to zero at the 1635th iteration, for a larger message length. The reason for this is the same as described for Figure 4 and hence the same inference can be drawn here as well.

6.6 Average Energy Consumption with varying node density

The plots in Figure 8 show the average energy consumption of the network with proposed algorithm run for two different message lengths. The simulation environment is set up with all the nodes equipped with a uniform initial energy of 2J. The node density is varied to account for scalability of the WSN and at the same time will aid in understanding the behaviour of the network especially in terms of energy management of the network for varying node densities. For comparatively lower value of node density, the average energy consumption of the network is smaller being a little less than 0.06 J for a smaller message length, increasing steadily to about 0.09 J for a node density of 100. In comparison, it is found that the energy consumption is relatively more for a larger sized message, varying from 0.078 J for 40 nodes reaching a value of 0.12 J for 100 nodes. This behavior is much the same as for a smaller message, the difference being that obviously more energy is consumed for a larger message size. As the number of nodes increase, the complexity of the network configuring phase also increases proportionately leading to an increased overhead on the sink to dynamically form load balanced hierarchical structures. The complexity of the data communication phase is no less, with more number of nodes being involved in data communications and with the complexity increasing with increasing nodes. The energy consumption of the network increases in proportion to the number of nodes and the same analogy holds good for different message lengths, the consumption being much more for larger sized messages.

7. Conclusions

A WSN is composed of tens to thousands of sensor nodes which communicate through a wireless channel for information sharing and processing. The sensors can be deployed on a large scale for environmental monitoring and habitat study, for military surveillance, in emergent environments for search and rescue, in buildings for infrastructure, health monitoring, in homes to realize a smart environment etc.. SLDHP manages to balance the load on the principal nodes and hence the sensor nodes are relieved from the energy intensive tasks such as formation of hierarchy and scheduling of slots to send their sensed data. This job is effectively accomplished by the high powered sink. The simulation results indicate that the network lifetime is elevated to a large extent when compared to other hierarchical routing protocols. The future work includes applying our protocol to a distributed wireless sensor network and hence to improve the network performance as in present scenario.

8. References

E. Shin, S.H. Cho, N. Ickes, R. Min, A. Sinha, A. Wang and A. Chandrakasan. Physical Layer Driven Protocol and Algorithm Design for Energy-Efficient Wireless Sensor Networks. *Seventh Annual ACM SIGMOBILE Conference on Mobile Computing and Networking,* July 2001.

J. Ibriq and I. Mahgoub. Cluster-Based Routing in Wireless Sensor Networks: Issues and Challenges. *SPECTS,* pp. 759-766, 2004.

I.F. Akylidiz, Weilian Su, Yogesh Sankarasubramaniam and Erdal Cayirci. Wireless Sensor Network: A Survey on Sensor Networks. *IEEE Communications Magazine,* 40(8); pp. 102-114, August 2002.

M. Bhardwaj and A.P. Chandrakasan. Bounding the Lifetime of Sensor Networks Via Optimal Role Assignments. *Twenty-First Annual Joint Conference of the IEEE Computer and Communications Society, INFOCOMM,* 2002.

J. Agre and L. Clare . An integrated architecture for co-operative sensing networks. *IEEE Computer Magazine,* pp. 106-108, May 2000.

W.B. Heinzelman, A.P. Chandrakasan and H. Balakrishnan. Energy-Efficient Communication Protocol for Wireless Microsensor Networks. *Proc. 33rd Hawaii Int'l. Conf. Sys. Sci.,* January 2000.

W.B. Heizelman, A.P. Chandrakasan and H. Balakrishnan. An Application-Specific Protocol Architecture for Wireless Microsensor Networks. *IEEE Transactions on Wireless Communications,* 1(4); pp. 660-670, October 2002.

S. Lindsey, C. Raghavendra and K.M. Sivalingam. Data Gathering Algorithms in Sensor Networks using Energy Metrics. *IEEE Trans. Parallel and Distrib. Sys.,* (9); pp. 924-935, September 2002.

S.D. Muruganathan, Daniel C.F. Ma, R.I. Bhasin and A.O. Fapojuwo. A Centralized Energy-Efficient Routing Protocol for Wireless Sensor Networks. *IEEE Communications Magazine,* 43; pp. 8-13, March 2005.

G. Huang, X. Li and J. He. Energy-efficiency Analysis of Cluster-Based Routing Protocols in Wireless Sensor Networks. *IEEE Aerospace Conference,* March 2006.

Y. Yu, R. Govindan and D. Estrin. Geographical and Energy Aware Routing: A Recursive Data Dissemination Protocol for Wireless Sensor Networks. *UCLA Computer Science Department Technical Report UCLA/CSD-TR-01-0023,* pp. 159-169, May 2001.

A. Depedri, A. Zanella and R. Verdone. An Energy Efficient Protocol for Wireless Sensor Networks. December 2003.

V. Mhatre and C. Rosenberg. Homogeneous vs Heterogeneous Sensor Networks: A Comparative Study. *Proceedings of International Conference on Communications (ICC 2004),* June 2004.

O. Younis and S. Fahmy. HEED: A Hybrid, Energy-Efficient, Distributed Clustering Approach for Ad Hoc Sensor Networks. *IEEE Transcations on Mobile Computing,* 3(4), December 2004.

A.D. Amis and R. Prakash. Load-Balancing Clusters in Wireless Ad Hoc Networks. *Proceedings of the 3rd IEEE Symposium on Application-Specific Systems and Software Engineering Technology (ASSET'00)*, 2000.

Z. Zhang and G. Zheng. A Cluster Based Query Protocol for Wireless Sensor Networks. *The 8th International Conference on Advanced Communication Technology*, 1; pp. 140-145, February 2006.

G. Smaragdakis, I. Matta and A. Bestavros. SEP: A Stable Election Protocol for Clustered Heterogenous Wireless Sensor Networks. *The 8th International Conference on Advanced Communication Technology*, 2004.

S. Ghiasi, A. Srivastava, X. Yang and M. Sarrafzadeh. Optimal Energy Aware Clustering in Sensor Networks. *Sensors*, 2; pp. 258-269, July 2002.

Uk-Pyo Han, Sang-Eon Park, Seung-Nam Kim and Young-Jun Chung. An Enhanced Cluster Based Routing Algorithm for Wireless Sensor Networks. *International Conference on Parallel and Distributed Processing Techniques and Applications*, 1; June 2006.

Y. Chen and N. Nasser. Energy-Balancing Multipath Routing Protocol for Wireless Sensor Networks. *Proceedings of the 3rd international conference on Quality of service in heterogeneous wired/wireless networks*, 191; 2006.

L. Lin, N.B. Shroff and R. Srikant. Energy-Aware Routing in Sensor Networks: A Large Systems Approach. *WONS 2006 : Third Annual Conference on Wireless On-demand Network Systems and Services*, pp. 159-169, January 2006.

R.C. Shah and J. Rabaey. Energy Aware Routing for Low Energy Ad Hoc Sensor Networks. *WCNC 2002 Conference*, March 2002.

11

Mobile Wireless Sensor Networks: Architects for Pervasive Computing

Saad Ahmed Munir* ,Xie Dongliang*,
Chen Canfeng µ and Jian Ma µ
**Beijing University of Posts & Telecommunications, China*
µNokia Research Center, China

1. Introduction

A mobile wireless sensor network owes its name to the presence of mobile sink or sensor nodes within the network. The advantages of mobile wireless sensor network over static wireless sensor network are better energy efficiency, improved coverage, enhanced target tracking and superior channel capacity. In this chapter we will present and discuss different classifications of mobile wireless sensor network as well as hierarchical multi-tiered architectures for such networks. This architecture makes basis for the future pervasive computing age. The importance of mobility in traditional wireless sensor network (WSN) is highlighted in this chapter along with the impact of mobility on different performance metrics in mobile WSN. A study of some of the possible application scenarios for pervasive computing involving mobile WSN is also presented. These application scenarios will be discussed in their implementation context. While discussing the possible applications, we will also study related technologies that appear promising to be integrated with mobile WSN in the ubiquitous computing. With an enormous growth in number of cellular subscribers, we therefore place the mobile phone as the key element in future ubiquitous wireless networks. With the powerful computing, communicating and storage capacities of these mobile devices, the network performance can benefit from the architecture in terms of scalability, energy efficiency and packet delay, etc.

For mobile wireless sensor networks, there are basically two sensing modes, local sensing and remote sensing. By allowing and leveraging sink mobility and sink coordination, mobile WSN can achieve the goal of lower and balanced energy consumption with the following features:

- Single-hop clustering. By allowing only single hop transmission between sensor and sink node, most previous multi hop relaying sensor nodes may become unnecessary. In fact, sensor nodes can enter sleep mode until the sink approaches. Therefore, the original energy budget for multi hop relaying can be saved.
- Sink mobility and coordination. For a delay tolerant application, single mobile sink in fact equals virtually multiple static sinks at different positions. Multi-sink deployment can bring more uniform energy dissipation, therefore the possibility of energy hole will be reduced and network coverage will be improved.

- Mobility-assisted positioning and identification. Sensor nodes can estimate their position by learning mobile sink's position, which can be periodically broadcasted. If each sensor node can be geographically identified, then it is feasible to use more energy-efficient routing method, such as the geographic based routing.
- Improved network scalability: This merit is achieved by lowering overhead of MAC/routing protocols at the vast majority of resource constrained sensor nodes, especially for high-density networks. Complexity of other network maintenance functions such as topology and connectivity control may also get reduced.
- Adaptive network configuration: This feature is achieved through adaptive network re-organization and varying-scale observation based on the observed dynamics of targets being sensed, both in spatial and temporal domains.
- Sacrificed message delay: This defect can mainly be attributed to the increased sensor-sink meeting delay. Methods such as increasing the density of sink nodes and controlling the trajectory of mobile sinks can offset relinquished performance. In these tiered networks, one shared design rationale is to keep the logics of sensor nodes as simple as possible, and move complex functions to the overlaying mobile elements with richer resources. We also notice that, some more recent work has commenced on applying methods including Delay Tolerant Networking (DTN) and peer-to-peer (P2P) information sharing for asynchronous message switching in challenged wireless sensor networks. Unfortunately, we have not found efficient inter-tier communication methods for such cross-tier optimization approaches.

At the same time, the delay performance cannot be improved by simply increasing sink velocity. When mobile sinks are moving too fast through the effective communication region of static sensors, there may not be sufficient long dialogue durations for the sensor nodes to successfully deliver potentially long packets to the mobile sink. In other words, with the increase of sink velocity, the outage probability of packet transmission will rise. We further address the issue of optimal multi hop forwarding strategy under predictable sink mobility, which includes the distance characteristics both in the case of multi hop and single hop communication model.

Energy conservation is regarded as one of the most significant challenges in wireless sensor networks (WSN) due to the severe resource limitations of sensor nodes [1]. In addition, the peculiar non uniform traffic pattern in wireless sensor networks can lead to increased traffic for those sensor nodes close to the sink node. Therefore an unbalanced energy dissipation pattern will be inevitable, and those critical sensor nodes close to the sink node will withdraw from the network earlier due to faster energy depletion. The withdrawal of sensor nodes around sink node has lead to the known "energy hole" problem. The network may consequently lose sufficient connectivity and coverage, if there is no supplementary sensor deployment. Methods such as in-network processing and deploying multiple sinks can only partly tackle this problem by sacrificing the information accuracy and increasing the infrastructure cost.

Different from these approaches for flat networks, we have addressed this problem by leveraging mobility and multi-radio heterogeneity to create a cellular-sensor hybrid system with clustered and tiered network architecture. By combing the rationales in previous approaches such as Data MULE [3] and TTDD [2], the mobile enabled WSN (mWSN [29]) enables both local and remote sensing by mobile phones extracting information of interest from the sensory environment. As illustrated in Figure. 1, there are three tiers in the mWSN architecture: sensor tier, mobile sink tier, and base station tier. At the sensor tier, sensor nodes as well as various RFID tags may be organized in a clustered fashion with mobile sinks as the cluster heads. At mobile sink tier, mobile sinks may coordinate locally or remotely to exploit the redundancy via short-range or long-range radios equipped with each mobile phone.

At the base station tier, gathered sensory information can be stored and forwarded to Internet by the base stations of cellular networks, which serves as the access points to Internet. mWSN will enhance the performance of network connectivity and coverage by connecting isolated "islands" of wireless sensor networks designed for different applications. As illustrated in Figure. 2, there are basically two sensing modes in mWSN. In the case of local sensing, after mobile sink sends the information query command, sensory information collected by fixed sensors will be firstly forwarded by mobile sink to the base station for information fusion, where the digital information can be parsed and translated into meaningful interpretations.

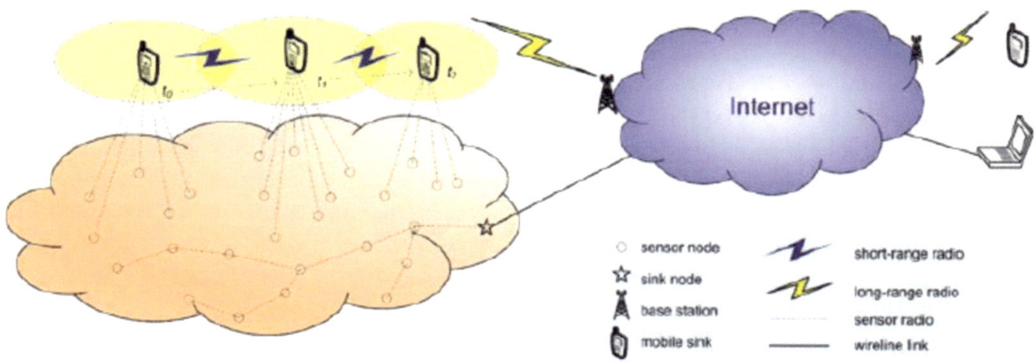

Fig. 1. Architecture Overview of Mobile enabled Wireless Sensor Network (mWSN).

In the case of remote sensing, the mobile sink will send the information query command to the base station, which will assign the sensing task to another mobile sink or fetch the information from a database of sensory information. The differentiation between local sensing and remote sensing may be based on the location information of sensors and mobile sink: if the location of a querying mobile sink is same with those of reporting sensors, it can be decided as a local sensing; otherwise, it should be a remote sensing.

Furthermore, with the knowledge of the remaining energy left at each sensor node, mobile sinks can choose the optimal path by circumventing the least energy sensor nodes [28]. The direct benefit of energy reduction is the lengthened network lifetime. As the route length can be reduced to one in mWSN, the scalability performance can also benefit from the hierarchical architecture of mWSN. However, the performance of packet delivery delay may be compromised, because packets have to be buffered before mobile sink approaches the sensor nodes.

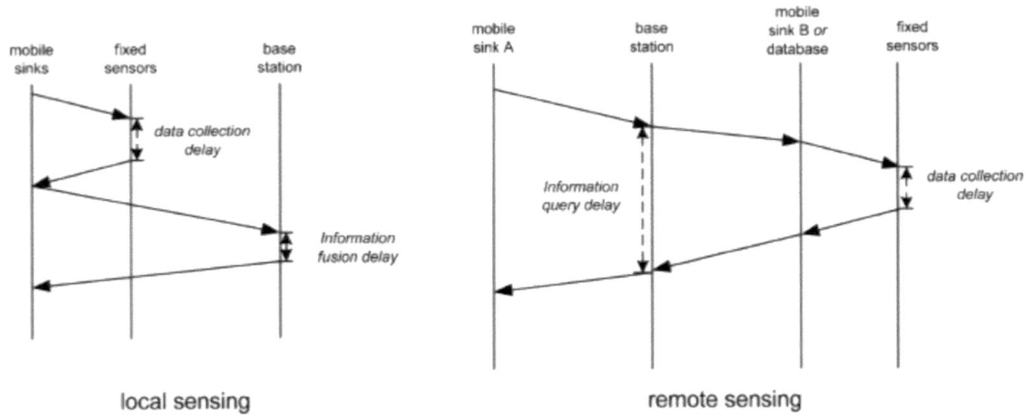

Fig. 2. Local Sensing and Remote Sensing in mWSN.

The concept of mobile wireless sensor network in the context of pervasive ubiquitous networks has emerged in recent years, although the genius of Marc Weiser envisaged this concept as early as in 1991 [31]. Along with the evolution of Wireless Sensor Networks (WSN) surfaced a new concept of presence of mobile sink or agents. Now, mobility in WSN is considered to be a blessing as opposed to problem. Their results confirm that mobility not only improves the overall network lifetime but also the data capacity of the network. Mobility can further address delay and latency problems. Most of the fundamental characteristics of mobile wireless sensor network are the same as that of normal static WSN. Some major differences, however, are as follows;

- Due to the mobility, mobile WSN has a much more dynamic topology as compared to the static WSN. It is often assumed that sink will move continuously in a random fashion, thus making the whole network a very dynamic topology. This dynamic nature of mWSN is reflected in the choice of other characteristic properties, such as routing and MAC level protocols and physical hardware.
- In most of the cases, it can be reasonably assumed that gateway sink has an infinite energy, computational and storage resources. The depleted batteries of mobile sinks can be recharged or changed with fresh ones and similarly mobile sink has access to computational and storage devices.
- The increased mobility in the case of mobile WSN imposes some restrictions on the already proposed routing and MAC level protocols for WSN. Most of the efficient protocols in static WSN perform poorly in case of mobile wireless sensor network.
- Due to the dynamic topology of mobile wireless sensor networks, communication links can often become unreliable. This is especially the case in hostile, remote areas where availability of constant communication channel for minimum QoS becomes a challenge.
- Because of the mobility involved, location estimation plays an important role so as to have an accurate knowledge of the location of the sink or node. Mobile wireless sensor networks have been shown to demonstrate enhanced performance over static wireless sensor networks. Because of the mobility of the sink, in general, much work can be shared by the mobile sink. Some of the advantages gained through mobile wireless sensor network over traditional sensor network are presented herewith.

Mobile Wireless sensor networks (Figure 1) are believed to have more channel capacity as compared to static WSN. The capacity gain has been calculated in case of mobile sink within WSN and has come out to be 3-5 times more than static WSN, provided the number of mobile sinks increases linearly with the growth of sensor nodes [32]. The other advantage of mWSN is its better targeting. Because, mostly the sensors are deployed randomly, as opposed to precisely, therefore there is often a requirement to move the sensor node for better sight or for close proximity. Also mobility helps in better quality of communication between sensor nodes. In a sparse or disconnected network, this property is especially helpful to maintain efficient network connectivity. Another advantage comes in the form of data fidelity. It is well known that the probability of error increases with increasing number of hops that a data packet has to travel. If we reduce the number of hops, this immediately reduces the probability of error. This not only increase the quality of data received but further reduces the energy spent at the static nodes by reducing the retransmissions required due to errors. Based on communication type, two kinds of mobile wireless sensor network exist at present. One is known as the infrastructure network in which the mobile unit is connected with the nearest base station that is within its communication radius to contact; as in the current mobile telephone system. The other one is called infrastructure-less mobile network, also knows as ad hoc network. No fixed routers are needed and all mobile units are capable of movement and still being able to self-organize and establish communication in an arbitrary manner. In this chapter we discuss the evolution of mobile Wireless Sensor Network and some of its characteristics which make the underlying network design considerations different from those of Wireless Sensor Networks. We present multi-tiered architectures for mobile wireless sensor networks, with an analysis on impact of mobility on delay and network connectivity. We then discuss the possible exciting technologies which could be integrated in future to work with WSN for ubiquitous computing. The chapter is organized in five sections. In the next section, we will present and discuss the multi-tiered architectures for mobile wireless sensor network based on overlay approach. In the third section, we will discuss the impact of mobility on the performance metrics of mobile wireless sensor networks. Fourth section will elaborate on some application scenarios of mobile wireless sensor networks in future pervasive world. The fifth section will discuss some of the existing enabling technologies for possible integration into mobile wireless sensor network for ubiquitous computing, followed by conclusion and references.

2. Hierarchical Architectures for Mobile Wireless Sensor Networks

Multi-tier architecture for traditional wireless sensor networks has been proposed in literature. We however present the multi-tiererd architecture for mobile wireless sensor network. A description of the ordinary planar wireless is presented, followed by the discussion of multi-tiered architecture for mobile wireless sensor network.

Planar Wireless Sensor Network: Typically, a Wireless Sensor Network (WSN) is composed of a large number of static nodes scattered throughout a certain geographical region. The sensory data is routed from the originator sensors to a remote sink in a multi-hop ad hoc fashion. In general, these sensor nodes have approximate energy conservation and transmission, sensing and caching capabilities, that is, they are homogeneous. A general example of planar wireless sensor networks is illustrated in Figure below.

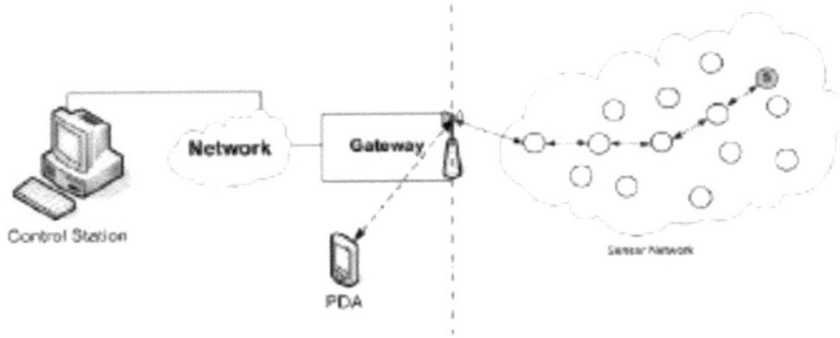

Fig 3. Typical planar Wireless Sensor Network

Using the ad hoc model, planar WSN would inherently pose some disadvantages on network performance. The throughput per node falls asymptotically with increasing nodes as $\theta\left(1/\sqrt{n}\right)$. When data is sent from one node to the next in a multi-hop network, there's a chance that a particular packet may be lost, and the odds grow worse as the size of the network increases. When a node sends a packet to a neighboring node, and the neighbor has to forward it; that takes energy. The bigger the network, the more nodes must forward data, and the more energy is consumed. The end result is: as the network grows, performance degrades.

Two-Tiered Sensor Network Architecture: In a two-tiered mobile-enabled wireless sensor network overlay, WSN utilizes mobile devices as the elements to construct the upper overlay. This owes to the development of microelectronic and wireless communication technologies resulting in the form of mobile phone, laptop and PDA.

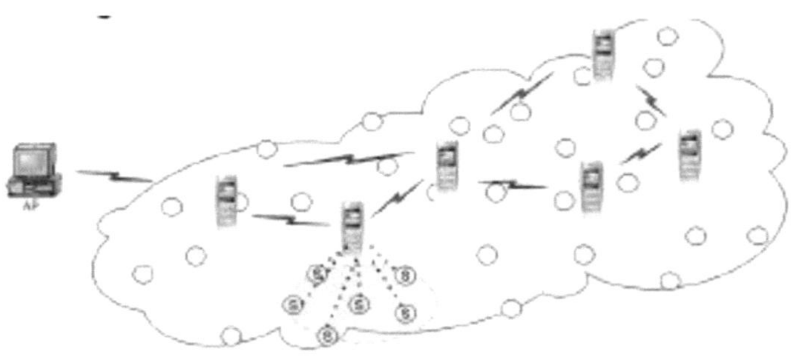

Fig. 4. Two-Tiered Sensor Network with Ad Hoc Configuration

Besides the basic ability of mobility, majority of these have the functions of processing complicated computing, caching and transmitting large number of information packet. These features enable these to be used in heterogeneous WSN and act as the elements of overlay structure. Based on the motivation mentioned above, we conceive the two-tier heterogeneous WSN architectures coupled with mobile overlay. Two brief illustrations are shown below in Figure 4 and Figure 5.

One major difference between the two architectures described in Figure 4 and Figure 5 is the topology of the overlay networking. In the first structure, all the mobile agents, represented by the mobile phones, are self-organized into an ad hoc network. The topology of mobile

overlay, which is random and temporary, has to depend on the relative positions of mobile agents, so it is possible only when the density of mobile devices is high enough. And the slower the mobile agents move, the more stably the overlay can be persisted. Some advanced wireless techniques, such as IEEE 802.11 and Bluetooth are suitable for constructing the wireless interconnected network.
But when the number of mobile phones is small, or the overlay belongs to a sparse network, the architecture mentioned above may not be viable. In this case, an alternative architecture presented in Figure 5 is more suitable. When each mobile phone gathers some data from the sensor nodes in its neighborhood, it doesn't forward it to the access point or other peers simultaneously, but only caches the data in its available memory. In order to avoid the data loss, the size of the memory in mobile agent should be kept large. The data loss is also decided by the average data generating rate from sensor nodes and the round trip time of the mobile agents, namely the maximum allowable delay of data delivery.

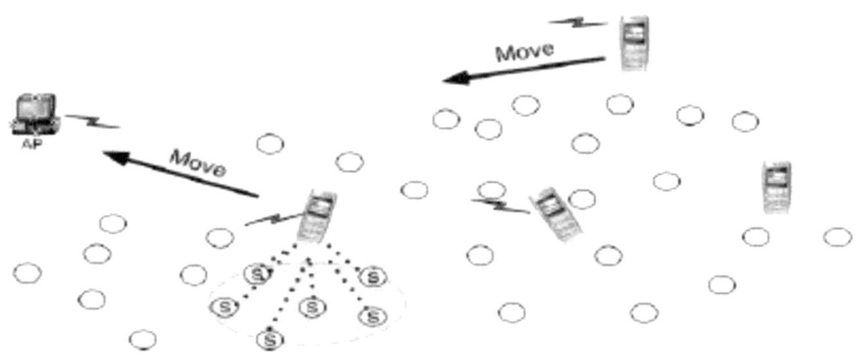

Fig. 5. Two-Tiered Sensor Network without ad hoc overlay

Three-Tiered Sensor Network Architecture: Combining the advantages of fixed access points and mobile agent serving as overlay elements together, we consider a three-tier heterogeneous mobile WSN that results from the introduction of ad hoc overlay and non-ad hoc overlay described above. It can be illustrated that if the *n* sensor nodes have a one-hop link to the nearest mobile agent, and forwarding is limited to the agent, to first order throughput scaling is achieved when the number of fixed access points exceeds $\sqrt{r}$, where $r < n$ is now the number of agents. This is because now the agents forward data for all the sensor nodes, therefore requirement on the number of access points relative to the two-tier hybrid network decreases. Thus, while agents may not reverse the scaling behavior, they can help reduce the number of access points and also lower the power consumption of the sensor nodes, both valuable resources in a variety of sensor applications. In addition to these general and theoretic networking issues, specifically for sensing applications, there are operational advantages to hierarchical heterogeneous layering that cannot be achieved with a "flat", homogeneous network of sensors, with its inherent limitations on power and processing capabilities. For instance, the mobile agents help preserve limited battery resources of sensors by eliminating the need for sensors to monitor communications from their neighbors. In data gathering networks, the medium layer offers the advantage of caching and forwarding compressed data to the destination. Thus for a variety of applications, it appears that a relatively small number of higher-level network elements with access to more power and better computing and communication capabilities could greatly improve the performance of the overall system in terms of throughput, reliability, longevity, and flexibility.

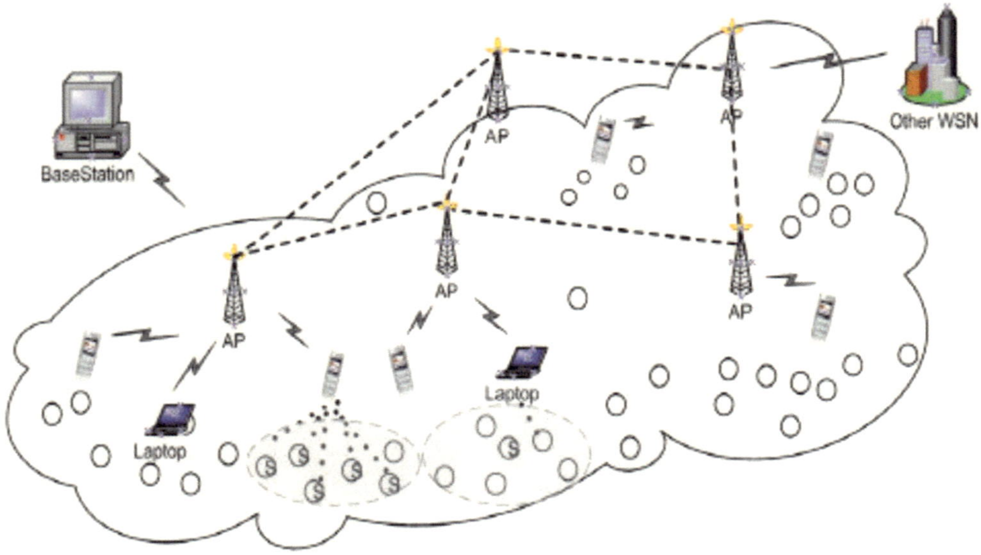

Fig. 6. A typical Three-Tiered Sensor Network Architecture

An example of such a three-tier hierarchical network is shown in Figure. 6. The lowest layer is a random deployed network composed of sensor nodes. These nodes are able to communicate immediately with upper layer agent in near range. They can also form an ad hoc network by communicating with each other, but it is not necessary. The most notable feature of medium layer is its mobility. The mobile agents move to anywhere at anytime if needed. They are responsible for gathering data from lower layer and then forwarding to upper layer. The behavior and collaboration of the mobile agents should be researched deeply so that the WSN performs well and achieves the best performance. The highest layer represents generally the fixed network consisted of some specific number of access points. This networking can be based on wired or wireless and can deploy some kinds of network models such as Mesh or ad hoc. The node of access points may be implemented by IEEE 802.11 AP, the base stations of cellular networks and so on. Meanwhile, this layer provides the network with the possibility of forming a large inter-city wireless sensor networks broadly. The relationship among the three layers is also described in Figure. 7 in which we can see that, many entries including mobile phones, vehicles, men, laptop and even animal, can act as mobile agents. But they have distinctive characteristics respectively, for instance, some are mobile controllable, some are mobile-predicted and some are random.

Two fashions to gather data from sensor nodes: At the lowest layer in the two or three-tier architecture mentioned above, there are two fashions to gather data from sensor nodes, as shown in Figure.8 and Figure.9. Here we will make some conclusions. The first fashion is that every sensor node is isolated. It doesn't communicate with its neighbors always and doesn't deliver received data until some mobile agent come close to it. This fashion does not depend on the topology of underlying network and it provides much energy conservation.

But the biggest disadvantage of this fashion lies in the poor-guaranteed latency of data delivery. The illustration is shown in Figure 7.
On the contrary to the isolated nodes fashion, sensor nodes can organize into an ad-hoc network at the initial phase. Data delivery and gathering can be done all the time. By this fashion, network performance is tightly dependant on the topology of underlying network. A perfect low latency can be achieved but maintaining and updating the network topology will consume much energy. The illustration of second kind of fashion is shown in Figure 8.

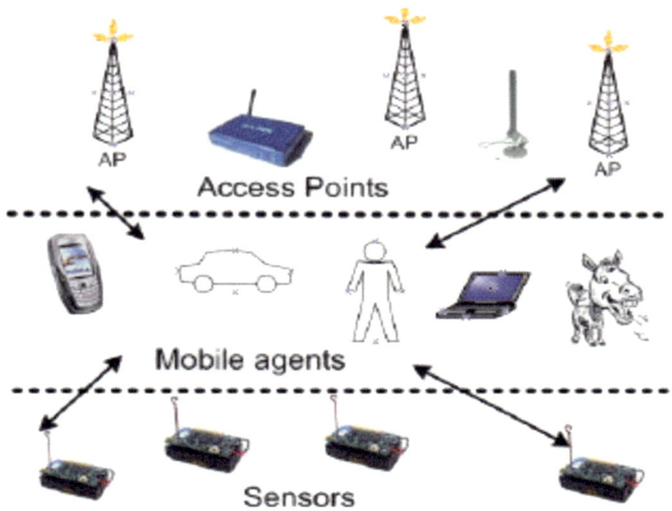

Fig. 7. A planar illustration of Three-tiered sensor network architecture

By observing the above architectures, one can predict that the ubiquitous environment of the future will comprise of both public and privately deployed sensor networks, which will enable the deployment of "smart services", accessible through advanced infrastructures, for instance, the capability-rich 3G mobile networks or open services gateways. In the following we discuss how such a sensor-based services model, which combines (mobile) telecommunication technologies and WSNs, can be realized. Figure 10 depicts the general architecture of a Sensor-Based Service (SBS) solution based on a 2.5G. or 3G network.

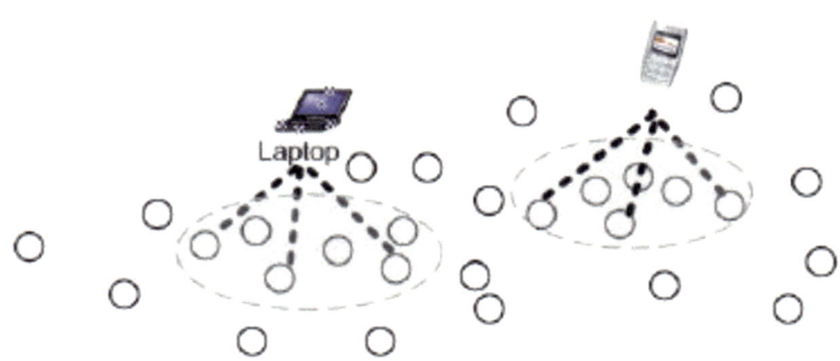

Fig. 8. Gathering data from isolated sensor nodes

The main modifications to the traditional network architecture will be:

The deployment of sensor networks in the monitored fields: A typical modern sensor network consists of sensor nodes and a gateway node. The gateway node has more processing, energy and communication capabilities than the other nodes and is a connecting link between the sensor network and other networks (e.g. mobile network or Internet). Hence, this node should be capable of communicating with the mobile network infrastructure, either directly (GSM/GPRS modem) or indirectly (Internet modem and Gateway GPRS Support Node - GGSN).

An application platform for the lifecycle management of the services and the handling of the remote sensor networks:

This platform is responsible for the creation, deployment, and management of the SBS and the WSN. Furthermore, this platform, as a central component of the overall architecture, can handle all the relevant charging and payment issues. Such platform should be open and support the largest possible number of underlying sensor network technologies.

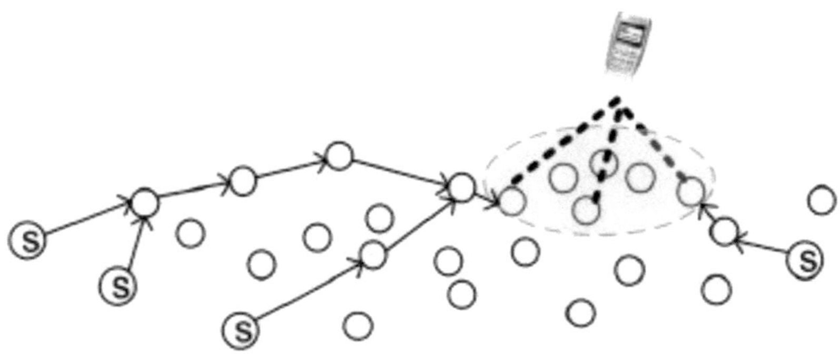

Fig. 9. Gathering data from connected sensor nodes

An open Sensor API for the communication between the WSN and the platform:

The accumulated experience from the wireless networking applications dictates the standardization and adherence to open public APIs for the interaction between the applications and the network elements.

Since the WSN can be regarded as a network element, the support of de facto standards for the WSN handling (i.e., sensor data retrieval) by the platform seems both crucial and feasible. Such an API could be the Parlay/OSA (Open Services Access). The only extension that would be required is the addition of a Sensor SCS (Service Capability Server) to the Parlay/OSA specification. Surely, the ongoing WSN research activity and the diversity in the WSN implementations introduce problems on such SCS standardization, but we can expect that the specific domain will be more clear and stable in a few years. The aforementioned enhancements to the network are not so extensive and have limited impact, because they do not make any unrealistic assumptions on the capabilities of the existing core network and do not imply any alterations to the existing mobile terminal equipment. Thus, the value added services based on such a solution, could experience a fast market penetration with minimal infrastructure investment.

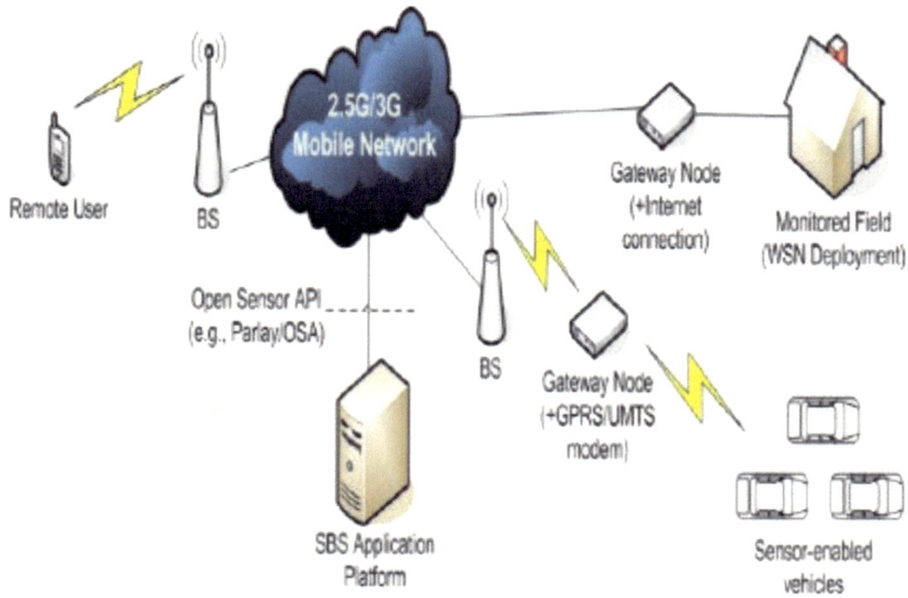

Fig. 10. Enabling Sensor-Based Services in mobile networks

3. Performance Influence from Sink Mobility in Single-hop mWSN

The existing approaches exploiting sink mobility can be categorized with respect to the property of sink mobility, communication/routing pattern, and sink amount. According to the obtainable knowledge about sink mobility, there are basically three kinds of sink mobility: random, predictable, and controlled sink mobility. In terms of the hop-count between sensors and sink, there are mainly two communication/routing patterns: single-hop and multi hop forwarding. The hop-count between sensors and sink has also defined the cluster radius in clustered wireless sensor networks. Majority of related work studied the mobility of single sink. However, a joint optimization is possible if coordination among multiple sinks is feasible. Table 1 lists the related work by comparing different approaches of leveraging sink mobility. Note here Mobile Base Station (MBS) and Mobile Data Collector (MDC) in [12] are with the same meanings as multihop and single-hop forwarding, respectively. For random sink mobility [2–10, 18], sensors can only choose to immediately deliver data to approaching mobile sinks, which leads to significant packet dropping due to insufficient sensor-sink communication duration. For predictable sink mobility [16–17, 19–21], sensors can learn the trajectory pattern of mobile sinks in spatial and temporal domains, based on which sensor topology can be adaptively reorganized. For instances, sensors can decide the transmission schedule which can maximize the opportunity of successful data transmission, and we can design routing strategies for more balanced load among sensors. For controlled sink mobility [11–15, 22–27], the optimization problem can be generally classified into two categories: finding the optimal sink trajectory, i.e. the rendezvous based solution or traveling salesman problem that aims to minimize mobile sink visiting time for all the sensor nodes; mWSN for Large Scale Mobile Sensing finding the optimal sink location, i.e. to optimally place multiple sinks or relays in order to minimize the energy consumption and maximize network lifetime.

It is well known that the traditional definition for a wireless sensor network is a homogeneous network with flat architecture, where all nodes are with identical battery capacity and hardware complexity, except the sink node as the gateway to communicate with end users across Internet. However, such flat network architecture inevitably leads to several challenges in terms of MAC/routing design, energy conservation and network management. In fact, as a kind of heterogeneity, mobility can create network hierarchy, and clustering is beneficial to improve network scalability and lifetime.

	References	Random, predictable, or controlled sink mobility	Single-hop or multihop forwarding	Single-sink or multiple sinks
Data MULEs, SENMA, DFT-MSN	[3–6]	Random	Single-hop	Multiple
CarTel, Message Ferry	[7–10]	Random	Multihop	Multiple
Mobile Element Scheduling	[11, 12]	Controlled	Single-hop	Single
AIMMS	[13–15]	Controlled	Multihop	Single/multiple
Predictable Mobile Observer	[16]	Predictable	Single-hop	Single
SEAD	[17]	Predictable	Multihop	Multiple
TTDD, EARM	[2, 18]	Random	Multihop	Single
HLETDR, Joint Mobility and Routing	[19, 21]	Predictable	Multihop	Single
Base Station Relocation, Maneuverable Relays	[22–27]	Controlled	Multihop	Multiple

Table 1. Comparison of Leveraging Sink Mobility in Wireless Sensor Network

Intuitively, increasing the sink velocity v will improve the system efficiency, since in unit time interval the mobile sink can meet more sensors and gather more information throughout the sensor field. However, we should carefully choose this parameter as explained below. On one hand, the higher the mobile sink velocity, the higher the probability for static sensors is to meet mobile sinks. On the other hand, when mobile sinks are moving too fast across the effective communication region of static sensors, there may not be a sufficient long session interval for the sensor and sink to successfully exchange one potentially long packet. In other words, with the increase of sink velocity, the "outage probability" of packet transmission will rise. Therefore, finding a proper value for sink velocity must be a tradeoff between minimizing the sensor-sink meeting latency and minimizing the outage probability.

3.1. Sensor-sink Meeting Delay

Suppose the network consists of m mobile sinks and n static sensors in a disk of unit size. Both sink and sensor nodes operate with transmission range of r. The mobility pattern of the mobile sinks $M_i(i=1,....m)$ is according to "Random Direction Mobility Model", however, with a constant velocity v. The sink's trajectory is a sequence of epochs and during each epoch the moving speed v of M_i is invariant and the moving direction of M_i over the disk is uniform and independent of its position. Denote Q_i as the epoch duration of M_i, which is measured as the time interval between M_i 's starting and finishing points. Q_i is an exponentially distributed random variable, and the distributions of different Q_i (i=1, ..., m) are independent and identically-distributed (i.i.d) random variables with common average of $\overline{Q}$. Consequently the epoch length of different L_i's are also i.i.d random variables, sharing the same average of $\overline{L} = \overline{Q}v$.

Assume a stationary distribution of mobile sinks, in other words, the probabilities of independent mobile sinks approaching a certain static sensor from different directions are equal. Specifically, the meeting of one static sensor N_j (j=1, ..., n) and one mobile sink M_i is defined as *Mi* covers *Nj* during an epoch. Since M_i will cover an area of size $\pi r^2 + 2rL_{i,k}$ during the k-th epoch, then the number of epochs X_i needed till the first sensor-sink meeting is geometrically distributed with average of (Theorem 3.1 of [30]), with the cumulative density function (cdf) as

$$F_{x_i}(x) = \sum_{x_k \le x} p(1-p)^{k-1}$$

In the case of multiple mobile sinks, the sensor sink meeting delay should be calculated as the delay when the first sensor-sink meeting occurs. Thus the number of epochs *X* needed should be the minimum of all X_i (i=1, ..., m), with the cdf as

$$F_x(x) = 1 - \left[1 - F_{x_i}(x)\right]^m \cong \sum_{x \le x} mp(1-p)^{k-1}$$

Denote $\overline{X}$ as the average of X, the expected sensor sink meeting delay will be

$$\overline{D_1} = \overline{X}.\frac{\overline{L}}{v}$$

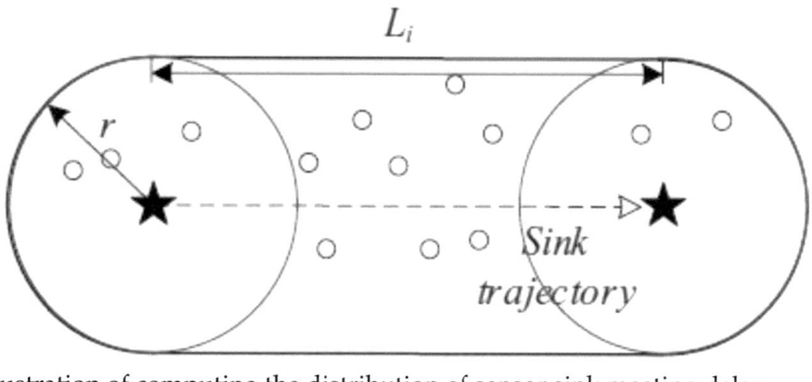

Fig. 11. Illustration of computing the distribution of sensor sink meeting delay.

This result gives us some hints on choosing the parameters to minimize the sensor-sink meeting delay. If we increase the radio transmission range *r*, or increase the number of mobile sinks *m*, or increase the sink velocity *v*, the sensor-sink meeting delay can get reduced. However, the above analysis has implicitly neglected the time consumed by packet transmission during each sensor-sink encounters. If the message length is not negligible, the message has to be split into several segments and deliver to multiple sinks.

3.2. Large Message Delivery Delay

In case of packet segmentations, the split packets are assumed to be sent to different mobile sinks and reassembled. Message delivery delay can be mainly attributed to the packet transmission time, while the packet re-sequencing delay is out of the scope of our study. Assume each sensor will alternate between two states, active and sleep, whose durations will be exponential distributed with a mean of $1/\lambda$. Thus the message arrival is a Poisson process with arrival rate λ. For constant message length of *L*, constant channel bandwidth *w*, the number of time slots required to transmit a message is $T=L/w$. Then with a service probability $p = m\pi r^2$, the service time of the message is a random variable with Pascal distribution (Lemma 1 of [6]). That is, the probability that the message can be transmitted within no more than *x* time slots, is

$$F_x(x) = \sum_{i=0}^{x-T} \binom{T+i-1}{T-1} p^T (1-p)^i$$

Such a Pascal distribution with mean value of $T/p = L/\pi m w r^2$. Under an average Poisson arrival rate λ and a Pascal service time with $\mu = p/T = \pi m w r^2 / L$, data generation and transmission can be modeled as an M/G/1 queue. Then the average message delivery delay can be expressed as follows:

$$\overline{D_2} = \frac{1}{\lambda}\left[\rho + \frac{\rho^2 + \lambda^2 \rho^2}{2(1-\rho)}\right]$$

where $\rho = \lambda/\mu$. For simplicity, we neglect the impact of arrival rate and set λ =1, thus

$$\overline{D_2} = \frac{1}{\mu - 1} = \frac{1}{\frac{\pi m w r^2}{L} - 1}$$

This result shows that, by decreasing message length *L*, or increasing transmission range *r* and number of mobile sinks *m*, the message delivery delay can be reduced. We have designed simulations to verify our analysis. One thousand five hundred sensor nodes have been deployed in a 10,000x10,000-m region. The data generation of each sensor nodes follows a Poisson process with an average arrival interval of 1s. By varying the ratio of sink velocity against transmission radius, and by varying the number of mobile sinks, we can evaluate the performance of average message delivery delay and energy consumption, as illustrated in Figure. 12 and Figure. 13.

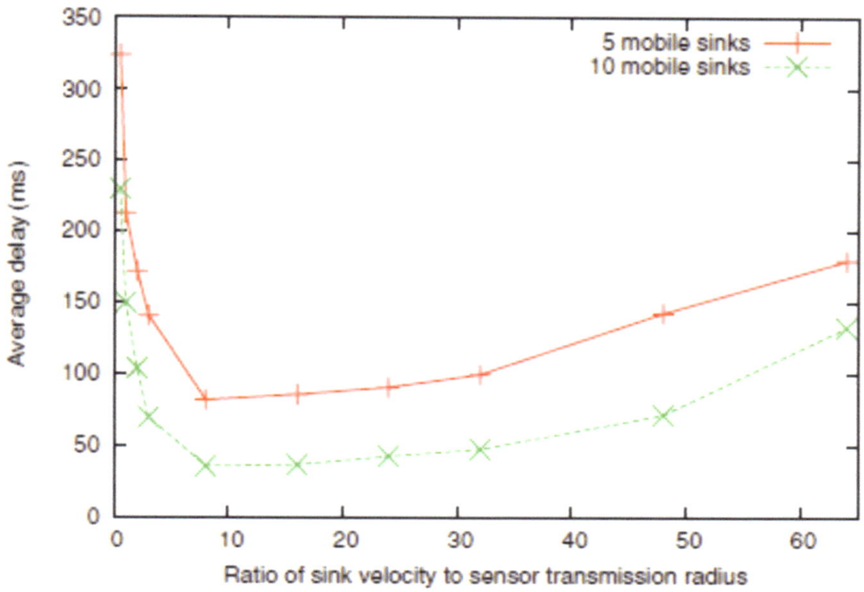

Fig. 12. Average message delivery delay under different scenarios by varying the number and velocity of mobile sinks.

As can be found in Figure. 12, it coincides with our expectation that the more mobile sinks deployed the less delay for message delivery between sensors and sinks. Besides, the simulation results are identical with our analysis on choosing the proper speed for mobile sinks. When the sink mobility is low, the sensors have to wait for a long time before encountering the sink and delivering the message. When the sink moves too fast, however, although the sensors meet the sink more frequently, they have to have the long messages sent successfully in several successive transmissions. In fact, there exists an optimal velocity under which the message delivery delay will be minimized. Average energy consumption is illustrated in Figure. 13. By different cluster size, we mean the maximal hop count between the sensor and mobile sink. It is worthy noting that when the cluster size is small (1 or 2), the average energy consumption will almost remain constant irrespective of the number of mobile sinks.

In other words, more deployed mobile sinks will not lead to further reduced energy consumption. However, when messages can be delivered to a mobile sink multiple hops away then the number of mobile sinks will have influence on the energy consumption: the more mobile sinks, the less energy will be consumed. In fact, the energy consumption in mWSN is more balanced compared with static WSN, which means the remaining energy of each sensor node is almost equal. It is easily understood that more balanced energy consumption will lead to more robust network connectivity and longer network lifetime.

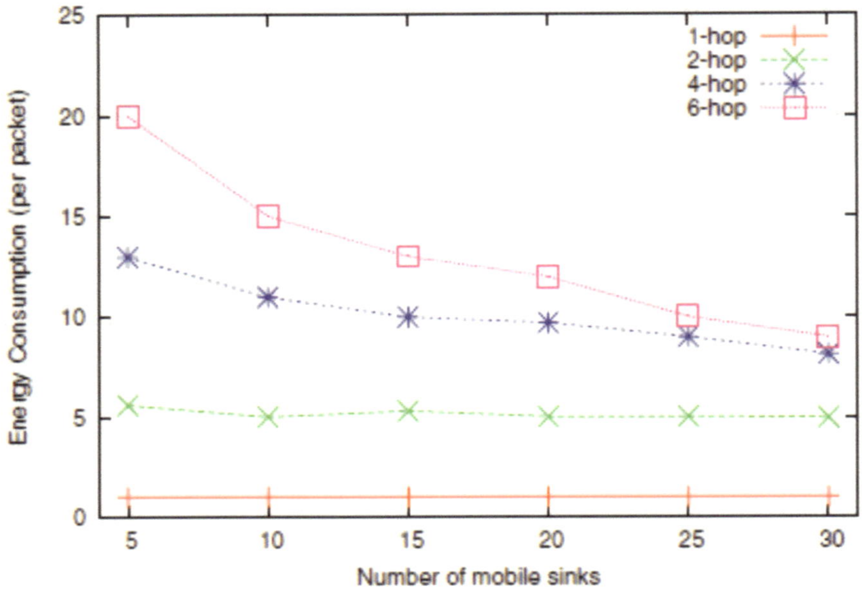

Fig. 13. Average message delivery delay under different scenarios by varying the cluster size and member of mobile sinks.

3.3. Outage Probability

In the above subsection, we have calculated the service time distribution for one sensor node (with multiple mobile sinks). However, while moving along predefined trajectory one mobile sink may potentially communicate with several sensor nodes simultaneously. In order for a successful packet delivery, we are interested in finding the relationship between such parameters as packet length *L* (number of time slot required is T=l/*w*), transmission range *r*, sink velocity *v*, and outage probability p_{outage}. Here we only qualitatively describe the relationship between p_{outage} and *r, v, T*. To guarantee the packet transmission completed in duration *T*, we first defined a zero-outage zone, as illustrated by the shaded region H in Figure 14. Nodes lying in H will be guaranteed with zero outage probability, because the link between sensor & sink remains stable for duration of *T* with probability 1. Intuitively, if H is viewed as a queuing system, then the larger the area of H, the higher the service rate, thus the lower the average outage probability. The border arc of H is the intersected area of two circles with radius *r*, and the width of H is determined by *(2r-vT)*. Therefore, the goal of enlarging the area of H can be achieved via increasing *r*, or decreasing *v* or *T*. With constant packet length (i.e. constant *T*), we can choose to increase *r* or to decrease *v*. However, increased *r* will require for larger transmission power, therefore, it is more energy efficient by decreasing sink velocity *v*. Some preliminary simulation results can verify the expectations on the parameter tuning methods. With 3,000 sensor nodes and one mobile sink in a 10,000x10,000-m region, when the sink velocity is 15 m/s and transmission range is 80 m, the outage percentage statistics have been shown in Figure. 15. One can find that, as analyzed above, the larger the transmission range r is, or the shorter the packet length T, is, the lower the outage percentage will be.

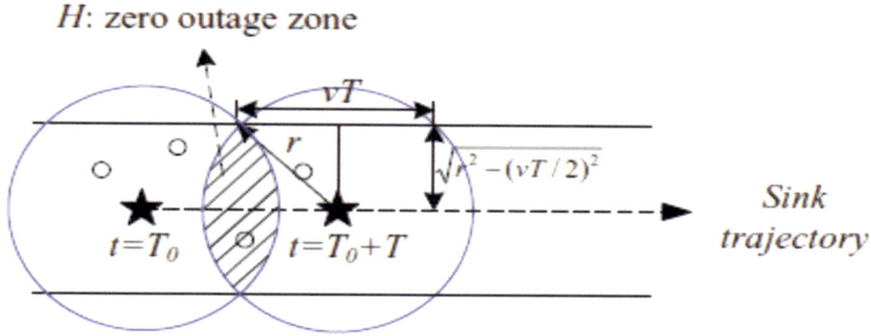

Fig. 14. Illustration for computing the relationship between zero-outage probability and r

It has been shown by Biao et. al. in [29] that with high probability, the average duration d until which a mobile sink first enters the field of sensor node S is given by,

$$d \leq \frac{4\log m}{crv}\sqrt{\frac{1}{m}}$$

where, the constant $c(c<1)$ is a scaling factor defined in [33,34], r is the communication radius of the sensor node, v is the velocity of the mobile sink, m is the number of mobile sinks present in the network Likewise, to calculate the impact of velocity of mobile sink on message delay an equation is

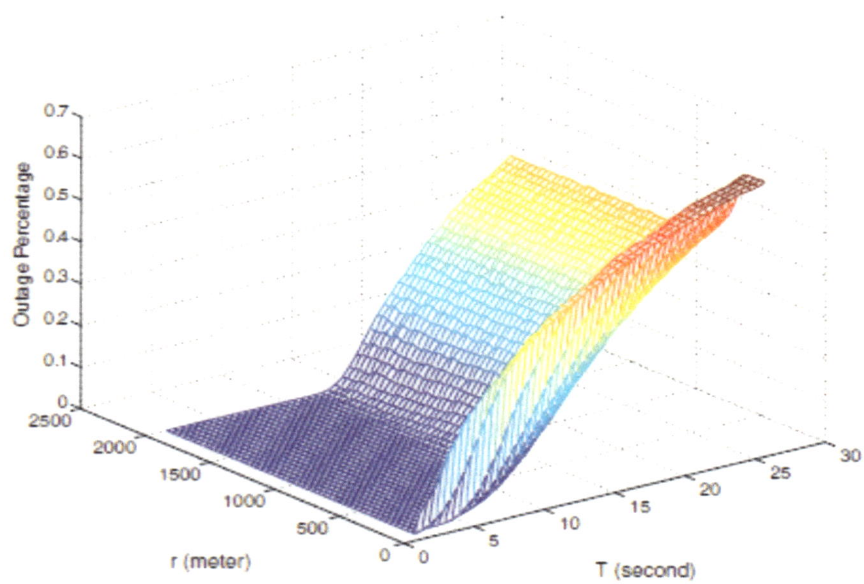

Fig. 15. Outage probability vs. r and T

derived as a Pascal distribution with Poisson arrival rate λ, and a Pascal service time $\mu = p/s$, where s is the number of time slots required to transmit a message of length L within a channel bandwidth of w. Another term p, is the service probability of a sensor node within the coverage of at least one mobile sink, and is given by,

$$p = \frac{crv\sqrt{m}}{4\log m}$$

we define the ratio of the packet arrival rate to the service time as $\rho = \lambda/\mu$, and similarly replace the value of pascal service time to study the impact of sink mobility on delay; the equation is given by,

$$\mu = \frac{wr\sqrt{\pi}\,p}{Lv}$$

The average message delivery delay can then be expressed as,

$$D = \frac{1}{\lambda}\left(\rho + \frac{\rho^2 + \lambda^2\rho^2}{2(1-\rho)}\right)$$

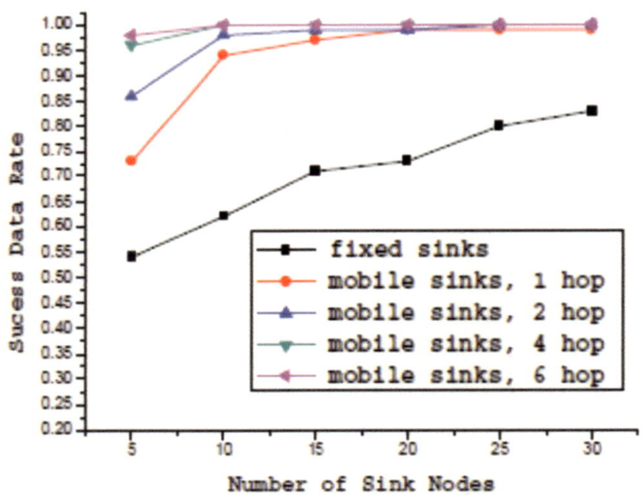

Fig. 16. Data success rates in loose-connectivity network

For simplicity, we neglect the impact of arrival rate and set $\lambda = 1$, thus

$$D = \frac{1}{\mu - 1}$$

The above equation therefore implies that on one hand, large v can improve the service probability p, on the other hand it increases the required times of mobiles sinks reaching it in order to finish a message transmission. Both sides of the impacts should be considered when choosing the appropriate velocity value of mobile sinks. The impact of mobility of the sink on the performance metrics of network connectivity is further highlighted in Figure 16. A comparison of data success rates between fixed sinks and mobile sinks in spare network is also presented herewith. In this case, the data success rate produced by mobile sinks is much better than that by fixed sinks. One of the advantages of mobile sinks is that they can move to such sensor nodes that are disconnected from others.

4. Future Application Scenarios

The possible application scenarios for traditional wireless sensor networks, which are envisaged at the moment, include environmental monitoring, military surveillance digitally equipped homes, health monitoring, manufacturing monitoring, conference, vehicle tracking and detection (telematics) and monitoring inventory control. Since, mobile wireless sensor networks are a relatively new concept; its specific, unique application areas are yet to be clearly defined. Most of its application scenarios are the same as that of traditional wireless sensor networks, with the only difference of mobility of mobile sink, preferably in the form of mobile phones. We, however, envisage a space where sensors will be placed everywhere around us, a concept of ubiquitous network, where different promising technologies will work together to help realize the dream of late Marc Weiser. We propose that with these sensors placed everywhere, a single individual mobile phone can enter into a "session" with the "current sensor network" in which he or she is present. A mobile phone will have the necessary interfaces available to allow it to communicate with the heterogeneous world. In most of the cases, this mobile phone will "enter" into the network as one of the mobile sinks. This way, a mobile phone can enter into the session anywhere at any time; at airport, railway station, commercial buildings, library, parks, buses, home etc. We will now discuss some of the possible application scenarios in ubiquitous computing age as a motivation for future work. This follows that we need to develop smart sensors and mobile phones to be able to take part in these applications. Mobile phones will be expected to have multiple radios to support multiple, heterogeneous technologies existing today. We believe that mobile WSN will be able to address multitude of applications, once the "world" gets smart.

Smart Transport System: One way in which mobile wireless sensor networks can help is through implementing an intelligent traffic system. With the sensors placed frequently around the city, these sensors can monitor and analyze the current traffic system at these areas at a given time. This information is delivered back to a central gateway or sink, having a link to different mobile phone operators, which in turn can provide this "traffic help" service to its customers, on demand.

Security: Similarly, with these sensors placed everywhere in and around the city, these very sensors can be used to implement security system in daily life. On an individual basis, a mobile phone of a person can enter into a "session" with the already present sensors in the area. In this way, it can keep a track of his belongings, car and even kids. Mobile enabled wireless sensor networks can help to monitor the environment, both external and internal. For internal environment monitoring, buildings can be made "smart building" to constantly monitor and analyze the environmental situation.

Social Interaction: One other possible scenario in ubiquitous computing is that of social interaction. There is a rapid increase in number of mobile subscribers in the world. We believe that with the possible integration of RFID tags and WSN, mobile phones can act as sinks to have a "social interaction" among peers who share the common interest. People can place their digital tags at their places of choice, or among their friends. Similarly, this combination of RFID tags and WSN can help mobile phones users in using their mobile phones as "single" tool to carry out all their tasks, be it shopping, billing, information gathering, guidance, social interaction, etc. By entering into a "session" with existing sensors or WSN in a particular area, the mobile phone user can get the necessary information on his mobile phone, like the location of his friends/relatives, the time table/schedule of the events taking place, environmental conditions etc. With the help of little initial information about the user, it is also possible to enter into any area, shop around, buy digital tickets and simply walk off, all with electronic billing. The same idea can be implemented in the form of e-voting in elections ranging from company elections to elections on much larger scale. "Context Aware" computation will be a significant key player in helping mobile WSN in social areas. Coupled with the superior image recognition techniques built in, people can interact with each other and with the environment. This single advancement in technology can have an enormous application potential, more than what we can imagine at the moment.

Health: One area which is already showing such signs of applications of ubiquitous computing is health monitoring. Emerging developments in this area are providing the means for people to increase their level of care and independence with specific applications in heart monitoring and retirement care. In recent years, one area of increasing interest is the adaptation of "micro grid" technology to operate in and around the human body, connected via a wireless body area network (WBAN). There are many potential applications that will be based on WBAN technology, including medical sensing and control, wearable computing, location awareness and identification. However, we consider only a WBAN formed from implanted medical sensors. Such devices are being and will be used to monitor and control medical conditions such as coronary care, diabetes, optical aids, bladder control, muscle stimulants etc. The advantages of networking medical sensors will be to spread the memory load, processing load, and improving the access to data. One of the crucial areas in implanting sensors is the battery lifetime. Batteries cannot be replaced or recharged without employing a serious medical procedure so it is expected that battery powered medical devices placed inside the body should last for ten to fifteen years. Networking places an extra demand on the transceiver and processing operations of the sensor resulting in increased power consumption. A network placed under a hard energy constraint must therefore ensure that all sensors are powered down or in sleep mode when not in active use, yet still provide communications without significant latency when required.

Miscellaneous Scenarios: We focus to concentrate on creating a smart world where a single user mobile phone can perform a multitude of applications. We envisage a scenario, where wireless sensor networks will be placed every where around the "smart" city and a person's mobile phone can just enter and leave the network as humans. Suppose a person goes into the shopping mall. With the already installed sensors and RFID tags installed over there, his mobile phone can interact with the environment. A user looks for his product of choice and is concerned about the price; he can just inquire through his mobile phone the price of the same item in other stores, at internet or even from the manufacturer. This will be made possible by having subscribed service from other retailers, distributors, internet sites or manufacturers. With the enormous growth of RFID, it is very much expected that every single item will have its own unique RFID tag, and with the help of grid computing and advanced database systems, it is not unreasonable to think of a data repository of this magnitude. For the huge number of sensor data collection, XML, which is good for firewalls and human readable, will help make sense of complex, huge senor data. We believe that sensor networks will populate the world as the present Internet does. For example, think of buildings covered with small, near invisible networked computers, which continually monitor the temperature of the building and modify it in relation to the amount of people in the building, thus saving energy. Or sensors buried in the ground, monitoring areas prone to earthquakes and landslides and providing vital feedback, which could prevent human loss and mass destruction.

5. Related Technologies for Ubiquitous Computing

In this section, we will highlight some of the existing enabling technologies which are believed to function along with WSN for the ubiquitous computing paradigm. Some of the exciting combinations are Mobile IPv6, RFID, P2P and grid technology. P2P and Grid technology are already believed to play a significant part in realizing the ubiquitous network dream. Grid and P2P systems share a number of common characteristics and it is now considered that they are both converging towards creating overlay infrastructures that support sharing resources among virtual communities that will also reduce the maintenance cost. We believe that the grid technology will be especially helpful in handling and managing the huge amount of sensor data that these future ubiquitous heterogeneous sensor networks will produce. However, a lot of issues remain to be solved to truly integrate these technologies, the biggest of which is mobility. On the other hand, a number of network owners will be ready to share information gathered by their networks (for example traffic status at their current location) for mutual benefit of all involved parties or will deploy networks with the sole intention of providing services to interested users and charging for them. In such environment where sensor networks come and go in an ad-hoc manner, deployed by numerous unrelated service operators, it will be impossible to establish a long lasting subscriber operator relationship between sensor networks and their users. Users will not know about the existence of sensor networks in a certain area in advance nor will know what type of services discovered networks provide. Instead, depending on their current requirements and needs, users will have to use ad hoc mechanisms to search for required services and available networks. Obviously, as variety of sensors and network types is enormous, both service discovery and communication protocols have to be very flexible and capable of supporting different types and formats of sensor data and services. A description of different related enabling technologies is now presented.

Mobile IPv6: There exist some characteristics of IPV6 which are attractive to WSN in its possible integration. We believe that the advantages that we will accrue from IPv6 are enormous and include some of the followings:

Enlarge address space: This means IP can increasingly mount up without considering short of addressing resource. With the possible integration of different technologies, Mobile IPv6 will help solve the addressing problem.
Identification and security: This improvement makes IPV6 more fit to those commercial applications that need sensitive information and resources.
Access Control: We can make identification and add some access control according to different username. IPV6 also proposes force management about consistency that can prevent the data from modifying during the transmission and resist the rebroadcast aggression. IPV6 also protect the aggression by other services like encryption, ideograph etc.
Auto-configuration: IPV6 supports plug and play network connection. Although we have seen the common issues about IPV6 and WSN, there still exist some questions to be solved. Embedded applications are not considered in IPV6 initially, so if we want to realize IPV6 in WSN we must do effort to the size of the protocol stack. We do not need to realize high layer stack in each wireless sensor node from the aspect of OSI. Power consumption is another issue. But if we want to apply IPV6 in WSN, we must reduce its power consumption. This can be realized through using duty-cycle model.
RFID Technology: RFID tag is the key device for the actualization of "context awareness", which is essence of ubiquitous computing and can recognize "data carriers" by electronic wave without physical contact. Auto-ID lab's EPC (Electronic Product Code) numbering code is based on 96-bit system, which is believed to be large enough to put electronic tag for every grain of rice on this planet earth. Contact-less chips in RFID do not have batteries; they operate using the energy they receive from signals sent by a reader. In context of integration of RFID technology into wireless sensor networks, probably, the most prominent integration application will be in the field of retail business. RFID, already, has been making a major breakthrough in the retail business, with giants like Wal-Mart beginning to embrace it. Although, RFID can be incorporated on its own in different application areas, it has some disadvantages, which are the main reasons for research community to pursue research in integration of RFID with WSN. Some of the disadvantages which make room for integration of RFID with WSN are

- Inability of RFID to successfully track the target object (customer) within a specified working space (department floor, exhibition etc.).
- Deployment of RFID systems on already existed working spaces. For example, if we have to deploy RFID on a department floor, it will be prohibitivel y expensive to do so.

In this regard one scheme is to implement the combined RFID and WSN technologies in enhancing the customer relationship management for a retail business. Mobile RFID has already started getting attention with Nokia incorporating it into its mobile phone, thus creating the first GSM phone with RFID capabilities. The kit uses the 13.56MHz radio frequency range at the very short range of typically 2-3cm using the ISO-14443A standard, and has 2 Xpresson RFID reader shells, 20 RFID tags, and the software for the phone (Nokia 5140) tag reading. The kit is best suited for applications with 1-20 users.
GRID Technology: Grid Computing delivers on the potential in the growth and abundance of network connected systems and bandwidth: computation, collaboration and communication over the advanced web. At the heart of Grid Computing is a computing infrastructure that provides dependable, consistent, pervasive and inexpensive access to computational capabilities. The main driving force behind grid computing is the desire to take advantage of idle resources in a network and use these in intensive computations. With

a grid, networked resources -- desktops, servers, storage, databases, even scientific instruments - can be combined to deploy massive computing power wherever and whenever it is needed most. We believe that with the huge amount of sensor data that future heterogeneous wireless sensor networks will produce, grid technology can be efficiently used to manage and store this magnitude of data. Technicalities at software and hardware level remain to be solved. Grid computing, at the moment, is not thought to be directly integrated into the WSN, but works as a third party in touch with the network base station or gateway. Playing a direct role can be wireless grid; technology to support less data intensive applications. Wireless grid technology has already got boost by some good progress in availability of compatible hardware. Wi-Fi technology and WLAN are supposed to play a key role in making wireless grid a reality. The wireless grid architecture represents a combination of high-performance WLAN switches with structured WLAN distribution systems and is believed to be a key development for the industry. One of the possible architecture is to employ densely deployed Wi-Fi radios with powerful centralized control to deliver predictable wired-LAN-like performance with the flexibility of WLANs. As the current wireless grid, with the help of WLAN standards already can support high data rate of 54 Mbps, it is therefore well set to integrate into the future densely deployed wireless sensor networks.

Mobile P2P: Mobile P2P can be simply defined as transferring data from one mobile phone to another. Some of the limitations that become challenges for mobile P2P to be implemented are low efficiency (in terms of CPU and Memory), low power, low bandwidth and billing issues. This concept basically presents the peer-to-peer networking concept that is widely in use today in fixed communication networks, but mapped to mobile environment. Each sensor network presents a peer node capable of working and providing information independently of other peers, but also of communicating with other nodes and sharing available information with them. Collaboration of completely uncoordinated and nomadic networks on execution of a common task in a mobile environment is obviously not easy to implement. Different types of information and services, various data formats and application requirements, connectivity of and ability to discover sensor networks connected to different mobile networks are some of the most interesting issues. An idea can be to expose the WSN to a P2P network and enable the UPnP (Universal Plug n Play) Gateway to discover remote sensor nodes through the P2P substrate and to instantiate UPnP proxies for them to ensure client connectivity.

6. Conclusion

Mobile enabled Wireless Sensor Network (mWSN) has been proposed to realize large-scale information gathering via wireless networking and mobile sinks. Through theoretical analysis it is established that by learning the mobility pattern of mobile sinks, d_{char} based multi hop clustering scheme can forward the packets to the estimated sink positions in a timely and most energy-efficient way. Besides, the less strict the packet deadline is, the more energy saving is achieved. In addition, the mobility's influence on the performance of single-hop clustering has been investigated. It is found that sink mobility can reduce the energy consumption level, and further lengthen the network lifetime. However, its side effects are the increased message delivery delay and outage probability. The same problems

will remain by tuning the sink density or coverage (i.e. sink amount and transmission range), so the conclusion is that sink mobility and sink density are permutable, since sink mobility increase its spatial redundancy similar with deploying multiple sinks.
In this chapter, we have further presented multi-tier architecture for the mobile wireless sensor network as a key element of future ubiquitous computing paradigm. The multi-tier architecture has been discussed in previous research for traditional wireless sensor network; however we consider the multi-tier architecture in mobile WSN, with a special emphasis on integration into a pervasive network. The detailed architectural implementation is presented in this chapter, followed by an analysis of the impact of mobility on performance related issues in WSN. The hierarchical multi tiered architecture is believed to perform efficiently and is also scalable to large network size. We have further discussed some of the future application scenarios for this ubiquitous computing age along with a description of some of the related existing technologies which play a significant role in the proposed architecture.

7. References

[1] I. F. Akyildiz, Weilian Su, Yogesh Sankarasubramaniam, and Erdal Cayirci, "A Survey on Sensor Networks", *IEEE Communications Magazine*, 2002 (August).
[2] F. Ye, H. Luo, J. Cheng, and S.L.L. Zhang, "A Two Tier Data Dissemination Model for Large scale Wireless Sensor Networks" Proc. of MOBICOM_02, Atlanta, Georgia, USA, 2002, pp.148-159 (September 23-26).
[3] R.C. Shah et al., "Data MULEs: Modeling a Three-tier Architecture for Sparse Sensor Networks" Proc. of IEEE SPNA workshop, 2003, pp. 30-41 (May 11).
[4] S. Jain et al., "Exploiting Mobility for Energy Efficient Data Collection in Wireless Sensor Networks" Mobile Networks and Applications, vol. 11, no. 3, 2006, pp. 327- 339 (June).
[5] L. Tong, Q. Zhao, and S. Adireddy, "Sensor networks with mobile agents" Proc. of IEEE MILCOM 2003, Boston, MA, USA, 2003, pp.688-693 (October 13-16).
[6] Y. Wang and H. Wu, "DFT-MSN: The Delay Fault Tolerant Mobile Sensor Network for Pervasive Information Gathering" Proc. of IEEE INFOCOM_06, Barcelona, Spain, 2006 (April 23-29).
[7] B. Hull et al., "CarTel: A Distributed Mobile Sensor Computing System" Proc. of ACM SenSys_06 (October 31-November 3).
[8] W. Zhao, et al., "A Message Ferrying Approach for Data Delivery in Sparse Mobile Ad Hoc Networks" Proc. of MobiHoc 2004, Roppongi, Japan, 2004, pp. 187-198 (May 24-26).
[9] W. Zhao et al., "Controlling the Mobility of Multiple Data Transport Ferries in a Delay-Tolerant Network" Proc. Of INFOCOM 2005, Miami, Florida, USA, 2005, pp. 1407-1418 (March 13-17).
[10] M.M.B. Tariq, et al., "Message Ferry Route Design for Sparse Ad hoc Networks with Mobile Nodes" Proc. of MobiHoc 2006, Florence, Italy, 2006, pp. 37-48 (May 22-25).
[11] A.A. Somasundara, A. Ramamoorthy, and M. B. Srivastava, "Mobile Element Scheduling for Efficient Data Collection in Wireless Sensor Networks with Dynamic Deadlines" Proc. Of IEEE int_l Real-Time Sys. Symp., Lisbon, Portugal, 2004, pp. 296-305 (December 5-8).

[12] E. Ekici et al., "Mobility-Based Communication in Wireless Sensor Networks" IEEE Communications Magazine, vol. 44, no. 7, 2006, pp. 56-62 (July).
[13] A. Kansal et al., "Intelligent Fluid Infrastructure for Embedded Networks,^ Proc. of MobiSys 2004" Boston, Massachusetts, USA, 2004 pp. 111-124 (June 6-9).
[14] D. Jea et al., "Multiple Controlled Mobile Elements (DataMules) for Data Collection in Sensor Networks" Proc. of DCOSS 2005, Marina del Rey, CA, USA, 2005 (June 30-July 1).
[15] A. Somasundara et al., "Controllably Mobile Infrastructure for Low Energy Embedded Networks" IEEE Transactions on Mobile Computing, vol. 5, no. 8, 2006, pp. 958-973 (August).
[16] A. Chakrabarti et al., "Using Predictable Observer Mobility for Power Efficient Design of Sensor Networks" Proc. Of IPSN 2003, Palo Alto, California, USA, 2003 (April 22-23).
[17] H.S. Kim et al., "Minimum-Energy Asynchronous Dissemination to Mobile Sinks in Wireless Sensor Networks" Proc. Of SenSys 2003, Los Angeles, California, USA, 2003, pp. 193-204 (November 5-7).
[18] K. Akkaya and M. Younis, "Energy-aware Routing to a Mobile Gateway in Wireless Sensor Networks" Proc. of Globecom 2004, Dallas, Texas, USA, 2004, pp.16-21 (November 29- December 3).
[19] P. Baruah & R. Urgaonkar, "Learning-Enforced Time Domain Routing to Mobile Sinks in Wireless Sensor Fields" Proc. of LCN 2004, Tampa, Florida, 2004 (November 16-18).
[20] J. Luo and J.-P. Hubaux, "Joint Mobility and Routing for Lifetime Elongation in Wireless Sensor Networks" Proc. Of INFOCOM 2005, Miami, USA, 2005, pp. 1735-1746 (March 13-17).
[21] J. Luo et al., "MobiRoute: Routing towards a Mobile Sink for Improving Lifetime in Sensor Networks" Proc. of DCOSS 2006, San Francisco, CA, USA, 2006 (June 18-20).
[22] S.R.Grandham et al., "Energy Efficient Schemes for Wireless Sensor Networks with Multiple Mobile Base Stations" Proc. of IEEE Globecom 2003, San Francisco, CA, USA, 2003, pp. 377-381 (December 1-5).
[23] Z.M. Wang et al., "Exploiting Sink Mobility for Maximizing Sensor Networks Lifetime" Proc. of HICSS 2005, Hawaii, 2005 (January 3-6).
[24] W.Wang et al., "UsingMobile Relays to Prolong the Lifetime of Wireless Sensor Networks" Proc. of MobiCom 2005, Cologne, Germany, 2005, pp. 270-283 (August 28 - September 2).
[25] Y.T. Hou et al., "Prolonging Sensor Network Lifetime with Energy Provisioning and Relay Node Placement" Proc. Of SECON 2005, Santa Clara, California, USA, 2005 (September 26-29).
[26] S. Eidenbenz et al., "Maneuverable Relays to Improve Energy Efficiency in Sensor Networks" Proc. of PerCom 2005 Workshops, Hawaii, 2005, pp. 411-417, (March 8-12).
[27] Z. Vincze et al., "Adaptive Sink Mobility in Event-driven Multihop Wireless Sensor Networks" Proc. of InterSense 2006, Nice, France, 2006 (May 30-31).

[28] L. Sun, Y. Bi, and J. Ma, "A Moving Strategy for Mobile Sinks in Wireless Sensor Networks" Proc. of IEEE SECON_06 (poster), Reston, VA, USA, 2006 (September 25–28).

[29] B. Ren, J.Ma, and C. Chen, "The Hybrid Mobile Wireless Sensor Networks for Data Gathering" Proc. of International Wireless Communications and Mobile Computing Conference (IWCMC_06), Vancouver, Canada, 2006 pp. 1085–1090 (July 3–6).

[30] Liu, B., Brass, P., Dousse, O., Nain, P., Towsley, D. "Mobility Improve Coverage of Sensor Networks," *Proceedings of ACM MobiHoc 2005*

[31] Mark Weiser, "The Computer for the Twenty-First Century," *Scientific American, September 1991*

[32] Canfeng Chen, Jian Ma, "MEMOSEN: Multi-radio enabled Mobile Wireless Sensor Network" *In Proc. of AINA 2006*

[33] Y. Wang and H. Wu, "DFT-MSN: The Delay Fault Tolerant Mobile Sensor Network for Pervasive Information Gathering", *In Proc. Of IEEE InfoCOM'06, Barcelona, Spain, April 23-29, 2006*

[34] N. Bansal and Z. Liu, "Capacity, delay, and mobility in wireless ad hoc networks", *In Proc. of IEEE InfoCOM, Vol. 3, 2003*

Enabling Compression in Tiny Wireless Sensor Nodes

Francesco Marcelloni[1] and Massimo Vecchio[2]
[1]*Dipartimento di Ingegneria dell'Informazione, University of Pisa, Via Diotisalvi 2, 56122 Pisa, Italy, e-mail: f.marcelloni@ing.unipi.it*
[2]*INRIA Saclay, Ile de France sud, France, e-mail: massimo.vecchio@inria.fr*

1. Introduction

A Wireless Sensor Network (WSN) is a network composed of sensor nodes communicating among themselves and deployed in large scale (from tens to thousands) for applications such as environmental, habitat and structural monitoring, disaster management, equipment diagnostic, alarm detection, and target classification. In WSNs, typically, sensor nodes are randomly distributed over the area under observation with very high density. Each node is a small device able to collect information from the surrounding environment through one or more sensors, to elaborate this information locally and to communicate it to a data collection centre called *sink* or *base station*. WSNs are currently an active research area mainly due to the potential of their applications. However, the deployment of a large scale WSN still requires solutions to a number of technical challenges that stem primarily from the features of the sensor nodes such as limited computational power, reduced communication bandwidth and small storage capacity. Further, since sensor nodes are typically powered by batteries with a limited capacity, energy is a primary constraint in the design and deployment of WSNs.

Datasheets of commercial sensor nodes show that data communication is very expensive in terms of energy consumption, whereas data processing consumes significantly less: the energy cost of receiving or transmitting a single bit of information is approximately the same as that required by the processing unit for executing a thousand operations. On the other hand, the energy consumption of the sensing unit depends on the specific sensor type. In several cases, however, it is negligible with respect to the energy consumed by the communication unit and sometimes also by the processing unit. Thus, to extend the lifetime of a WSN, most of the energy conservation schemes proposed in the literature aim to minimize the energy consumption of the communication unit (Croce et al., 2008). To achieve this objective, two main approaches have been followed: power saving through duty cycling and in-network processing. Duty cycling schemes define coordinated sleep/wakeup schedules among nodes in the network. A detailed description of these techniques applied to WSNs can be found in (Anastasi et al., 2009). On the other hand, in-network processing consists in reducing the amount of information to be transmitted by means of aggregation (Boulis et al., 2003) (Croce et al., 2008) (Di Bacco et al., 2004) (Fan et al., 2007)

(Intanagonwiwat et al., 2003) (Lindsey et al., 2002) (Madden et al., 2002) and/or compression techniques. In this chapter, we do not consider aggregation: the interested reader can refer to (Fasolo et al., 2007) for a brief discussion and classification of aggregation approaches.

Data compression algorithms fall into two broad classes: lossless and lossy algorithms. Lossless algorithms guarantee the integrity of data during the compression/decompression process. On the contrary, lossy algorithms generate a loss of information, but generally ensure a higher compression ratio.

Due to the limited resources available in sensor nodes, to apply data compression in WSNs requires specifically designed algorithms. Two approaches have been followed:

1. to distribute the computational cost on the overall network (Chen et al., 2004) (Ciancio & Ortega, 2005) (Ciancio et al., 2006) (Deligiannakis et al., 2004) (Ganesan et al., 2003) (Gastpar et al., 2006) (Girod et al., 2005) (Guestrin et al., 2004) (Lin et al., 2006) (Pradhan et al., 2002) (Rebollo-Monedero, 2007) (Tang & Raghavendra, 2004), (Wagner et al., 2007) (Zixiang et al., 2004);
2. to exploit the statistical features of the data under monitoring so as to adapt some existing algorithms to the constraints imposed by the limited resources available on the sensor nodes (Ganesan et al., 2003) (Lynch et al., 2004) (Sadler & Martonosi, 2006).

The first approach is natural in cooperative and dense WSNs, where data measured by neighbouring nodes are correlated both in space and in time. Thus, we can apply distributed transforms or estimate distributed models which allow decorrelating the data measured by sensors, and, therefore, representing these data by using fewer bits. Obviously, the models are generally only approximations of the data. Thus, distributed compression algorithms are intrinsically lossy.

To the best of our knowledge, only a few papers have discussed the second approach. Examples of compression techniques applied to the single node adapt some existing dictionary-based compression algorithms to the constraints imposed by the limited resources available on the sensor nodes. For instance, in (Sadler & Martonosi, 2006) and (LZO, 2008), the authors have introduced two lossless compression algorithms, namely S-LZW and miniLZO, which are purposely adapted versions of LZW (Welch, 1984) and LZ77 (Ziv & Lempel, 1977), respectively. Since S-LZW outperforms miniLZO, as shown in (Sadler & Martonosi, 2006), we will consider only S-LZW as comparison in this chapter. The Lightweight Temporal Compression (LTC) algorithm proposed in (Schoellhammer et al., 2004) is an efficient and simple lossy compression technique for the context of habitat monitoring. LTC introduces a small amount of error into each reading bounded by a control knob: the larger the bound on this error, the greater the saving by compression.

The choice of the algorithm type (lossless or lossy) depends on the specific application domain. Typically, applications, which are not particularly critical, tolerate the use of sensors that, though very cheap, collect data affected by a non-negligible noise. In this context, lossy compression algorithms can provide a double advantage: to reduce noise and to compress data (Ganesan et al., 2003). On the other hand, the criticality of some application domains demands sensors with high accuracy and cannot tolerate that measures, are corrupted by the compression process. In Body Area Networks, for instance, sensor nodes permanently monitor and log vital signs: each small variation of these signs have to be captured because it might provide crucial information to make a diagnosis. Thus, we believe that both lossless and lossy compression algorithms suitable to WSNs have to be deeply investigated. Since sensor nodes are typically equipped with a few kilobytes of memory and a 4-8MHz microprocessor, embedding classical data compression schemes in these tiny nodes is practically infeasible (Barr & Asanović, 2006) (Kimura & Latifi, 2005).

To overcome these problems, in a previous paper (Marcelloni & Vecchio, 2009), we have proposed a Lossless Entropy Compression algorithm (LEC), which exploits the natural correlation that exists in data typically collected by WSNs and the principles of entropy compression. We have shown how its low complexity and the small amount of memory required for its execution make the algorithm particularly suited to be used on available commercial sensor nodes. Other important features of LEC are i) its ability to compute a compressed version of each value on the fly and ii) to exploit a very short fixed dictionary, whose size depends on the precision of the analog-to-digital converter (ADC).

The LEC algorithm follows a scheme similar to the one used in the baseline JPEG algorithm for compressing the so-called DC coefficients of a digital image: the basic idea is to divide the alphabet of values into groups whose sizes increase exponentially and consequently to implement the codewords as a hybrid of entropy and binary codes (Pennebaker & Mitchell, 1992). In particular, the entropy code (a variable-length code) specifies the group, while the binary code (a fixed-length code) represents the index within the group. In (Marcelloni & Vecchio, 2009), we have adopted the Huffman table proposed in JPEG to entropy encoding the groups.

In this chapter, first we briefly introduce the LEC algorithm and, by using two real datasets, we discuss how the LEC algorithm outperforms S-LZW and three well-known compression algorithms, namely gzip, bzip2 and rar. We used these three algorithms as benchmarks, but actually these algorithms are not embeddable in tiny sensor nodes. Second, we analyze how the correlation between consecutive samples affects the performance of LEC by downsampling the datasets with different downsampling factors and by evaluating how much the compression ratio decreases. Third, we investigate the use of semi-adaptive and adaptive Huffman coding to increase the performance in the case of reduced correlation between consecutive samples. Fourth, we discuss how LEC can be transformed into a lossy compression algorithm and show how the lossy version considerably outperforms the lossless version in terms of compression ratios without introducing a significant error. Finally, we compare the lossy version of LEC with LTC.

The Chapter is organized as follows. Section 2 introduces the LEC algorithm. In Section 3, we assess the performance of LEC in terms of compression ratios and complexity. Section 4 introduces the lossy version of LEC and shows some preliminary experimental results. Finally, Section 5 gives some conclusions.

2. The LEC Algorithm

Figure 1 shows the block scheme of the LEC algorithm. In the sensing unit of a sensor node, each measure m_i acquired by a sensor is converted by an ADC to a binary representation r_i on R bits, where R is the resolution of the ADC, that is, the number (2^R) of discrete values the ADC can produce over the range of analog values.

For each new acquisition m_i, LEC computes the difference $d_i = r_i - r_{i-1}$, which is input to an entropy encoder (in order to compute d_0 we assume that r_{-1} is equal to the central value among the 2^R possible discrete values). The entropy encoder performs compression losslessly by encoding differences d_i more compactly based on their statistical characteristics. LEC exploits a modified version of the Exponential-Golomb code (Exp-Golomb) of order 0 (Teuhola, 1978), which is a type of universal code. The basic idea is to divide the alphabet of numbers into groups whose sizes increase exponentially. Like in Golomb coding (Golomb, 1966) and Elias coding (Elias, 1975), a codeword is a hybrid of unary and binary codes: in particular, the unary code (a variable-length code) specifies the group, while the binary code (a fixed-length code) represents the index within the group. Indeed, each nonzero d_i value is represented as a bit sequence bs_i composed of two parts $s_i \mid a_i$, where s_i codifies the number n_i of bits needed to represent d_i (that is, the group to which d_i belongs) and a_i is the representation of d_i (that is, the index position in the group). When d_i is equal to 0, the corresponding group has size equal to 1 and therefore there is no need to codify the index position in the group: it follows that a_i is not represented.

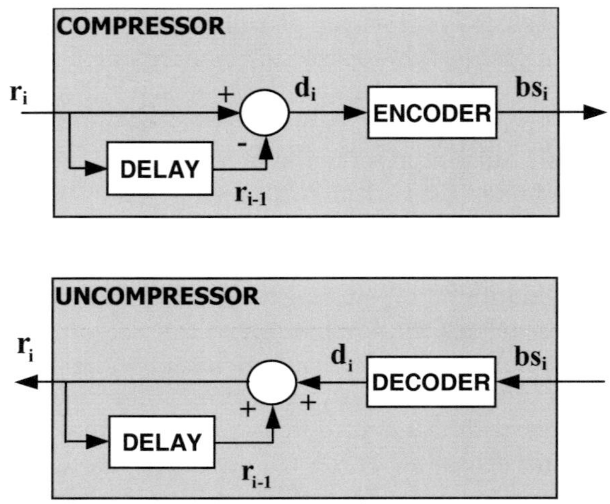

Fig. 1. Block diagram of the encoding/decoding schemes.

For any nonzero d_i, n_i is trivially computed as $\lceil \log_2(|d_i|) \rceil$: at most n_i is equal to R. Thus, in order to encode n_i a prefix-free table of R + 1 entries has to be specified. This table depends on the distribution of the differences d_i: more frequent differences have to be associated with shorter codes. From the observation that, in typical data collected by WSNs, the most frequent differences are those close to 0, in (Marcelloni & Vecchio, 2009) we adopted Table 1, where the first 11 lines coincide with the table used in the baseline JPEG algorithm for compressing the DC coefficients (Pennebaker & Mitchell, 1992). On the other hand, these coefficients have statistical characteristics similar to the measures acquired by the sensing unit. Of course, whether the resolution of the ADC is larger than 14 bits, the table has to be appropriately extended.

In order to manage negative d_i, LEC maps the input differences onto nonnegative indexes, using the following bijection:

$$index = \begin{cases} d_i, & d_i \geq 0 \\ 2^{n_i} - 1 - |d_i|, & d_i < 0 \end{cases} \tag{1}$$

Finally, s_i is equal to the value at entry n_i in the prefix-free table and a_i is the binary representation of *index* over n_i bits. Since d_i is typically represented in two's complement notation, when $d_i < 0$, a_i is equal to the n_i low-order bits of $d_i - 1$.

The procedure used to generate a_i guarantees that all possible values have different codes. Using Table 1, we have, for instance, that d_i = 0, d_i = +1, d_i = -1, d_i = +255 and d_i = -255 are encoded as 00, 010|1, 010|0, 111110|11111111 and 111110|00000000, respectively. Once bs_i is generated, it is appended to the bitstream which forms the compressed version of the sequence of measures m_i.
In the uncompressor, the bit sequence bs_i is analyzed by the decoder block which outputs difference d_i. Difference d_i is added to r_{i-1} to produce r_i.

n_i	s_i	d_i
0	00	0
1	010	-1,+1
2	011	-3,-2,+2,+3
3	100	-7,…,-4,+4,…,+7
4	101	-15,…,-8,+8,…,+15
5	110	-31,…,-16,+16,…,+31
6	1110	-63,…,-32,+32,…,+63
7	11110	-127,…,-64,+64,…,+127
8	111110	-255,…,-128,+128,…,+255
9	1111110	-511,…-256,+256,…,+511
10	11111110	-1023,…,-512,+512, …,+1023
11	111111110	-2047, …,-1024,+1024, …,+2047
12	1111111110	-4095, …,-2048,+2048, …,+4095
13	11111111110	-8191, …,-4096,+4096, …,+8191
14	111111111110	-16383, …,-8192,+8192, …,+16383

Table 1. The default dictionary table.

3. Performance Assessment Results

In our experiments, we have used the temperature and relative humidity measurements collected from a randomly extracted node (NODE ID= 84) of the WSN SensorScope LUCE deployment (SensorScope, 2009), within the time interval from 23/11/2006 to 17/12/2006. The resulting temperature and relative humidity datasets are composed by 64913 samples and we will refer to them as LU_ID84_T and LU_ID84_H, respectively. The WSN adopted in the deployment employs a TinyNode node type (TinyNode, 2009), which uses a TI MSP430 microcontroller, a Xemics XE1205 radio and a Sensirion SHT75 sensor module (Sensirion, 2009).
This module is a single chip which includes a capacitive polymer sensing element for relative humidity and a bandgap temperature sensor. Both the sensors are seamlessly coupled to a 14-bit ADC and a serial interface circuit on the same chip. The Sensirion SHT75 can sense air temperature in the $[-20°C,+60°C]$ range and relative humidity in the $[0\%, 100\%]$ range. The outputs *raw_t* and *raw_h* of the ADC for temperature and relative humidity are represented with resolutions of 14 and 12 bits, respectively. The outputs *raw_t*

and *raw_h* are converted into measures *t* and *h* expressed, respectively, in Celsius degrees (°C) and percentage (%) as described in (Sensirion, 2009). The datasets corresponding to the deployments store measures *t* and *h*. On the other hand, the LEC algorithm works on *raw_t* and *raw_h*. Thus, before applying the algorithm, we extracted *raw_t* and *raw_h* from *t* and *h*, respectively, by using the inverted versions of the conversion functions in (Sensirion, 2009).
Table 2 shows some statistical characteristics of the two datasets. In particular, we have computed the mean $\overline{s}$ and the standard deviation $\sigma_{\overline{s}}$ of the samples, the mean $\overline{d}$ and the standard deviation $\sigma_{\overline{d}}$ of the differences between consecutive samples, the information entropy $H = -\sum_{i=1}^{N} p(x_i) \cdot \log_2 p(x_i)$ of the original signal, where N is the number of possible values x_i (the output of the ADC) and $p(x_i)$ is the probability mass function of x_i, and the information entropy $H_d = -\sum_{i=1}^{N} p(d_i) \cdot \log_2 p(d_i)$ of the differentiated signal.

Dataset	Samples	$\overline{s} \pm \sigma_{\overline{s}}$	$\overline{d} \pm \sigma_{\overline{d}}$	H	H_d
LU_ID84_T	64913	7.21±3.16	$-2.87 \cdot 10^{-5} \pm 0.05$	10.07	4.05
LU_ID84_H	64913	87.04±8.04	$1.12 \cdot 10^{-4} \pm 0.55$	10.08	5.85

Table 2. Statistical characteristics of the datasets.

In the following, we first show the compression ratios achieved by LEC and compare them with the ones achieved by S-LZW and three well-known compression algorithms. We also discuss the complexity of LEC and S-LZW. Then, we investigate the dependence of the compression performance of LEC on the correlation between consecutive samples of the signal to be compressed and show how semi-adaptive and adaptive Huffman coding can help LEC to increase the compression ratios. Finally, we introduce a problem that affects LEC and in general all the differential compression algorithms and discuss how this problem can be solved without considerably penalizing the compression ratios achieved by LEC.

3.1 Compression ratios and complexity

The performance of a compression algorithm is usually computed by using the compression ratio (*CR*) defined as:

$$CR = 100 \cdot \left(1 - \frac{comp\_size}{orig\_size}\right) \quad (2)$$

where *comp_size* and *orig_size* are, respectively, the sizes in bits of the compressed and the uncompressed bitstreams. Considering that uncompressed samples are normally byte-aligned, both temperature and relative humidity samples are represented by 16-bit unsigned integers. Thus, from Table 2, it is easy to compute *orig_size* for the given datasets.
Moreover, assuming that all samples have to be transmitted to the sink by using the lowest number of messages so as to have power saving (Mainwaring et al., 2002) and supposing

that each packet can contain at most 29 bytes of payload (Croce et al., 2008), we can define the packet compression ratio as:

$$PCR = 100 \cdot (1 - \frac{comp\_pkt}{orig\_pkt}) \qquad (3)$$

where *comp_pkt* and *orig_pkt* represent the number of packets necessary to deliver the compressed and the uncompressed bitstreams, respectively.
Table 3 shows the results obtained by LEC in terms of CR and PCR on the two datasets. As expected, the LEC algorithm achieves higher compression ratios on the temperature dataset which is characterized by a lower entropy H_d and, in general, a low variability between consecutive samples (that is, low values of the mean and standard deviation of the differences between consecutive samples).

	Dataset	*orig_size*	*comp_size*	*CR(%)*	*orig_pkt*	*comp_pkt*	*PCR(%)*
LEC	LU_ID84_T	1038608	303194	70.81	4477	1307	70.81
	LU_ID84_H	1038608	396442	61.83	4477	1709	61.83

Table 3. Compression ratios obtained by LEC on the two datasets.

To assess the goodness of the results shown in Table 3, we have also applied the S-LZW algorithm and three well-known compression methods to the same datasets. S-LZW is a lossless compression algorithm purposely developed to be embedded in sensor nodes. S-LZW splits the uncompressed input bitstream into fixed size blocks and then compresses separately each block. During the block compression, for each new string, that is, a string which is not already in the dictionary, a new entry is added to the dictionary. For each new block, the dictionary used in the compression is re-initialized by using the 256 codes which represent the standard character set. Due to the poor storage resources of sensor nodes, the size of the dictionary has to be limited. Thus, since each new string in the input bitstream produces a new entry in the dictionary, the dictionary might become full. If this occurs, an appropriate strategy has to be adopted. For instance, the dictionary can be frozen and used as-is to compress the remainder of the data in the block (in the worst case, by using the code of each character), or it can be reset and started from scratch. To take advantage of the repetitive behaviour of sensor data, a mini-cache is added to S-LZW: the mini-cache is a hash-indexed dictionary of size N, where N is a power of 2, that stores recently used and created dictionary entries. Further, the repetitive behaviour can be used to pre-process the raw data so as to build appropriately structured datasets, which can perform better with the compression algorithm.
In (Sadler & Martonosi, 2006), the authors show that the use of structured datasets and the introduction of the mini-cache increase the compression ratios without introducing appreciable computational overhead. It follows that S-LZW has to balance four major inter-related parameters: the size (BLOCK_SIZE) of the data block, the maximum number (MAX_DICT_ENTRIES) of dictionary entries, the strategy (DICTIONARY_STRATEGY) to follow when the dictionary is full and the number (MINI-CACHE_ENTRIES) of mini-cache

entries. With the aim of putting S-LZW in its best situation, we adopted the values suggested in (Sadler & Martonosi, 2006): a block size of 528 bytes, a dictionary of 512 entries that is maintained once full and a mini-cache of 32 entries.
As regards the well-known compression methods, we have considered gzip, bzip2 and rar. These methods have a parameter which allows setting the compression level. This parameter is between 1 and 9 (default 6) for gzip and bzip2, and between 1 and 5 (default 3) for rar. We fixed this paramater to the maximum possible compression (9 for gzip and bzip2 and 5 for rar).
Table 4 shows the results obtained by the four algorithms. We can observe that LEC outperforms the other algorithms. In the table, we have not shown the PCRs for gzip, bzip2 e rar. Indeed, these algorithms have been used only as benchmarks to validate the compression ratios obtained by applying LEC. Actually, as already observed in (Ganesan et al., 2003) (Kimura & Latifi, 2005) (Sadler & Martonosi, 2006), these algorithms cannot be executed in a sensor node, due to memory requirements and computational power needed for their execution. Indeed, the executable codes are too large to be embedded in tiny sensor nodes. Further, the compression ratios are obtained after collecting all the samples and therefore all the samples have to be stored in memory. This implies that large datasets cannot be managed. In addition, the compression cannot be performed on the fly. Finally, during their execution, these algorithms require a large memory to manage some step of the execution.

Dataset	**Algorithm**	***CR(%)***	***PCR(%)***
LU_ID84_T	S-LZW	48.99	48.98
	gzip	48.87	-
	bzip2	69.24	-
	rar	69.16	-
LU_ID84_H	S-LZW	31.24	31.22
	gzip	37.86	-
	bzip2	57.82	-
	rar	59.03	-

Table 4. Compression ratios obtained by S-LZW, gzip, bzip2 e rar algorithms on the two datasets.

As regards complexity of LEC and S-LZW, we have performed a comparative analysis on the number of instructions required by both the algorithms to compress data. To this aim, we have adopted the Sim-It Arm simulator (Sim-It, 2009), since there already exists a free available version of S-LZW implemented for this simulator by the same authors of this compression algorithm. Sim-It Arm is an instruction-set simulator that runs both system-level and user-level ARM programs. Since S-LZW compresses each dataset block by block, we executed the two algorithms on Sim-It Arm simulator to compress the first block of each

dataset. A block consists of 528 bytes (corresponding to 264 samples of 16 bits). Table 5 shows the number of instructions required for compressing one block, the number of saved bits and the number of instructions per saved bit for the temperature and relative humidity datasets, respectively.
We note that, though the LEC algorithm achieves a higher compression ratio than S-LZW, it requires a lower number of instructions. In particular, we observe that, the LEC algorithm executes, for instance, 15.33 instructions for each saved bit against 29.93 executed by S-LZW for compressing the first block of the temperature dataset.

	LEC		S-LZW	
	Temperature	**Relative Humidity**	**Temperature**	**Relative Humidity**
number of instructions	44784	62817	63207	63207
number of saved bits	2922	2086	2112	96
number of instructions per saved bit	15.33	30.11	29.93	658.41

Table 5. Complexity of LEC and S-LZW.

3.2 Compression ratio versus correlation between consecutive samples

In the previous section we have adopted the default Huffman table for compressing the two datasets so as to show the effectiveness of the LEC algorithm when dealing with high correlated datasets. In this section, we analyze the behaviour of LEC when the correlation between consecutive samples decreases. To this aim, we performed the following experiment: we simulated different lengths of the sampling interval by downsampling the sequence of data. Since in the original datasets, samples are obtained by measuring temperature and relative humidity at intervals of 30 seconds along 25 days, we considered downsampling factors of 2, 4, 8, 16, 60 and 120, which correspond, respectively, to consider time intervals of 1, 2, 4, 8, 30 and 60 minutes. We expect that, like for all compression algorithms based on differential coding, the sampling rate affects the achievable compression ratio: when the sampling interval is long, the correlation between consecutive samples typically decreases, thus reducing the performance of the LEC algorithm. The significance of this reduction depends on the variability of the signal.
Figure 2 shows the results obtained by compressing the downsampled temperature and relative humidity datasets. We can observe that the compression ratios decrease with the increase of the downsampling factors. For instance, for the temperature, we pass from a compression ratio of 70.81% with the original data (downsampling factor equal to 0) to a compression ratio of 41.38% with a downsampling factor equal to 120. As expected, the decrease of the compression ratios is therefore quite relevant. On the other hand, the Huffman table shown in Table 1 has been proposed for data with high correlation, where high probabilities are associated with differences between consecutive samples very close to 0. Actually, these differences are characterized by a high occurrence frequency. By downsampling the original signal with factors from 16 to 120, this assumption is not true

anymore. To be fair in the experiment, we should compute again the occurrence frequencies of the differences between consecutive samples (this approach is known in the literature as semi-adaptive Huffman coding (Salomon, 2007)) and modify appropriately the Huffman table used in the compression.

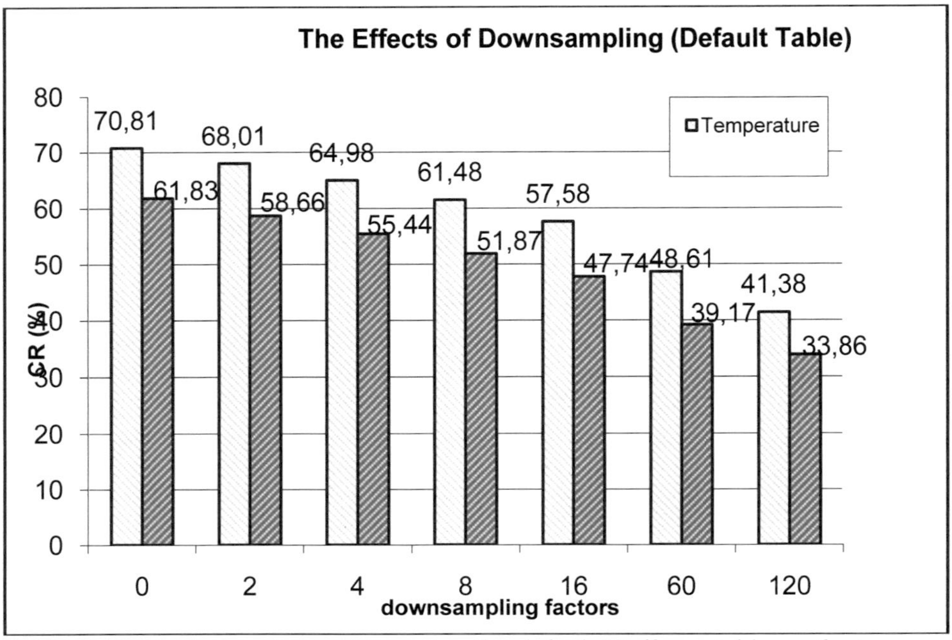

Fig. 2. Compression ratios obtained by using the default Huffman table on the temperature and humidity datasets sampled with different downsampling factors.

Figure 3 shows the results obtained by semi-adaptive Huffman coding. We can observe that the compression ratios still decrease with respect to increasing downsampling factors (on the other hand, the correlation between consecutive samples is lower), but now this decrease is less significant. To take on-line these variations of difference distributions into account, in the literature adaptive Huffman coding has been proposed. The method was originally developed by Faller (Faller, 1973) and Gallager (Gallager, 1978) with substantial improvements by Knuth (Knuth, 1985).

Figure 4 shows the results obtained by using the adaptive Huffman coding in LEC. Obviously, the use of the adaptive coding increases the compression ratios with respect to the use of a fixed table, but does not allow outperforming the use of the semi-adaptive Huffman coding. On the other hand, unlike fixed table and adaptive Huffman coding, semi-adaptive Huffman coding exploits the knowledge of all data. Obviously, this knowledge cannot be assumed in real applications. Thus, the compression ratios obtained by using the semi-adaptive Huffman coding can be considered as an upper limit. However, we observe that the compression ratios achieved with the adaptive Huffman coding are very close to the ones obtained with the semi-adaptive Huffman coding. On the other hand, we have to consider that the use of the adaptive Huffman coding increases the complexity of LEC. Further, since the adaptive coding/decoding scheme is symmetric, a possible loss of one packet makes the decompression process completely unreliable. Thus, in real applications of WSNs the use of a fixed table is certainly desirable and practically mandatory.

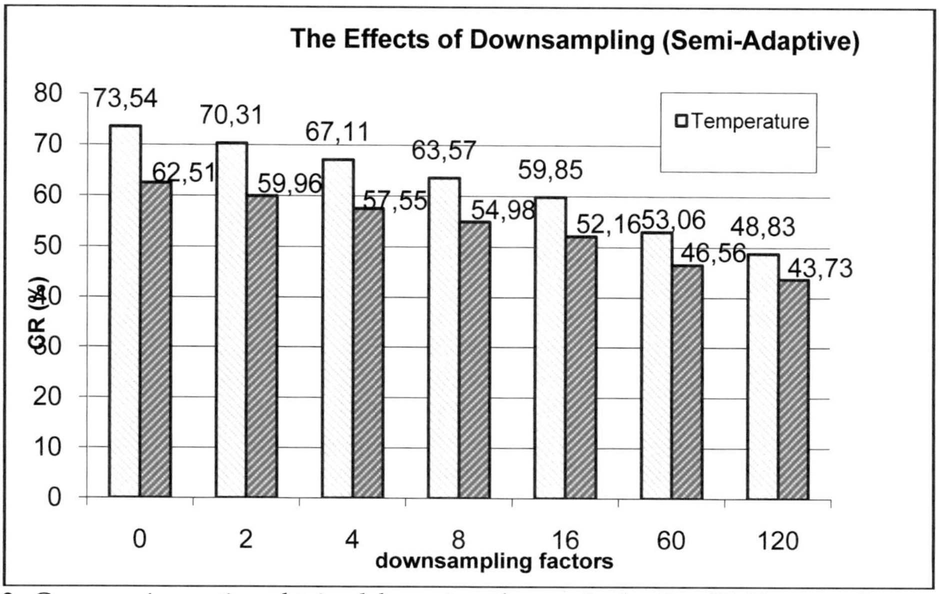

Fig. 3. Compression ratios obtained by using the semi-adaptive Huffman coding on the temperature and humidity datasets sampled with different downsampling factors.

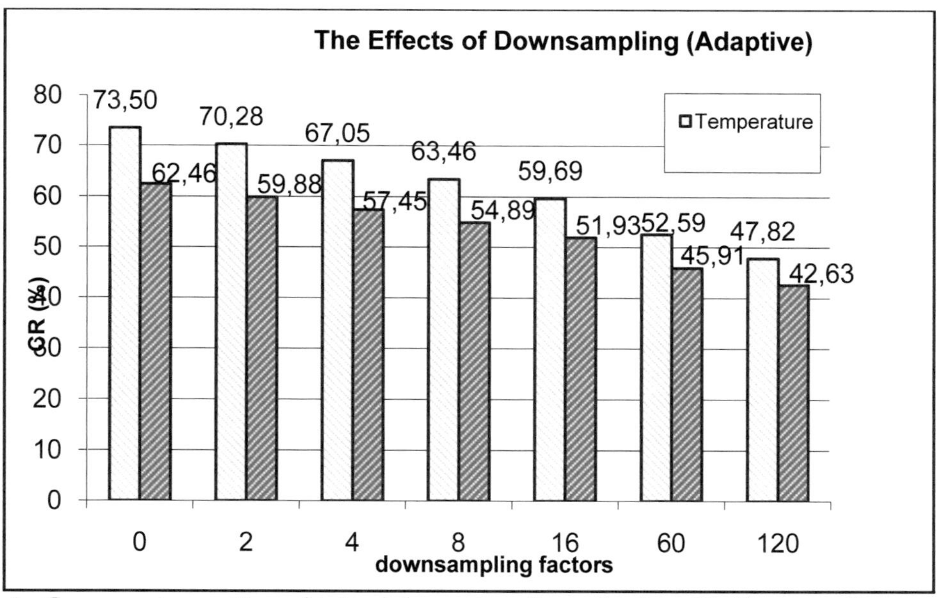

Fig. 4. Compression ratios obtained by using the adaptive Huffman coding on the temperature and humidity datasets sampled with different downsampling factors.

The fixed Huffman table used in the original version of LEC can guarantee satisfactory performance when the correlation between consecutive samples is high. However, when the correlation is not high, we can find a fixed Huffman table suitable for the specific application. Indeed, we would like to remark that, in real habitat monitoring applications, the sampling rate is a parameter of the application domain: once fixed, rarely it is modified. Since the trend of the environmental signals is generally known, this allows us to make quite reliable assumptions on the distributions of the differences, thus permitting us to generate fixed Huffman tables which guarantee high compression ratios. We could also consider to adopt a two-phase approach. In the first phase, we collect an appropriate number of samples so as to perform an analysis of occurrence frequency of the differences. Then, in the second phase, we use the fixed Huffman table generated by the analysis performed in the first phase to compress the data on the fly.
To highlight that the lack of sample correlation does not affect only LEC, but in general all the compression algorithms, we have also applied S-LZW to the temperature and humidity datasets sampled with downsampling factors of 2, 4, 8, 16, 60 and 120. Figure 5 compares the compression ratios obtained by S-LZW with the ones achieved by the LEC algorithm executed by using the default table. As expected, we can observe that also the performance of S-LZW are considerably affected by downsampling.

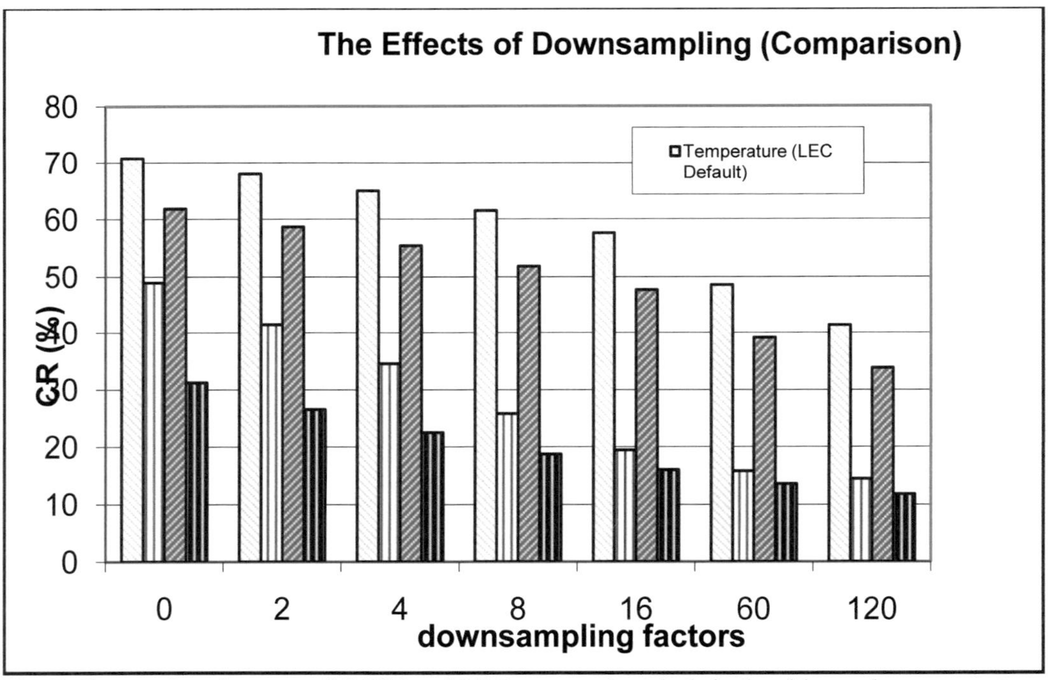

Fig. 5. Comparison between S-LZW and LEC executed with default table on the temperature and humidity datasets sampled with different downsampling factors.

4.3 The problem of the first sample

LEC, as all the differential compression algorithms, suffers from the following problem. In order to reconstruct the original samples, the decoder must know the value of the first sample: if the first sample has been lost or corrupted, all the other samples are not correctly decoded. In our case, the compressed bitstream is sent by wireless communication to the collector, which takes the decompression process in charge. Since the transmission can be non-reliable, the first packet could be lost and thus also the first value, making correct reconstruction of samples impossible.

To make communication reliable, a number of solutions have been proposed. In general, these solutions involve protocols based on acknowledgements which act at Transport layer. Obviously, these protocols require a higher number of message exchanges between nodes and this increases the power consumption. A review of these algorithms is out of the scope of this chapter. Anyway, a solution to this problem can be also provided at the application layer without modifying the protocols of the underlying layers: when we insert the first sample into the payload of a new packet, we do not insert the difference between the current and the previous sample, but rather the difference between the current sample and a reference value known to the decoder (for instance, the central value of the ADC). Thus, the decoding of each packet is independent of the reception of the previous packets. Table 6 compares the PCRs obtained by using this expedient (this PCR will be denoted as PCR*) with those shown in Table 3: we can note that the decrease of PCR is not high. Further, the PCR*s are still higher than those achieved by S-LZW. Thus, we can conclude that the LEC scheme can be made more robust without significantly affecting its performance.

Dataset	***PCR(%)***	***PCR*(%)***
LU_ID84_T	70.81	68.19
LU_ID84_H	61.83	58.21

Table 6. PCRs obtained without (PCR) and by (PCR*s) transmitting the first value in each packet.

5. From Lossless to Lossy

In some WSN applications, like environmental monitoring, the accurateness of the measures is less important than the sensor cheapness. Thus, often commercial wireless nodes are equipped with sensors which, though cheap, collect measures affected by considerable noise. In this context, the use of lossless compression algorithms can be penalising. Indeed, noise increases the entropy of the signal and therefore hinders the lossless compression algorithm to achieve considerable compression ratios. The ideal solution would be to adopt on the sensor node, a lossy compression algorithm in which the loss of information would be just the noise. Thus, we could achieve high compression ratios without losing relevant information. To this aim, we exploit the observation that data typically collected by WSNs are strongly correlated. Thus, differences between consecutive samples should be regular and generally very small. If this does not occur, it is likely that samples are affected by noise. To de-noise and simultaneously compress the samples, we introduce a lossy version of LEC. In this version, the difference $d_i = r_i - r_{i-1}$ is not directly encoded, but is first quantized and then encoded following the Differential Pulse Code Modulation (DPCM) scheme often used for digital audio signal compression. The schemes of the lossy versions of the compressor and uncompressor are shown in Fig. 6.

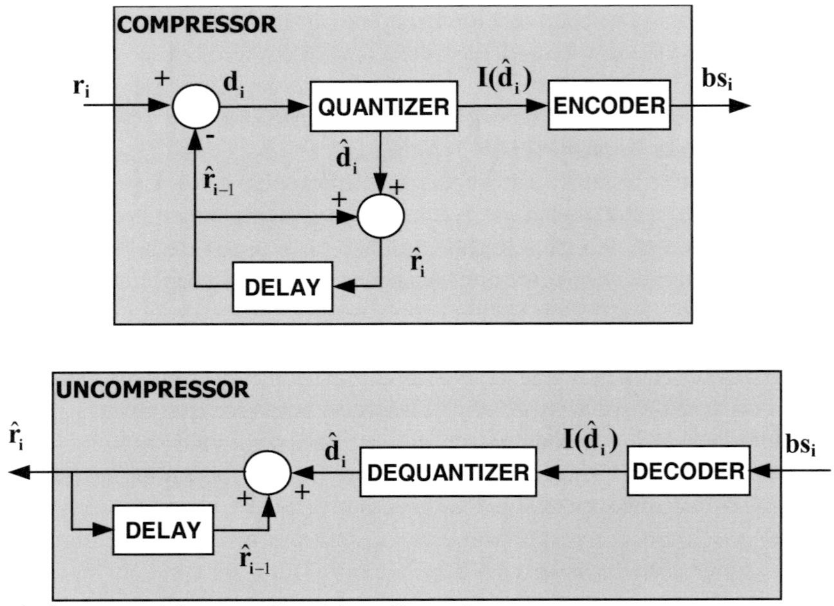

Fig. 6. Block diagram of the encoding/decoding schemes.

Actually to avoid the well-known problem of the *accumulation of the error* (Salomon, 2007), we quantize the difference between sample r_i and the most recent reconstructed value $\hat{r}_{i-1}$. The problem originates from the following consideration: the compressor can compute the exact differences d_i from the original data samples r_i and r_{i-1}, while the uncompressor can work only with quantized differences $\hat{d}_i$. The uncompressor uses $\hat{d}_i$ to generate the reconstructed samples $\hat{r}_i$ ($\hat{r}_i = \hat{r}_{i-1} + \hat{d}_i$) rather than the original samples r_i. The generic nth reconstructed sample $\hat{r}_n$ at the uncompressor will contain the sum of the quantization errors accumulated during the reconstruction of the previous n-1 samples plus the quantization error of the current sample:

$$\hat{r}_n = r_n + \sum_{i=1}^{n} q_i \quad (3)$$

where q_i is the quantization error.

To overcome this problem, the compressor is modified so as to compute the generic difference $d_i = r_i - \hat{r}_{i-1}$, that is, to calculate difference d_i by subtracting the most recent reconstructed value $\hat{r}_{i-1}$ (which both the compressor and the uncompressor have) from the current original sample r_i. Thus, the uncompressor first decodifies r_0. Then, when it receives the first quantized difference $\hat{d}_1$, it computes $\hat{r}_1 = r_0 + \hat{d}_1 = r_0 + d_1 + q_1 = r_1 + q_1$. When it receives the second quantized difference $\hat{d}_2$, it computes $\hat{r}_2 = \hat{r}_1 + \hat{d}_2 = \hat{r}_1 + d_2 + q_2 = \hat{r}_1 + r_2 - \hat{r}_1 + q_2 = r_2 + q_2$. The decoded value $\hat{r}_2$ contains just the single quantization error q_2, and in general, the decoded value $\hat{r}_i$ contains just the quantization error q_i.

Difference d_i is input to the block QUANTIZER that outputs the quantization level $\hat{d}_i$ assigned to d_i and the index $I(\hat{d}_i)$ of $\hat{d}_i$. The index $I(\hat{d}_i)$ is input to the ENCODER block, which generates the codeword bs_i using the same bijection defined in (1) for mapping integer inputs to natural values, and the same combination of unary and binary codes described in Section 2. The ENCODER block, therefore, encodes the quantization index corresponding to the quantized difference rather than the difference as in LEC. Again, the dictionary table used to produce the codes should be generated based on the occurrence frequency of the quantization indexes. In these preliminary experiments, we have decided to adopt the same dictionary used in Table 1, where in place of d_i, the reader should read $\hat{d}_i$. Since the number of quantization levels $\hat{d}_i$ is lower than the number of possible d_i, the table might have a lower number of entries.

In the uncompressor, the codeword bs_i is analyzed by the DECODER block which outputs the index $I(\hat{d}_i)$, exploiting the same dictionary table. This index is elaborated by the block DEQUANTIZER to produce $\hat{d}_i$ which is added to $\hat{r}_{i-1}$ to output $\hat{r}_i$.

Currently, we are simply adopting a uniform quantization. In this case, the unique parameter to be fixed is the difference *D* between two consecutive levels. This parameter is very important because it affects the value of the quantization error and indirectly the compression ratio. To show the performance of the lossy version of LEC, we set *D* to six different values: 10%, 20%, 30%, 40%, 50% and 60% of the Manufactured Error (ME) of the sensor used to collect data. In the case of the sensors (Sensirion SHT75) used in our experiments, ME = ± 0.3 °C and ME = ± 1.8% for temperature and relative humidity, respectively (Sensirion, 2009). Table 7 shows the compression ratios and the root mean squared errors (RMSEs) obtained on the temperature and relative humidity datasets.

RMSE is computed as:

$$RMSE = \sqrt{\frac{1}{N}\sum_{i=1}^{N}(r_i - \hat{r}_i)^2} \tag{5}$$

where r_i is the original sample, $\hat{r}_i$ is the reconstructed sample and *N* is the number of samples of the signal. We observe that, as expected, the compression ratios are higher than the ones obtained by the original version of LEC. On the other hand, the lossy version introduces an error on the reconstructed signal. Anyway, this error is lower than ME, which represents a sort of uncertainty of the measure.

To assess the results shown in Table 7, we have applied LTC to the same datasets. LTC is an efficient and simple lossy compression technique for the context of habitat monitoring. LTC generates a set of line segments which form a piecewise continuous function. This function approximates the original dataset in such a way that no original sample is farther than a fixed error *e* from the closest line segment. Thus, before executing the LTC algorithm, we have to set error *e*. To perform a fair comparison with the lossy version of LEC, we have set *e* to the 10%, 20% and 30% of the ME of the sensor. This allows obtaining RMSEs comparable with the ones obtained by the lossy version of LEC when *D* is equal to the 20%, 40% and 60% of the ME. Table 8 shows the compression ratios and the RMSEs obtained on the temperature and relative humidity datasets. We can observe that the lossy version of LEC outperforms LTC in terms of CR for comparable RMSEs, thus proving the good characteristics of the proposed lossy compression algorithm.

Dataset	**Algorithm**	*CR(%)*	*RMSE*
LU_ID84_T	0.1·ME	78.18	0.0082
	0.2·ME	81.26	0.0171
	0.3·ME	83.45	0.0256
	0.4·ME	83.46	0.0353
	0.5·ME	84.94	0.0428
	0.6·ME	86.14	0.0517
LU_ID84_H	0.1·ME	74.65	0.0450
	0.2·ME	78.83	0.0872
	0.3·ME	80.89	0.1296
	0.4·ME	82.13	0.1721
	0.5·ME	82.97	0.2166
	0.6·ME	83.61	0.2598

Table 7. Compression ratios obtained by the lossy version of LEC on the two datasets.

Dataset	**Algorithm**	*CR(%)*	*RMSE*
LU_ID84_T	0.1·ME	55.00	0.0190
	0.2·ME	77.53	0.0348
	0.3·ME	86.12	0.0502
LU_ID84_H	0.1·ME	26.49	0.0824
	0.2·ME	55.97	0.1681
	0.3·ME	70.99	0.2496

Table 8. Compression ratios obtained by the LTC algorithm on the two datasets.

6. Conclusions

In this chapter, we have discussed how enabling compression helps in wireless sensor nodes. First, we have briefly introduced LEC, a lossless compression algorithm we proposed in a previous paper. LEC divides the alphabet of differences between consecutive samples into groups whose sizes increase exponentially. Each codeword is a hybrid of unary and binary codes: in particular, the unary code (a variable-length code) specifies the group, while the binary code (a fixed-length code) represents the index within the group. In the original version, we used the Huffman table proposed in JPEG for coding the groups. Here, we have investigated semi-adaptive and adaptive Huffman coding and carried out a comparison in terms of compression ratios with the LEC algorithm with fixed Huffman table. We have shown that semi-adaptive and adaptive Huffman coding can increase the compression ratios when the correlation between consecutive samples decreases. We have compared all the approaches with S-LZW, a compression algorithm specifically proposed for sensor nodes, and with three classical compression algorithms, namely gzip, bzip2 and rar, though these algorithms are not embeddable in tiny sensor nodes. We have shown that the different versions of LEC can achieve considerable compression ratios in all the datasets considered in the experiments. Finally, we have discussed how LEC can be transformed into a lossy compression algorithm and have shown that this lossy version outperforms LTC, a lossy compression algorithm specifically designed for being embedded in tiny sensor nodes.

7. Acknowledgements

This work was supported by the Italian Ministry of University and Research (MIUR) under the PRIN project #2005090483_005 "Wireless sensor networks for monitoring natural phenomena" and the FIRB project "Adaptive Infrastructure for Decentralized Organization (ArtDecO)".

8. References

Anastasi, G., Conti, M., Di Francesco, M. & Passarella, A. (2009) Energy conservation in wireless sensor networks: A survey. *Ad Hoc Networks*, Vol. 7, 537-568.

Barr, K. C. and Asanović, K. (2006) Energy-aware lossless data compression. *ACM Trans. Comput. Syst.*, Vol. 24, 250-291.

Boulis, A., Ganeriwal, S. & Srivastava, M.B. (2003) Aggregation in sensor networks: an energy- trade-off. *Ad Hoc Networks*, Vol. 1, 317-331.

Chen, H., Li, J. & Mohapatra, P. (2004) RACE: time series compression with rate adaptivity and error bound for sensor networks. Proceedings of the First IEEE International Conference on Mobile Ad-hoc and Sensor Systems, pp. 124-133, Fort Lauderdale, FL, USA, 24-27 October,. IEEE, Piscataway, NJ, USA.

Ciancio, A. & Ortega, A. (2005) A distributed wavelet compression algorithm for wireless multihop sensor networks using lifting. Proceedings of IEEE International Conference on Acoustics, Speech, and Signal Processing, pp. 825-828, Philadelphia, PA, USA, 18-23 March,. IEEE, Piscataway, NJ, USA.

Ciancio, A., Pattem, S., Ortega, A. & Krishnamachari, B. (2006) Energy-efficient data representation and routing for wireless sensor networks based on a distributed wavelet compression algorithm. Proceedings of the Fifth international Conference on Information Processing in Sensor Networks, pp. 309-316, Nashville, TN, USA, 19-21 April, ACM, New York, NY, USA.

Croce, S., Marcelloni, F. & Vecchio, M. (2008) Reducing power consumption in wireless sensor networks using a novel approach to data aggregation. *The Computer Journal*, Vol. 51, No. 2, 227–239.

Deligiannakis, A., Kotidis, Y. & Roussopoulos, N. (2004) Compressing historical information in sensor networks. Proceedings of the 2004 ACM SIGMOD International Conference on Management of Data, pp. 527-538, Paris, France, 13-18 June, ACM, New York, NY, USA.

Di Bacco, G., Melodia, T. & Cuomo, F. (2004) A MAC protocol for delay-bounded applications in wireless sensor networks. Proceedings of the Third Annual Mediterranean Ad Hoc Networking Workshop, pp. 208-220, Bodrum, Turkey, 27-30 June, available on-line: http://www.ece.osu.edu/medhoc04/.

Elias, P. (1975) Universal codeword sets and representations of the integers. *IEEE Transaction on Information Theory*, Vol. 21, No. 2, 194–203.

Faller, N. (1973) An adaptive system for data compression. Proceedings of the 7th Asilomar Conference on Circuits, Systems, and Computers, pp. 593–597, Pacific Grove, CA, USA, November, IEEE Press, Piscataway, NJ, USA.

Fan, K-W, Liu, S. & Sinha, P. (2007) Structure-free data aggregation in sensor networks. *IEEE Transactions on Mobile Computing*, Vol. 6, 929-942.

Fasolo, E., Rossi, M., Widmer, J. & Zorzi, M. (2007) In-network aggregation techniques for wireless sensor networks: a survey. *Wireless Communications*, Vol. 14, 70-87.

Gallager, R. G. (1978) Variations on a theme by Huffman. *IEEE Transactions on Information Theory*, Vol. 24, No. 6, 668-674.

Ganesan, D., Estrin, D. & Heidemann, J. (2003) DIMENSIONS: why do we need a new data handling architecture for sensor networks?, *SIGCOMM Comput. Commun. Rev.*, Vol. 33, 143-148.

Gastpar, M., Dragotti, P. L. & Vetterli, M. (2006) The distributed Karhunen-Loève transform. *IEEE Transactions on Information Theory*, Vol. 52, No. 12, 5177-5196.

Girod, B., Aaron, A., Rane, S. & Rebollo-Monedero, D. (2005) Distributed video coding. *Proceedings of the IEEE*, Special Issue Advances Video Coding, Delivery, Vol. 93, No. 1, 71–83.

Golomb, S. W. (1966) Run-length encodings. *IEEE Transactions on Information Theory*, Vol. 12, No. 3, 399-401.

Guestrin, C., Bodi, P., Thibau, R., Paski, M. & Madden, S. (2004) Distributed regression: an efficient framework for modelling sensor network data. Proceedings of the Third International Symposium on Information Processing in Sensor Networks, pp.1-10, Berkeley, CA, USA, 26-27 April, ACM, New York, NY, USA.

Intanagonwiwat, C., Govindan, R., Estrin, D., Heidemann, J. & Silva, F. (2003) Directed diffusion for wireless sensor networking. *IEEE/ACM Trans. Netw.*, Vol. 11, 2-16.

Kimura, N. & Latifi, S. (2005) A survey on data compression in wireless sensor networks. Proceedings of the International Conference on Information Technology: Coding and Computing, pp. 8-13, Las Vegas, NV, USA, 4-6 April, IEEE Computer Society, Washington, DC, USA.

Knuth, D. E. (1985) Dynamic Huffman Coding. *Journal of Algorithms*, Vol. 6, 163-180.

Lin, S., Kalogeraki, V., Gunopulos, D. & Lonardi, S. (2006) Online information compression in sensor networks. Proceedings of the IEEE International Conference on Communications, pp. 3371-3376, Istanbul, Turkey, 11-15 June, IEEE Press, Piscataway, NJ, USA.

Lindsey, S., Raghavendra, C. & Sivalingam, K. M. (2002) Data gathering algorithms in sensor networks using energy metrics. *IEEE Trans. Parallel Distrib. Syst.*, Vol. 13, 924-935.

Lynch, J. P., Wang, Y., Sundararajan, A., Law, K. H. & Kiremidjian, A. S. (2004) Wireless sensing for structural health monitoring of civil structures. Proceedings of the International Workshop on Integrated Life-Cycle Management of Infrastructures, Hong Kong, 9-11 December.

LZO (2008) http://www.oberhumer.com/opensource/lzo/

Madden, S., Franklin, M. J., Hellerstein, J. M. & Hong, W. (2002) TAG: a Tiny AGgregation service for ad-hoc sensor networks. *SIGOPS Oper. Syst. Rev.*, Vol. 36, 131-146.

Mainwaring, A., Culler, D., Polastre, J., Szewczyk, R. & Anderson, J. (2002) Wireless sensor networks for habitat monitoring. Proceedings of the First ACM international Workshop on Wireless Sensor Networks and Applications, pp. 88-97, Atlanta, GA, USA, 28 September, ACM, New York, NY, USA.

Marcelloni, F. & Vecchio, M. (2009) An efficient lossless compression algorithm for tiny nodes of monitoring wireless sensor networks, *The Computer Journal*, Section B: Networks and Computer Systems, Advance Access, doi:10.1093/comjnl/bxp035.

Pennebaker, W. B. & Mitchell, J. L. (1992) *JPEG still image data compression standard*. Kluwer Academic Publishers, Norwell, MA, USA.

Pradhan, S. Kusuma, J. & Ramchandran, K. (2002) Distributed compression in a dense microsensor network, *IEEE Signal Processing Mag.*, Vol. 19, 51–60.

Rebollo-Monedero, D. (2007) Quantization and transforms for distributed source coding. PhD thesis The Department of Electrical Engineering and the Committee on Graduate Studies of Stanford University.

Sadler, C. M. & Martonosi, M. (2006) Data compression algorithms for energy-constrained devices in delay tolerant networks. Proceedings of the 4th ACM International Conference on Embedded networked sensor systems, pp. 265-278, Boulder, Colorado, USA, October 31 - November 3, ACM, New York, NY, USA.

Salomon, D. (2007) *Data Compression: The Complete Reference*, Springer Verlag, London, UK.

Schoellhammer, T., Osterweil, E., Greenstein, B., Wimbrow, M. & Estrin, D. (2004) Lightweight temporal compression of microclimate datasets. Proceedings of the 29th Annual IEEE International Conference on Local Computer Networks, pp. 516-524, Tampa, FL, USA, 16-18 November, IEEE Computer Society, Washington, DC, USA.

Sensirion (2009) Sensirion homepage, www.sensirion.com

SensorScope (2009), SensorScope deployments homepage http://sensorscope.epfl.ch/index.php/Main_Page

Sim-It (2009) Sim-It Simulator homepage, http://simit-arm.sourceforge.net/

Tang, C. & Raghavendra, C. S. (2004) Compression techniques for wireless sensor networks. In: *Wireless Sensor Networks*, Raghavendra, C. S., Sivalingam, K. M. and Znati, T. (Ed.), Kluwer Academic Publishers, Norwell, MA, USA.

Teuhola, J. (1978) A Compression Method for Clustered Bit-Vectors. *Information Processing Letters*, Vol. 7, 308-311.

TinyNode (2009) TinyNode homepage, http://www.tinynode.com

Wagner, R. S., Baraniuk, R. G., Du, S., Johnson, D. B. & Cohen, A. (2006) An architecture for distributed wavelet analysis and processing in sensor networks. Proceedings of the Fifth International Conference on Information Processing in Sensor Networks, pp. 243-250, Nashville, TN, USA, 19-21 April, ACM, New York, NY, USA.

Welch, T.A. (1984) A technique for high-performance data compression, *Computer*, Vol. 17, 8-19.

Ziv, J. & Lempel, A. (1977) A universal algorithm for sequential data compression, *IEEE Transactions on Information Theory*, Vol. 23, 337-343.

Zixiang, X., Liveris, A.D. & Cheng, S. (2004) Distributed source coding for sensor networks. *IEEE Signal Processing Magazine*, Vol. 21, No. 5, 80-94.

13

Implementation of Accelerometer Sensor Module and Fall Detection Monitoring System based on Wireless Sensor Network

Youngbum Lee and Myoungho Lee
*Yonsei University, Department of Electrical and Electronic Engineering
Republic of Korea*

1. Introduction

ADL means 'Activity of Daily Living' and literally the activity from everyday living. In the early days, the activity measurement system using accelerometer measures in one direction at one part. This method has an advantage that easy and quantitative measurement is possible using one sensor. But that is so simple method that precise activity assessment for various posture classifications in daily living is impossible [2]. For the study about the correlation between the human's movement and energy consumption, the method that measures 3 direction activity data using 3-axis accelerometer sensor is used. This method is better than using many sensors, but the classification for various human's movement is still impossible [5]. In this study, using accelerometer sensor module, we develop the algorithm that classify the wearer's posture and activity. And we implement the monitoring system based on wireless sensor network. For the performance assessment of developed accelerometer module, algorithm and monitoring system, the experiment for 30 subjects is executed.

This research implements wireless accelerometer sensor module and algorithm to determine wearer's posture, activity and fall. Wireless accelerometer sensor module uses ADXL202, 2-axis accelerometer sensor (Analog Device). And using wireless RF module, this module measures accelerometer signal and shows the signal at 'Acceloger' viewer program in PC. ADL algorithm determines posture, activity and fall that activity is determined by AC component of accelerometer signal and posture is determined by DC component of accelerometer signal. Those activity and posture include standing, sitting, lying, walking, running, etc. By the experiment for 30 subjects, the performance of implemented algorithm was assessed, and detection rate for postures, motions and subjects was calculated. Lastly, using wireless sensor network in experimental space, subject's postures, motions and fall monitoring system was implemented. By the simulation experiment for 30 subjects, 4 kinds of activity, 3 times, fall detection rate was calculated. In conclusion, this system can be application to patients and elders for activity monitoring and fall detection and also sports athletes' exercise measurement and pattern analysis. And it can be expected to common person's exercise training and just plaything for entertainment.

2. Wireless Accelerometer Sensor Module Design and Implementation

In this part, we describe the design and implementation of wireless accelerometer sensor module. The system consists of wireless accelerometer sensor module and base station module. In case of wireless accelerometer sensor module, that consists of accelerometer sensor part, MCU (Micro Controller Unit) part and RF part. In case of base station module, that consists of wireless receiver part and USB interface part. Lastly, we describe the monitoring software in PC.

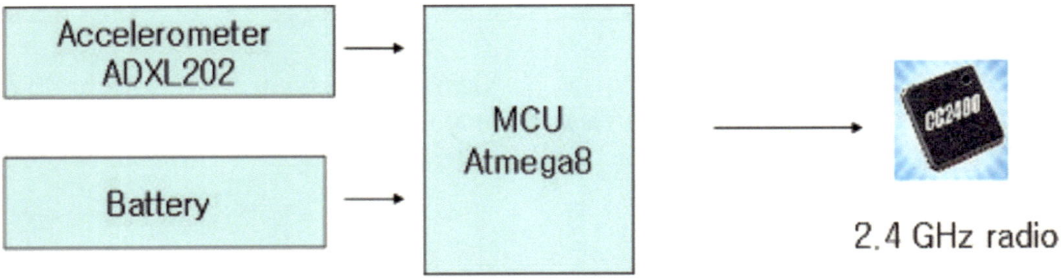

Fig. 1. Block diagram of wireless accelerometer sensor module

2.1 Accelerometer sensor part

We use ADXL 202 (Analog Device, USA), 2-axis accelerometer sensor that measures +/-2g acceleration and the output is PWM type digital signal. The module receive this signal by interrupt and using timer, the pulse width is calculated and sent to receiver by wireless. The receiver sends this data to USB driver and the 'Acceloger' viewer program collects this data and show the graph in display.

Fig. 2. The size comparison of wireless accelerometer sensor module

2.2 MCU module

We use ATmega8 (ATMEL, USA), and SPI port is used for firmware writing and SD card interface. Using embedded ADC, MCU read the output of accelerometer sensor. MCU give the serial clock at wireless module and read the packet data from wireless module. ATmega series have advantage to develop firmware more easily using efficient GCC and Tool-chain.

2.3 RF wireless module

2.4 GHz wireless radio chip has advantage of its chip size and transmission speed. So, it is good for embedded application, but its directivity is high, so if there are some obstacles, the communication doesn't work well. This problem can be solved using wireless sensor network. We use wireless radio chip nRF2401 (nVLSI, Norway). This chip is connected to MCU by 8 pin connector. This chip has double independent transceiver, but we use only one transceiver. Transceiver uses 76 channels from 2.4-2.5GHz frequency band. We set up that the channel can be used by any users. The communication protocol in link layer use Shock Burst embedded in nRF2401 chip. In this mode, 32 byte data can be transmitted with 256 Kbps or 1 Mbps speed. One wireless data packet is 256 bit (32 byte) that consists of 40 bit receiver address, 40 bit sender address, 20 byte data and 2 byte CRC field. Transceiver treats transmission to receiver and CRC check task. Antenna is located in PCB board as pattern type.

2.4 Wireless receiver

Wireless receiver is small dongle type device connected to USB port in PC to deliver the acceleration signal to PC. Wireless receiver has also ATmega microcontroller and nRF2401 radio chip. ATmega microcontroller uses firmware to implement USB packet processor for USB Slave. We develop this using AVR-GCC in window's virtual Linux environment (CygWin). And this has wireless chip control function such as wireless packet validation, wireless packet rearrangement and wireless packet error correction. In case of USB Slave, we implement firmware for relatively simple low speed (1.1Mbps) control transfer. This process is described below.

- When inserted at USB port that is worked as low speed USB mode delivers various descriptors to host and finish the setup process.
- In host's control packet's user function definition, lamp blinking, RF packet read and RF packet write function's service routine is embedded and these 3 routines can be executed using control packet's function number.

2.5 Acceleration signal viewer program

Figure 3 shows the signal when we take the wireless acceleration module in hand and shake. Upper graph is X axis information, lower graph is Y axis information. When the 'Cont' checkbox is pushed, the program received the data continuously. 'LedOn' and 'LedOff' buttons show the receiver's status and used when the receiver's LED is blinking. 'Open' button is used when connecting to device driver. 'GetIO', 'GetRF' and 'RXMODE' buttons are for wireless communication debugging and change the mode of wirless receiver's IO register dump, wireless packet data dump and receiver's wireless transceiver to receiving mode forcibly. Data transmission speed is controlled by changing the firmware.

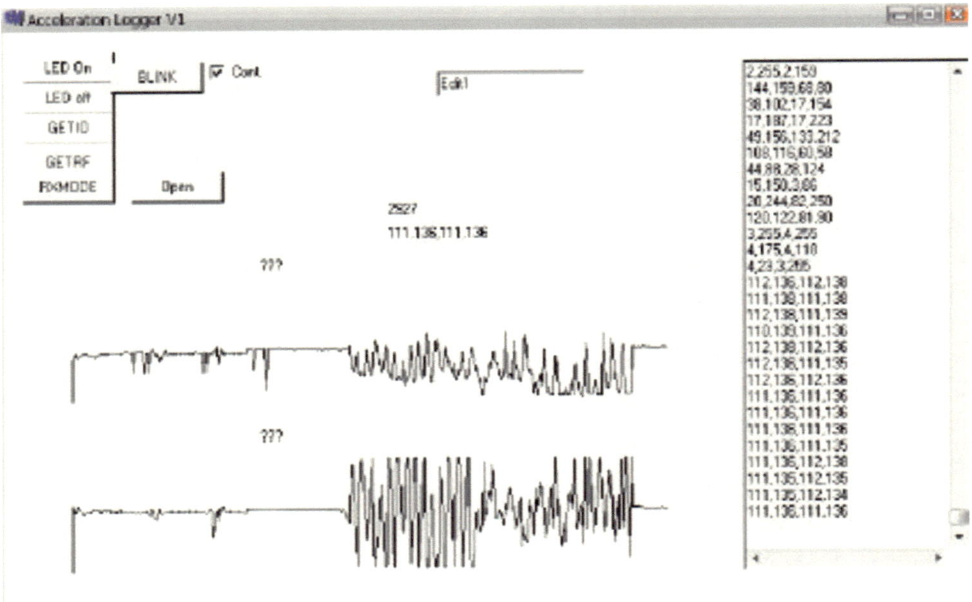

Fig. 3. 'Acceloger' viewer program

3. Implementation of Fall detection monitoring system based on Wireless Sensor Network

Wireless sensor network is currently almost standardized by 'Zigbee', but when there are specific purpose and limited space, it is better to have optimized wireless communication stack in wireless sensor network. In this case, there are max 8 relay-nodes in one base-station. And each relay-node can have max 32 mobile-nodes or fixed-nodes in topology. Every relay-node, fixed-node and mobile-node can be freely configured as master or slave. Fixed-node and mobile-node are not in specific relay-node but connected to voluntary one or many relay-nodes.

Fig. 4. Developed wireless sensor network RF module

3.1 Wireless sensor network design

First, relay-node has a function to repeat retransmitting the received wireless packet infinitely. But when retransmitting, relay-node turn on ID bit in packet's specific item and increase relay-node's counter number by 1. This function definition is the minimum condition for ad-hoc network and self organizing network. Fixed-node makes and transmits the wireless packet by constant time interval or specific event. The packet from fixed-node has logical serial number, relay-node's ID item, relay-node's counter number and sensor value. Mobile-node has mobility and other character is same as fixed-node. Wireless acceleration sensor can be modeled as mobile-node because that is taken by mobile object. The relay-node's situation is very non-deterministic that is typical feature of wireless sensor network. Relay-node is installed in fixed location and each relay-node's location must be considered carefully. Relay-node is basically located within other relay node's visibility range because 2.4 GHz radio wave has strong directivity. By relay-node's antenna sensitivity and transmission power, the distance between relay-node can be different but typically, when 0 dBm (1mW), 10m is the basis. This system uses 1 dBm output. The topology can be serial, star shape, circle or informal, but each relay-node must link to at least one relay-node or base-station.

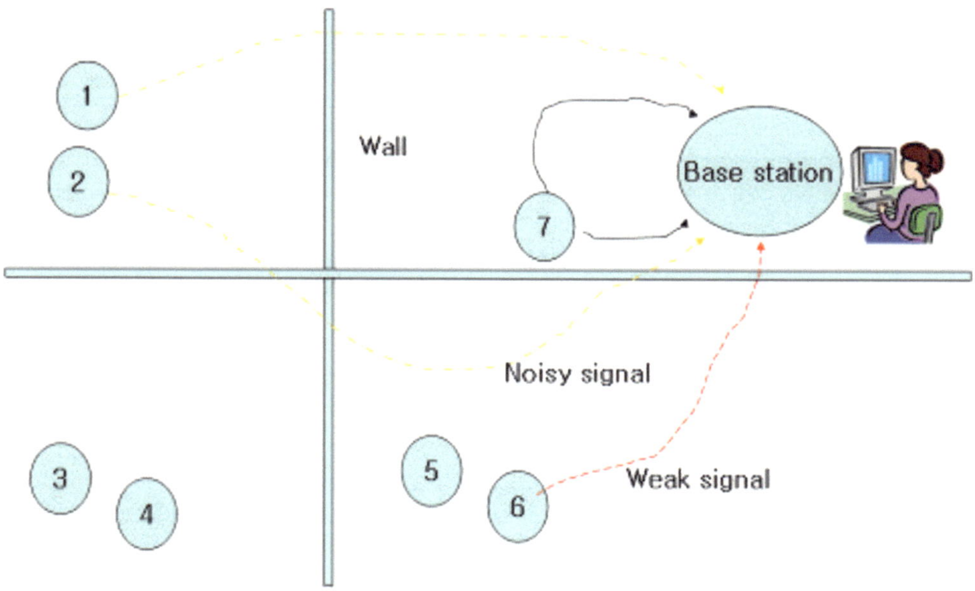

Fig. 5. Simple wireless sensor network without repeater

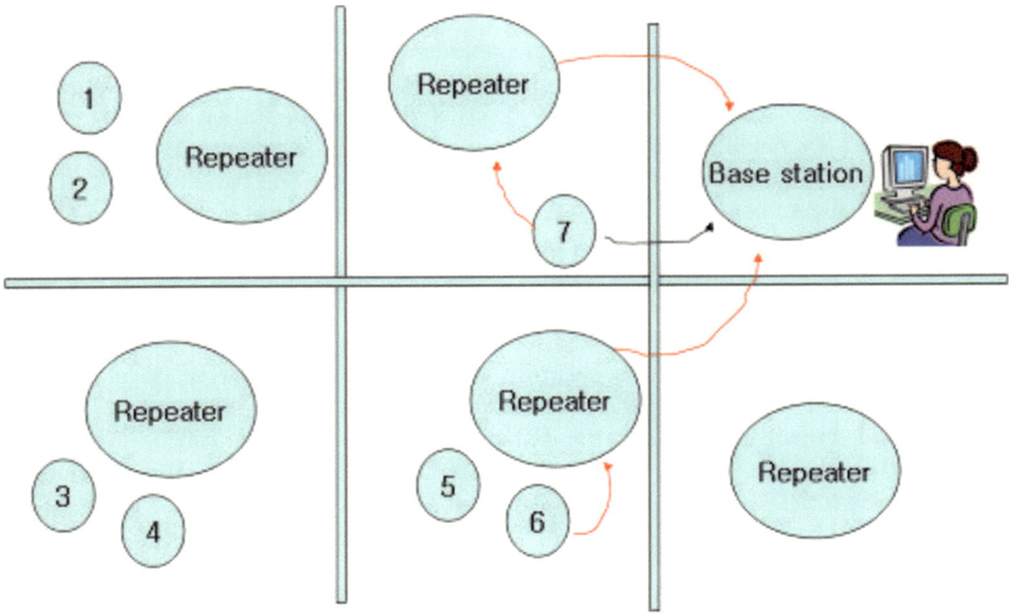

Fig. 6. The example of wireless sensor network construction using repeater

In above figure, the signal from wireless acceleration sensor module can go directly to base-station or go through relay-node. Relay-node inserts the information in wireless packet. Using this method, we install the relay-node in each room and make wireless sensor network. When the RF wave has a problem to go directly to base-station, it goes through relay-node. In this point, wireless sensor network algorithm must solve the complex problem that is infinitely repeatable stray packet detection between relay-node, unnecessary retransmission between relay-node, optimal shortest path finding problem between base-station and specific relay-node in very complex topology. To solve these 3 problems, the system typically becomes very complex. In this system, the design purpose is minimum power consumption, minimum hardware implementation, and optimized algorithm for small sensor network in limited space. So, we don't consider optimal path finding problem and redundant retransmission problem but detect and remove the critical stray packet for network management. Relay-node changes the counter value and prevents the transmitted packet from receiving. In this case, this algorithm doesn't relay any more that stop infinite repetitions.

3.2 Monitoring system development

Figure 7 shows implemented monitoring program based on wireless sensor network. The program reads the plain figure of rooms and we can configure the location of relay-node using mouse pointer. (point A, B, C, D in figure) each wireless station is appeared around relay-node by number character.

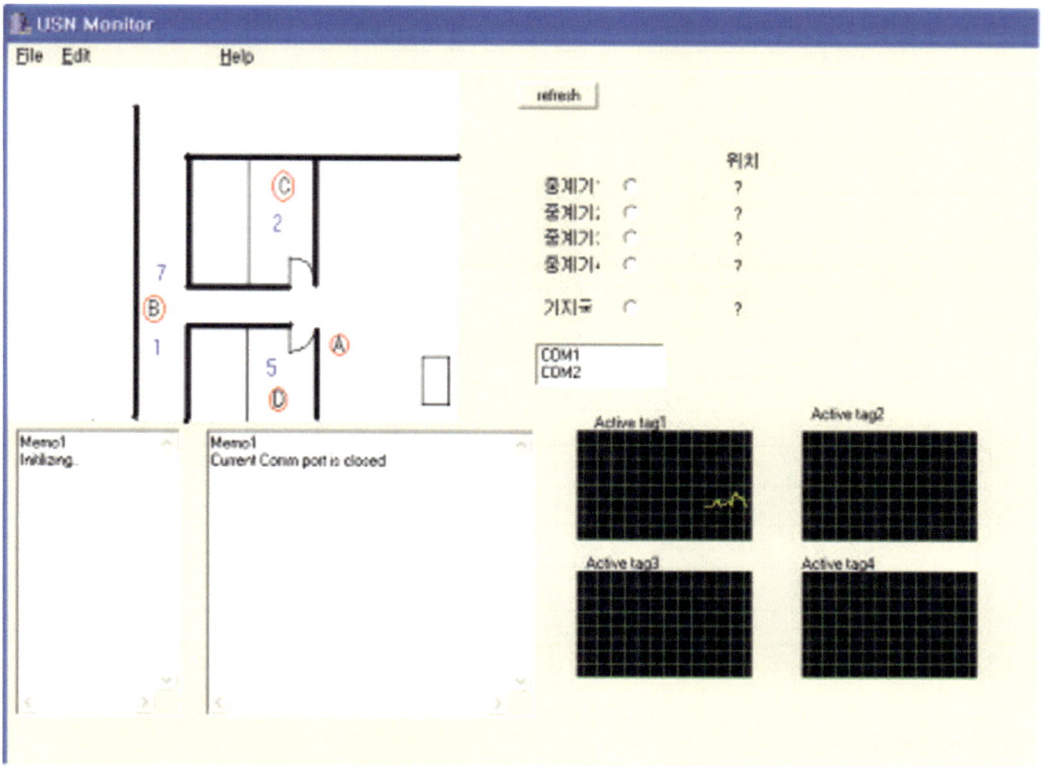

Fig. 7. Implemented monitoring system based on wireless sensor network

4. Experiment and Discussion

The implemented monitoring system based on wireless sensor network is installed in experimental space and as the result of the experiment; we obtain the posture, activity and fall detection rate for subjects. Specially, we assume this system can be application for patient's and elderly fall detection in sanatorium and hospital, and execute the simulation. In this experiment, we classify the fall into forward fall, backward fall, side fall and just sitting and standing. And the detected fall is marked as 'Success' and the undetected fall is 'Fail'. In case of just sitting and standing, when the fall is not detected, that means 'Success'. For 30 subjects, we repeat the above 4 kinds of activity by 3 times in experimental space. For all 360 fall simulation tries, the 337 falls are detected and 23 falls are not detected. The fall detection rate is 93.2%.

Gender, Number	Item	Avg±SD	Min	Max
Male, 20	Age	26.4±3.67	20	32
	Height	175.8±4.20	168	185
	Weight	70.3±6.64	58.1	85

Female, 10	Age	28±3.26	21	31
	Height	161.1±5.30	152	171
	Weight	55.6±6.72	46.2	67.4
Total 30	Age	26.9±3.61	20	32
	Height	170.9±8.31	152	185
	Weight	65.4±9.62	46.2	85

Table 1. General data of 30 subjects

Tries	Fall Detection							
	Forward		Backward		Side		Sit and Stand	
	S	F	S	F	S	F	S	F
1	28	2	27	3	29	1	30	0
2	28	2	28	2	27	3	29	1
3	27	3	27	3	28	2	29	1
Total	83	7	82	8	84	6	88	2
	Success : 337 Fail : 23							

Table 2. Fall detection rate using wireless sensor network monitoring system

5. Conclusion

In this study, using acceleration sensor, we implement wireless acceleration sensor module and algorithm to detect wearer's posture, activity and fall. To assess the performance of algorithm, in specific space, we develop wearer's posture, activity and fall detection monitoring system, and for 30 subjects, the fall simulation experiment is executed for 4 kinds of activity, 3 times and calculate fall detection rate. The result is 337 times detection success and 23 times fail among 360 tries. So, fall detection rate is 93.2%. The developed system can be used for patient or the senior people's activity monitoring and fall detection, also, sports athlete's activity measurement and pattern analysis, normal people's exercise learning and just plaything.

6. References

Henry J. Montoye, Han C. G. Kemper, Wim H. M. Saris, Richard A. Washburn, "Measuring physical activity and energy expenditure", Human Kinetics, pp.72-96, 1996.

Kim L. Coleman, Douglas G. Smith, David A. Boone, Aaron W. Joseph, Michael A. del Aguila, "Step activity monitor: long-term, continuous recording of ambulatory function", Journal of Rehabilitation Research and Development, Vol.36, NO.1, 1999.

F. Foerster, M. Smeja, J. Fahrenberg, "Detection of posture and motion by accelerometry: a validation study in ambulatory monitoring", Computers in Human Behavior, Vol.15, pp.571-583, 1999.

H. G. van Steenis, J. H. M. Tulen, "The effects of physical activities on cardivascular variability in ambulatory situations", Proceedings-19th International Conference-IEEE/EMBS, pp.105-108, 1997.

Carlijn V.C. Bouten, Karel T. M. Koekkoek, Maarten Verduin, Rens Kodde, Jan D. Janssen, "A triaxial accelerometer and portable data processing unit for the assessment of daily physical activity", IEEE Transactions on Biomedical Engineering, Vol.44, pp.136-147, 1997.

K. Aminian, Ph. Robert, E. E. Buchser, B. Rutschmann, D. Hayoz, M. Depairon, "Physical activity monitoring based on accelerometry: validation and comparison with video observation", Medical & Biological Engineering & Computing, Vol.37, pp.304-308, 1999.

B. Najafi, K. Aminian, F. Loew, Y. Blanc, "An ambulatory system for physical activity monitoring in elderly", IEEE-EMBS Special Topic Conference on Microtechnologies in Medicine & Biology, pp.562-566, 2000.

S. H. Lee, H. D. Park, H. R. Yoon, K. J. Lee, "Design of a Portable Activity Monitoring System", The Korean Institute of Electrical Engineer, Vol.51, pp.32-38, 2002

Bijan Najafi, "Ambulatory System for Human Motion Analysis Using a Kinematic Sensor: Monitoring of Daily Physical Activity in the Elderly", IEEE Trans. on biomedical Engineering, Vol. 50, No.6, June 2003

G.Williams, "A Smart Fall & Activity Monitor for Telecare Application", Proceeding of the 20th Annual International Conference of the IEEE Engineering n Medicine and Biology Society, 1998

J.Y.Hwang, "Development of Novel Algorithm and Real-time Monitoring Ambulatory System Using Bluetooth Module for Fall Detection in the Elderly", Proceedings of the 26th Annual International Conference of the IEEE EMBS San Francisco, CA, USA, September 1-5, 2004

14

Realizing a CMOS RF Transceiver for Wireless Sensor Networks

Hae-Moon Seo[1, 2]
[1]Korea Electronics Technology Institute (KETI), [2]Kyungpook National University (KNU)
South Korea

1. Introduction

The choice of the CMOS radio frequency (RF) transceiver architecture affects the design of the whole system and is thus a fundamental one. In order to make a good choice, several factors have to be considered, the most important ones being: performance, power consumption, die size, cost, integration level, and time-to-market. The minimum required performance is dictated by the IEEE802.15.4 standard approval. The relative weight of all other factors is determined by the wireless sensor network application at hand. As the RF transceiver developed here targets very small devices such as information-gathering nodes for sensor network applications, a small size and low power consumption are key requirements. In particular, as power consumption sets dimensions and type of the battery, it also has a major impact on size, weight, and cost of the system.

In this chapter, we explore the implementation and testing of a fully CMOS integrated RF transceiver for wireless sensor networks in sub-GHz ISM-band applications. A comprehensive description of the radio system architecture, RF transceiver circuits, and measurement results is described in this sub-chapter. At the end of this chapter, a fully CMOS RF transceiver chip is presented to give an impression of the possible die size and floor plan for a highly integrated transceiver chip.

1.1 Introduction of Wireless Sensor Networks

Recently, the desire for wireless connectivity has led an exponential growth in wireless communication. In particular, wireless sensor networks are potential wireless network applications for the following future ubiquitous computing system. Ubiquitous sensor networks are an emerging research area with potential applications in environmental monitoring, surveillance, military, health, and security (Y. K. Park et al. 2005), . The power dissipation of wireless sensor networks does require low power consumption for several years' operation. There has been a great deal of interest in realizing low power, low cost, compact RF transceiver IC for wireless sensor networks. Several technological trends that are driving the technical evolution of wireless technology include the process scaling of CMOS transistors and higher bandwidth available at ISM bands. Almost all of the license free bands propose both linear and nonlinear modulation standards for wireless applications, and thus requiring different design optimizations in the RF transceiver. Along with these issues, there exists the challenge to develop fully integrated wireless solutions in silicon-based substrates (S. Sarkar et al., 2003).

2. The Radio System Architecture for Wireless Sensor Networks

Conventional transceiver architectures as shown in Fig.1 include heterodyne, zero-IF (intermediate-frequency), and low-IF conversion structure (P. S. Choi et al. (2003), C. Cojocaru et al. (2003), M. Valla et al. (2005), Ilku-Nam et al. (2003)), each having their own advantages and disadvantages. However, it becomes further challenging to meet all the specifications of many applications while keeping more competitiveness than the others.

The super-heterodyne architecture is without any doubt the most often used transceiver topology and it has been in use for a long time already and its way of operating is very well known. It is the most widely used architecture, mainly because of its high performance. However, it usually requires image-reject and external channel selection filters and is therefore not well suited for fully integrated systems. Also, it has difficulties in the multi-band/mode transceiver and has problems of high power consumption, high cost.

The low-IF and zero-IF architectures can achieve much better performance at low-power consumption and are well suited for high integration.

The concept of the low-IF (P. S. Choi et al. (2003), C. Cojocaru et al. (2003)) starts from the survey that all information necessary to separate the mirror frequency from the wanted frequency is available in the two low frequencies after quadrature conversion. This scheme can avoid the DC offset problem and eliminate IF SAW and image RF filters. However, it suffers from impairments such as I/Q mismatching, even-order nonlinearity, and local oscillator (LO) pulling/leakage. Some calibration techniques for stringent image rejection may be used at the expense of additional complexity and power consumption.

Finally, zero-IF (M. Valla et al. (2005), Ilku-Nam et al. (2003)) architecture performs a direct down-conversion of the wanted frequency to the baseband. The consequence is that the mirror signal is equal to the wanted frequency. This does however not mean that there would not be a mirror signal problem in the zero-IF receiver. But, this architecture remains the most suitable solution for high integration, low power consumption, and low cost. Moreover, it has advantage in elimination of image rejection requirements. However, it may suffer from impairments of DC offset, flicker noise, and complication of LO frequency-planning to evade LO pulling/leakage.

The communication nodes for ubiquitous networks are required to be integrated in one die for low power consumption and low cost wireless sensor network applications. The overall wireless personal area networks (WPAN) system architecture is shown in Fig.2. It consists of the RF transceiver and a companion digital baseband (BB) processor, which implements both physical (PHY) and medium access control (MAC) layers of the IEEE 802.15.14 standard (IEEE Computer Society (2003)). Fig.2 shows the architecture of a radio chip, which consists of a receiver, a transmitter and a frequency synthesizer with on-chip voltage-controlled oscillator (VCO). Note that RF transceiver chip includes a 6-bit digital-to-analog converter (DAC) and 4-bit I/Q analog-to-digital converters (ADCs). The receiver adopts zero-IF architecture to have low power consumption, low cost and small size (M. Valla et al. (2005), Ilku- Nam et al. (2003), Kwang-Jin Koh et al. (2004), M. Zargari et al. (2004), S. F. R. Chang et al. (2005), W. Hioe et al. (2004)). The RF front-end circuits of receiver are shown in Fig.3. The sub-GHz RF signal is first amplified by a low noise amplifier (LNA)

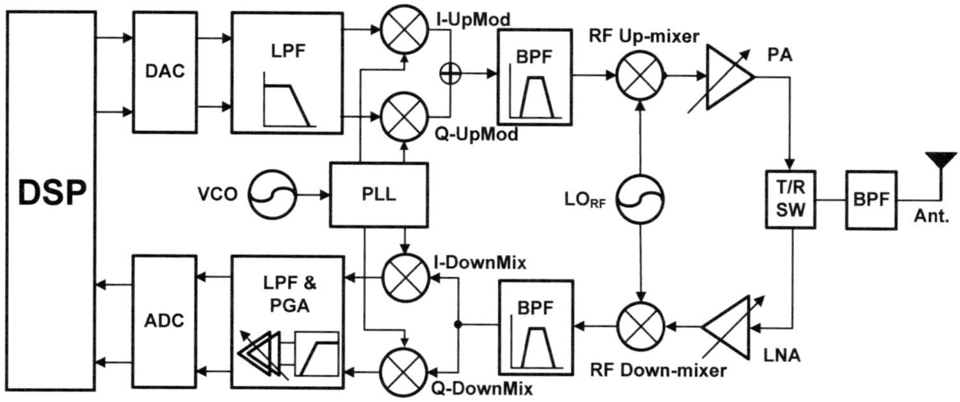

(a) Super-heterodyne architectre

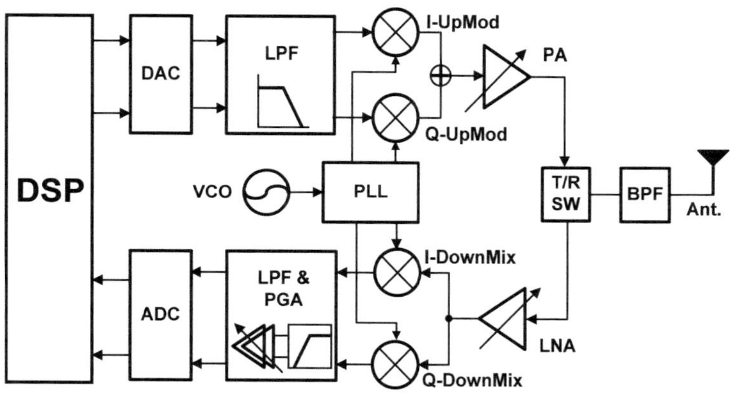

(b) Zero-IF architectre

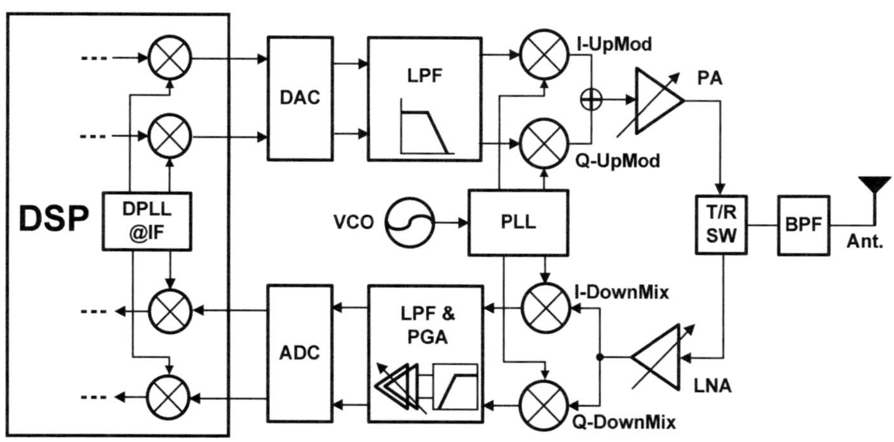

(a) Low-IF architectre

Fig. 1. Transceiver architectures

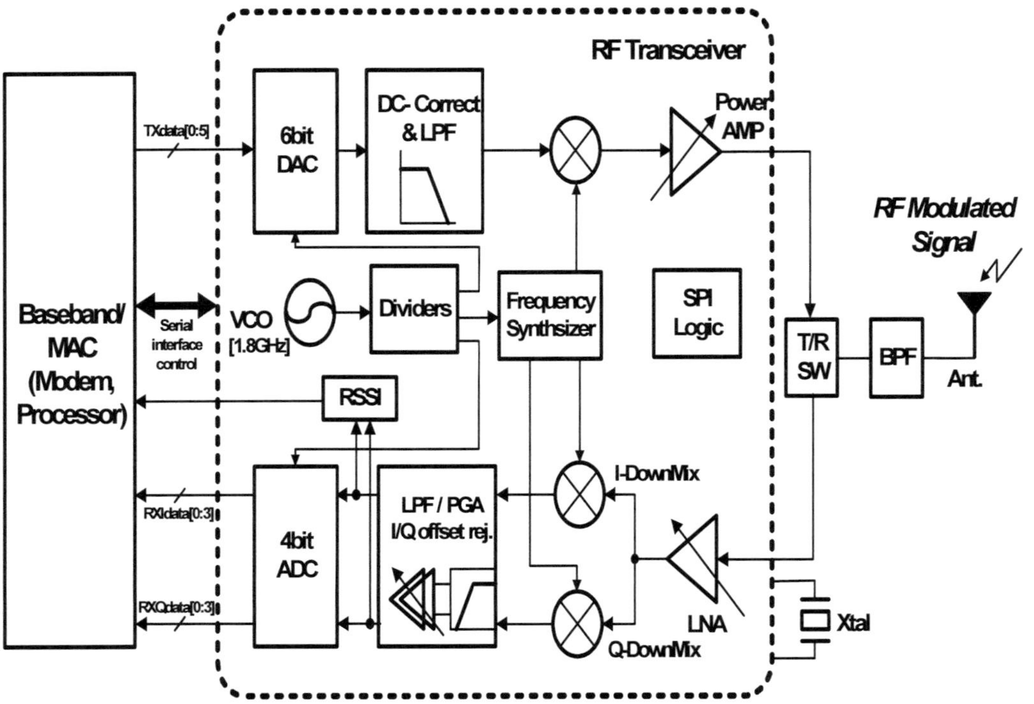

Fig. 2. Overall system architecture supporting wireless sensor networks in sub-GHz ISM-band: RF transceiver & Baseband Processor

and then down-converted to zero-IF I/Q signals by two identical mixers driven by quadrature LO signals from a frequency synthesizer. At the analog baseband stage, using a third-order RC filter and programmable gain amplifier simultaneously performs channel selection filtering, signal amplification, and dc-offset cancellation. In addition, I/Q 4-bit dual flash-ADCs are connected to interface of MODEM block. The transmitter adopts a zero-IF modulation with up-conversion mixer using a current mixing scheme. Baseband BPSK signals generated by digital modulator in MODEM block are followed by a 6-bit DAC. A mixer does directly up-convert the baseband signals directly 900-MHz, which is combined by RC low-pass filter. Since BPSK modulation is a constant envelop modulation, a nonlinear power amplifier with high efficiency can be used for high power emission. For generating 900-MHz LO signals with 2-MHz channel spacing, an integer-N frequency synthesizer derived from a 30-MHz crystal oscillator with 30-ppm accuracy is implemented. A 1.8GHz LO signals are generated by a VCO with a small area and high Q (quality factor) on-chip inductor. A divide-by-two circuit then produces the 900-MHz LO I/Q signals for frequency mixing of TX and RX mode. The frequency synthesizer is implemented in fully differential type, for immunity to common mode noise.
In consideration of RF transceiver IC implementation for WSN applications, the low power consumption is a key issue. To achieve this, adequate trade-offs are required for system power consumption, chip area, gain, noise figure, and linearity. Since the radio will operate with a very low-duty cycle for WSN applications, the sleep mode current and battery leakage current can be reduced with the optimization of current sources. Also, the use of

small devices with a small active area, regardless of system IC performance degradation, can be applied for the reduction of sleep mode current. The power dissipation in driving pad and trace parasitic capacitances for off-chip inductors is removed with an on-

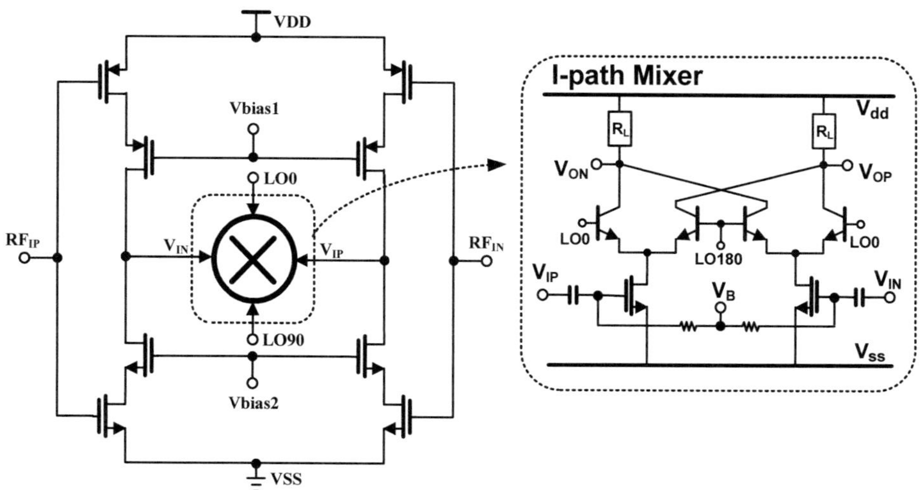

Fig. 3. RF front-end circuits of receiver: low-noise amplifier & I/Q down-conversion mixer

chip inductor. Since the transmit power is very low (max. -3 dBm) as compared with other standards, the transmit RF front-end can be implemented with low power consumption using a simpler current mixing scheme and resistive load.

3. RF Transceiver Circuit Implementation

RF transceiver chip is designed using 0.18-μm mixed-signal CMOS process including six metal layers with 2-μm thick top metal. This process provides high gain and good quality factor Q (8) for on-chip inductor, resulting in low power consumption in RF and analog circuits.

3.1 Receiver

The RF front-end (RFE) of a realized WPAN receiver chain consists of low-noise amplifier (LNA), quardrature down-conversion mixer. The fully balanced sub-GHz LNA shown in Fig. 2 uses current-reuse complementary technique (pMOS and nMOS) without inductor requiring large area. Input matching is realized by external passive LC components. The LNA features 2.6 dB noise figure (*NF*) and a third-order input intercept point (IIP_3) of 5.2 dBm at maximum gain. The differential outputs of LNA are down-converted directly into a common analog baseband path by a Gilbert-cell-based quadrature frequency demodulator. The selection of the vertical bipolar transistors in the switching quadrant decrease the gain of mixer, however, the average integrated noise floor in the direct-conversion receiver improves due to the reduced *1/f* noise (Ilku-Nam et al. (2003)). The large voltage headroom achieved by Gilbert multiplier type with source grounded topology helps maximize the contribution of linearity in the overall IIP_3. The estimated IIP_3 is 6 dBm.

The analog front-end (AFE) of a realized WPAN receiver consists of continuous-time low pass filters, highly linear programmable gain amplifier (PGA), filter tuning circuit, and DC

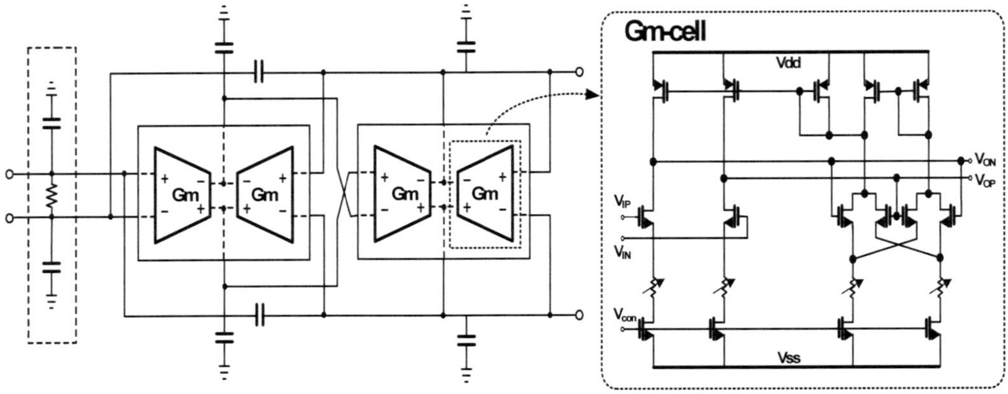

Fig. 4. Analog baseband circuits of receiver I: the channel selection filter with third-order Butterworth LPF using proposed transconductance cells (Gm-cell)

offset cancellation block. The third order Butterworth filter was implemented cascading a biquad cell and a single pole cell, and the programmable gain cell was stationed at the middle to improve the cascaded dynamic range. The AFE design is concentrated on optimizing the dynamic range and keeping the required die area small and low power consumption. The baseband noise is dominated by the thermal noise of the PMOS current sources at the quadrature mixer outputs. The flicker noise is not a significant problem at baseband since all transistors are designed with a long channel length for better matching. Moreover, the output of the DAC is DC blocked using a baseband modem control signal to minimize the effect of the internal DC offsets from limiting the dynamic range of the receiver.

The channel filter allows a signal of the desired band to pass and attenuates the adjacent channel and the alternate channel. The filter requirement in this chapter, is as follows. Since it is a direct-conversion receiver (DCR) structure, *1/f* noise should be reduced and the DC offset should be small. In addition, in order to alleviate the SFDR requirements of the PGA and the ADC, most of the interference is filtered in the first part (J. Silava-Martinez et al. (1992), Y. Palaskas e al. (2004)). Figure 4 shows the designed third order Butterworth LPF. Using the single pole of the passive RC at the output stage of the mixer reduces the interference that can affect the dynamic range at the baseband input stage, and using the overshoot of biquad compensates the in-band loss. Figure 4 shows the proposed Gm-cell with degeneration resistor. Two Gm-cells are used as one to reduce the area that LPF occupies. The lumped resistor and the size of MOS should be properly adjusted to improve the linearity of the Gm-cell.

The signal level of the RF input requires a minimum dynamic range of 78 dB, namely from -98 dBm to -20 dBm. The automatc gain-control (AGC) control signal receives the digital control signal from the baseband modem to control the gain of the receiver. The PGA of this receiver utilizes the three gain stages to control the gain of 0 ~ 65 dB with a 1-dB step. The resistor switching method was utilized in order not to lose the linearity of PGA. I/Q 4bit

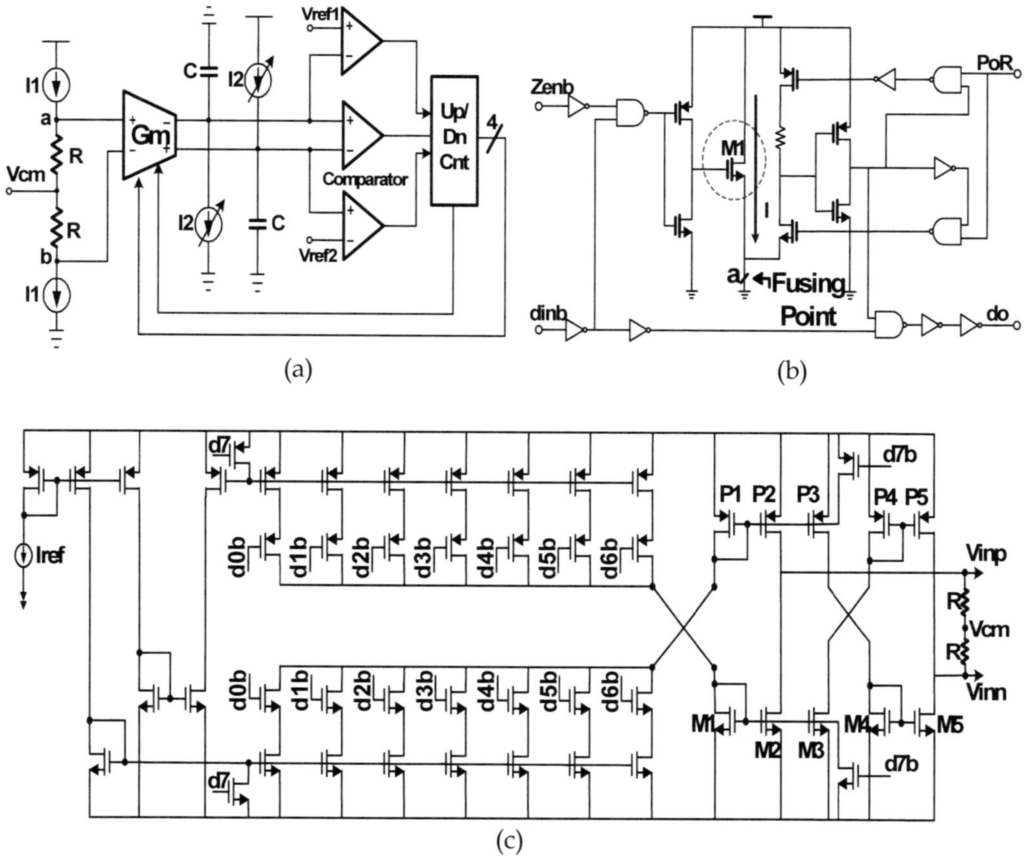

Fig. 5. Analog baseband circuits of receiver II: (a) The tuning circuit for channel selection filter, (b) The circuit of a fusing cell for filter-tuning, (c) DAC schematic for DC offset adjustment

dual flash-ADCs are designed for interface of baseband modem block. The simulated maximum DC current consumption of an overall receiver path is 6 mA.

Figure 5 shows the automatic-tuning circuit, which is based on indirect tuning method. Since the characteristics of the Gm-C filter are determined by the transconductance value, the *gm* has to be controlled to keep a fixed pole frequency. The *gm* value should not be changed even by process variations or outer environment changes. As shown in Fig. 5(a), it is important to keep a *gm* value and a ratio of *gm* output current to *gm* input voltage equal. And the required current for sinking or sourcing is designed to minimize changes of *gm* by reducing current change due to the temperature variation from bias block. The current *I1* in Fig. 5(a) offsets the MOS of the bias part as well as the temperature variation of resistance so as to minimize the changes of voltage *Vab* due to the temperature and to evenly maintain the input voltage of the *gm*-cell. The converging time of tuning circuit is designed to less than 100 msec. If the cut-off frequency differs from the designed value, as a parameter set up the first time it distorts the value of *gm* by the process variations, *gm* should be adjusted by changing current *I2* by fusing. Fusing is controlled by serial port

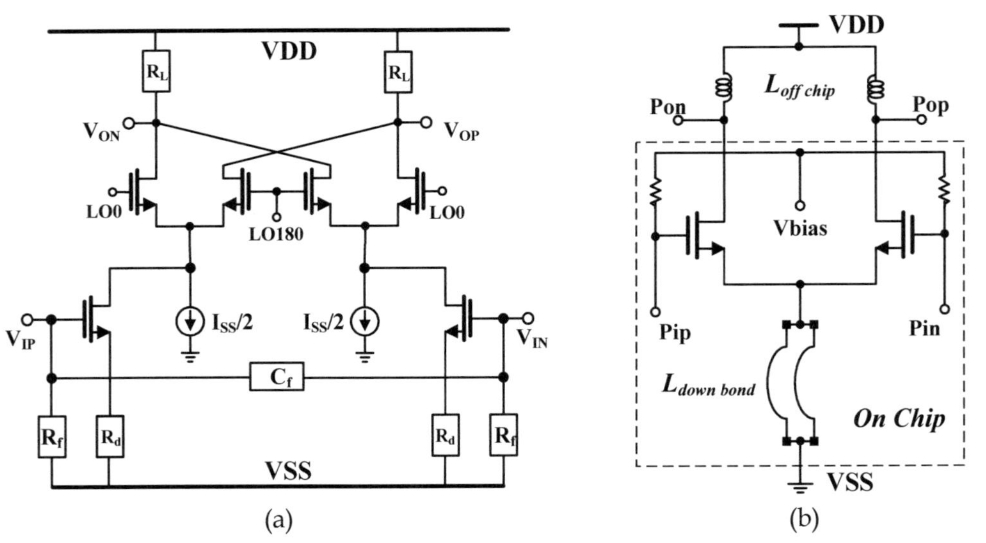

Fig. 6. Transmitter circuits: (a) Up-conversion I/Q-modulator using current-mixing scheme (b) Drive-amplifier with off-chip inductor

interface (SPI), and there is no change in value once it is put in. Figure 5(b) represents the circuit diagram of fusing cell. The fusing cell is a circuit which amplifies the voltage, which is set in ratio of PMOS channel resistance to NMOS channel resistance within the range of power on reset ('Low' *PoR* signal) at power-on. To inverting amplifier, the signal is latched and displays the latched value without change while normal operation ('High' *PoR* signal). The '*Zenb*' is a signal of 'fusing enable', '*dinb*' is a 'data input signal' controllable via SPI. The '*PoR*' is a signal for 'enable' at the mode of 'power on reset', while '*do*' is an output signal of fusing cell. Once the fusing signal turns to 'enable', the output signal of fusing cell is fixed regardless of the data input signal. The current capacity of *M1* should have more than 1 mA in order to disconnect the node of a fusing point at transmitting the fusing enable signal.

For DC offset adjustment, it is important for the cancellation of DC-offsets generated at the back side of PGA1 and to use the feedback loop to reduce the offset at the LPF output. Figure 5(c) shows the DAC to convert the 8-bit data into the input voltage of the PGA. The resolution for 1 bit is 5 mV, and the DC offset change at the LPF output is ±640 mV. The size of MOS (*P1~P5, M1~M5*) used, as a current mirror of the DAC circuit has to be appropriate in consideration of the current mismatch. The aspect ratio of the MOS is used by 20µm/2µm.

3.2 Transmitter

In the transmitter path, the BPSK modulated baseband signal is converted from digital to analog before being applied to frequency up-translation block. Fig.6 (a) shows the schematic of up-conversion mixer with RC low-pass filter. The baseband analog signal is filtered by second RC low-pass filter, and then is translated into RF frequency by up-conversion

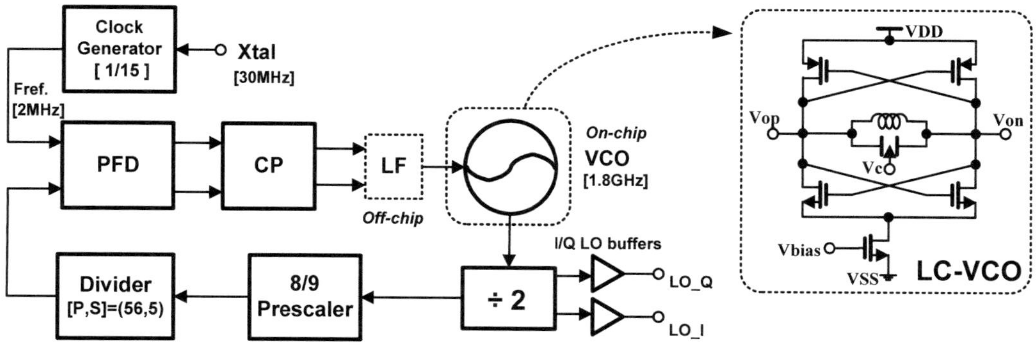

modulator with balanced Gilbert-cell using current-mixing scheme. The major advantage of current mixing relaxes a requirement of heavy linearity of modulator inputs from high

Fig. 7. Frequency synthesizer block-diagram with LC voltage-controlled oscillator

voltage-driving DAC output signal. In addition, this scheme for frequency-up modulation can produce satisfactory results for high modulation quality, low-power consumption, and good linearity. This balanced mixer converts baseband signal directly up to 900 MHz and deliver -20 dBm differential signal to power amplifier. LO emission is due to differential mismatch in the modulator circuit, while spectrum re-growth is due to LO (0/90-degree) quardrature imbalance and nonlinearity of the Gilbert-cell. Layout is fulfilled very carefully to maintain symmetry for differential and quardrature signals, which minimizes both LO emission and spectrum re-growth. Fig.6 (b) shows the driver amplifier of a differential common source topology with off-chip inductor having a high Q. The multiple down-bond wire inductors are applied for the minimization of spectrum re-growth. The simulated DC current consumption of an overall transmitter path is 7 mA.

3.3 Frequency Synthesizer

The integer-N frequency synthesizer, using a second-order passive loop filter, generates the LO signal for transmit/receive mode. A crystal reference of 30 MHz is internally divided. To minimize pulling, the 900-MHz LO signals are generated by 1.8 GHz voltage controlled oscillator (VCO), shown in Fig.7. The LC-resonator consists of four-turn spiral inductor and varactor. The negative-Gm core cell has nMOS/pMOS complementary topology for high power efficiency and gain.

$$f_{OSC} = \frac{1}{2\pi\sqrt{LC_{eff}}} \tag{1}$$

The oscillation frequency of VCO is shown as equation (1). The tuning frequency of VCO is simulated from 1.6 GHz to 2.2 GHz. The divider circuit for high frequency has a structure of negative-feedback type using two latches. The phase frequency detector (PFD) consists of two D-flip-flop (DFF), AND-gate, and delay-time block for locking speed and high linearity of phase transfer function. The charge-pump circuit has a structure of nMOS/pMOS cascade-type to minimize of up/down current mismatch and output switching noise. The clock generation block provides a reference clock of PLL and sampling-clocks of ADC/DAC

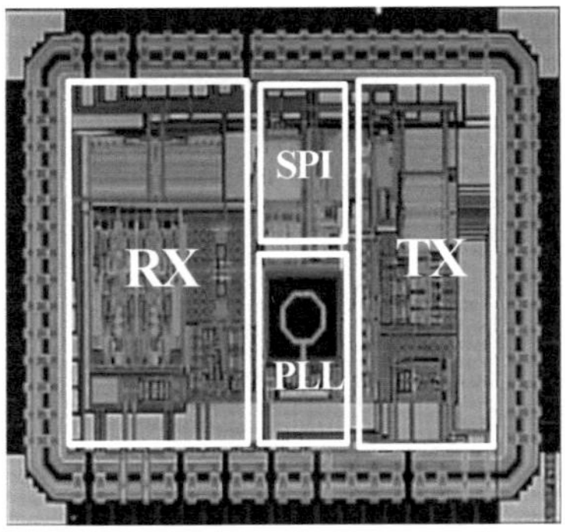

Fig. 8. Die microphotograph

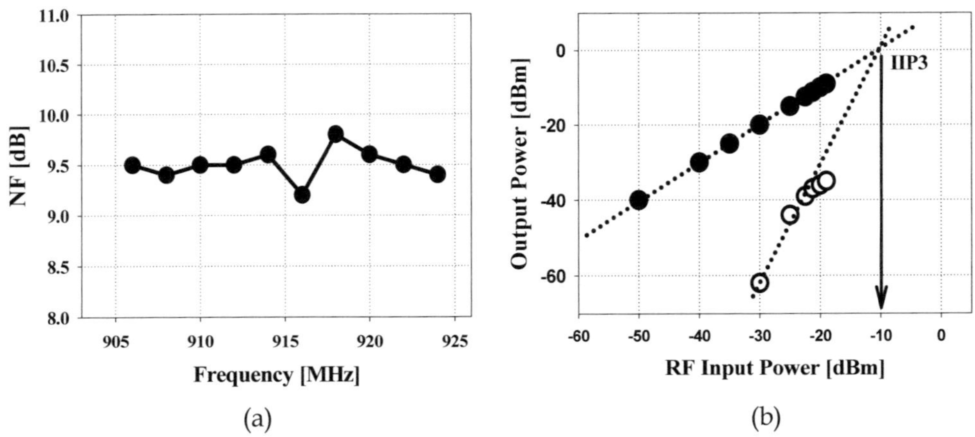

using an external 30-MHz crystal-oscillator. The simulated DC current consumption of an overall frequency synthesizer path is 8 mA.

Fig. 9. Measured results: (a) cascaded noise figure (NF), (b) cascaded IIP3 of overall receiver

4. Measured Results

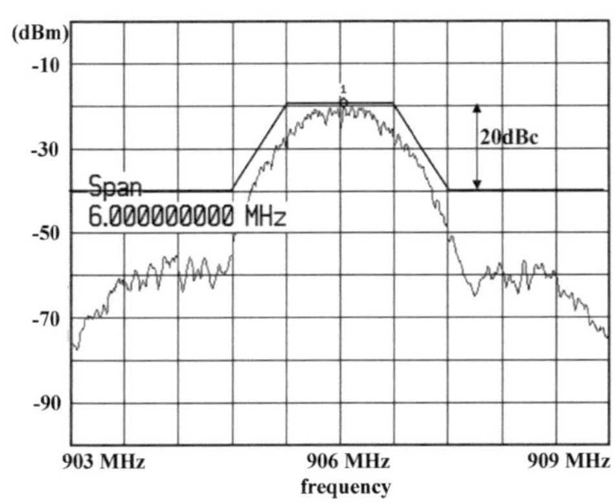

Fig. 10. Measured result of spectrum mask of transmitter

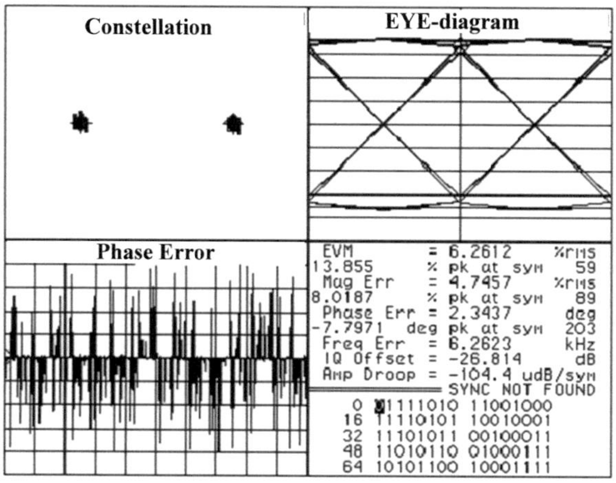

Fig. 11. Measured result of vector signal analysis of transmitter

A radio transceiver die microphotograph, which consists of transmitter, receiver, and frequency synthesizer with on-chip VCO, is shown in Fig. 8. The total die area is 1.8 × 2.2-mm^2 and it consumes only 29 mW in the transmit-mode, 25-mW in the receive-mode and a LPCC48 package is used. The overall receiver features a cascaded-*NF* of 9.5 dB for 900 MHz band as shown in Fig. 9(a). Overall receive cascaded- IIP_3 as shown in Fig. 9(b) is -10 dBm and the maximum gain of receiver is 88dB. The automatic gain control (AGC) of receiver is 86dB with 1dB step and selectivity is -48 dBc at 5 MHz offset frequency. The 40 kHz baseband single signal is up-converted by 906 MHz RF carrier signal and wanted-signals are 25dB higher than third-order harmonics. The spectrum density at the output of transmitter satisfies the required spectrum mask as shown in Fig. 10, which is above 28 dBc at the ±1.2-MHz offset frequency. Due to the low in-band integrated phase noise and the digital calibration that eliminates I/Q mismatch and baseband filter mismatch, transmitter EVM is dominated by nonlinearities (Behzad Razzavi (1997), I. Vassiliou et al. (2003), K. Vavelidis et al. (2004)). As shown in Fig. 11, a reference design achieves 6.3 % EVM for an output power

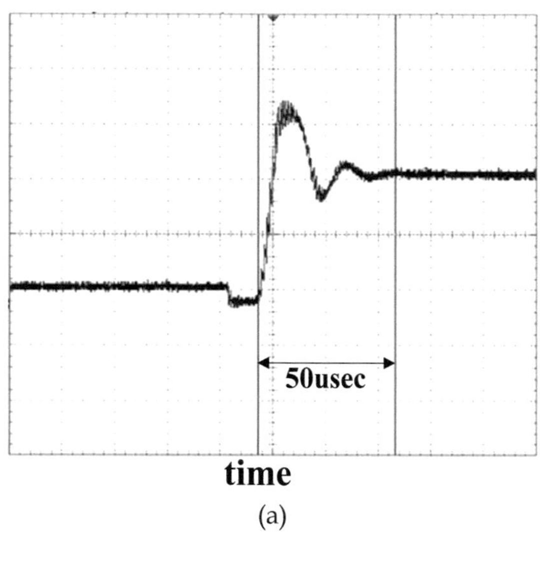

(a)

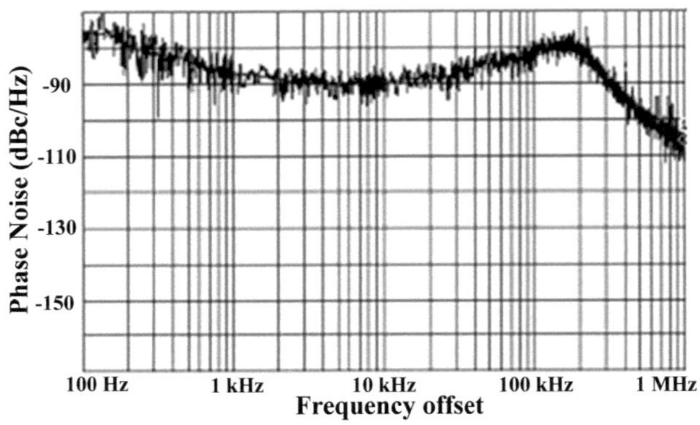

(b)

Fig. 12. Measured result of phase lock loop (PLL): (a) settling time, (b) phase noise

of -3dBm for sub-GHz ISM-band. Measured results of settling time and phase-noise plot of phase locked loop (PLL) are shown in Fig. 12. Table 1 summarizes the UHF RF transceiver's characteristics. The specifications of two RF transceivers (Walter Schucher et al. (2001)) and (Hiroshi Komurasaki et al. (2003)) for UHF applications are also shown for comparison in this table. The RX current is not the lowest; however, the power dissipation in RX mode is the smallest because of the 1.8 V supply voltage. Although the TX output power and RX IIP_3 are a little worse due to the antenna switch and the matching network, this work has great advantages.

Specification	This work	Walter Schucher et al. (2001)	Hiroshi Komurasaki et al. (2003)
VDD	1.8V	2.8V	1.8V
Current consum.	Rx./Tx.:14/16mA	Rx./Tx.: 11/20mA	Rx./Tx.: 34/26mA
Die size	3.96 mm^2		10 mm^2
NF	9.5dB	11.8dB	-76dBm
IIP_3	-10dBm	-23.2dBm	+3dBm
Max. Gain	88dB	-	-
AGC gain range	86	-	-
Selectivity	-48dBc (@5MHz)	-	-21dBc (@4MHz)
TX power	+0dBm	+10dBm	+0dBm
EVM	6.3%	-	-
OP1-dB	+1dBm	-	-
LO PN. (@1MHz)	-108dBc	-115dBc	-

Table 1. The Measured Results of UHF Transceivers

5. Conclusion

A low power fully CMOS integrated RF transceiver chip for wireless sensor networks in sub-GHz ISM-band applications is implemented and measured. The IC is fabricated in 0.18-µm mixed-signal CMOS process and packaged in LPCC package. The fully monolithic transceiver consists of a receiver, a transmitter and a RF synthesizer with on-chip VCO. The overall receiver cascaded noise-figure, and cascade IIP_3 are 9.5 dB, and -10 dBm, respectively. The overall transmitter achieves less than 6.3 % error vector magnitude (EVM) for 40kbps mode. The chip uses 1.8V power supply and the current consumption is 25 mW for reception mode and 29 mW for transmission mode. This chip fully supports the IEEE 802.15.4 WPAN standard in sub-GHz mode.

6. References

Behzad Razavi (1997). Design Considerations for Direct-Conversion, IEEE Transactions on circuit and systems-II, 14, 251-260, June.

C. Cojocaru, T. Pamir, F. Balteanu, A. Namdar, D. Payer, I. Gheorghe, T. Lipan, K. Sheikh, J. Pingot, H. Paananen, M. Littow, M. Cloutier, and E. MacRobbie (2003). A 43mW Bluetooth transceiver with -91dBm sensitivity, ISSCC Dig. Tech. Papers, 90-91.

Hiroshi Komurasaki, Tomohiro Sano, Tetsuya Heima, Kazuya Yamamoto, Hideyuki Wakada, Ikuo Yasui, Masayoshi Ono, Takahiro Miki, and Naoyuki Kato (2003). A 1.8 V Operation RF CMOS Transceiver for 2.4 GHz Band GFSK Applications, IEEE Journal of Solid-State Circuit, 38, May.

IEEE Computer Society (2003). IEEE Standard for Part 15.4: Wireless Medium Access Control (MAC) and Physical Layer (PHY) specifications for Low Rate Wireless Personal Area Networks (LR-WPANs), IEEE Standard 802.15.4TM.

Ilku Nam, Young Jin Kim, and Kwyro Lee (2003). Low 1/f Noise and DC offset RF mixer for direct conversion receiver using parasitic vertical NPN bipolar transistor in deep N-well CMOS Technology, IEEE symposium on VLSI circuits digest of technical.

I. Vassiliou, K. Vavelidis, T. Georgantas, S. Plevridis, N. Haralabidis, G. Kamoulakos, C. Kapnistis, S. Kavadias, Y. Kokolakis, P. Merakos, J.C. Rudell, A. Yamanaka, S. Bouras, and I. Bouras (2003). A single-chip digitally calibrated 5.15 GHz-5.825 GHz 0.18 μm CMOS transceiver for 802.11a wireless LAN, IEEE J. Solid-State Circuits, 38, 2221–2231, December.

J. Bouras, S. Bouras, T. Georgantas, N. Haralabidis, G. Kamoulakos, C. Kapnistis, S. Kavadias, Y. Kokolakis, P. Merakos, J. Rudell, S. Plevridis, I. Vassiliou, K. Vavelidis, and A. Yamanaka (2003). A digitally calibrated 5.15– 5.825 GHz transceiver for 802.11a wireless LANS in 0.18 μm CMOS, IEEE Int. Solid-State Conf. Dig.Tech. Papers, February.

J. Silva-Martinez, M.S.J. Steyaert, and W. Sansen (1992). A 10.7 MHz, 68 dB SNR CMOS Continuous-Time Filter with On-Chip Automatic Tunig, IEEE J. Solid-State Circuits, 27, 1843-1853, December.

Kwang-Jin Koh, Mun-Yang Park, Cheon-Soo Kim, and Hyun-Kyu Yu (2004). Subharmonically Pumped CMOS Frequency Conversion (Up and Down) Circuits For 2 GHz WCDMA Direct-Conversion Transceiver, IEEE J. Solid-State Circuits, 39, 871-884, June.

K. Vavelidis, I. Vassiliou, T. Georgantas, A. Yamanaka, S. Kavadias, G. Kamoulakos, C. Kapnistis, Y. Kokolakis, A. Kyranas, P. Merakos, I. Bouras, S. Bouras, S. Plevridis, and N. Haralabidis (2004). A dual- band 5.15-5.35 GHz, 2.4-2.5 GHz 0.18 μm CMOS Transceiver for 802.11a/b/g wireless LAN, IEEE J. Solid-State Circuits, 39, 1180-1185, July.

M. Zargari, M. Terrovitis, S.H.M. Jen, B.J. Kaczynski, MeeLan Lee, M.P. Mack, S.S. Mehta, S. Mendis, K. Onodera, H. Samavati, W.W. Si, K. Singh, A. Tabatabaei, D. Weber, D.K. Su, and B.A. Wooley (2004). A Single-Chip Dual-Band Tri-Mode CMOS Transceiver for IEEE 802.11a/b/g Wireless LAN", IEEE J. Solid-State Circuits, 39, 2239-2249, December.

M. Valla, G. Montagna, R. Castello, R. Tonietto, and I. Bietti (2005). A 72 mW CMOS 802.11a Direct Conversion Front-End with 3.5 dB NF and 200 kHz 1/f Noise Corner, IEEE J. Solid-State Circuits, 40, 970-977, April.

Pengfei Zhang, T. Nguyen, C. Lam, D. Gambetta, T. Soorapanth, Baohong Cheng, S. Hart, I. Sever, T. Bourdi, A. Tham, and B. Razavi (2003). "A 5 GHz Direct-Conversion CMOS Transceiver" IEEE Journal of Solid-State Circuit, 38, December.

P. S. Choi, H. C. Park, S. Y. Kim, S. C. Park, I. K. Nam, T. W. Kim, S. J. Park, S. H. Shin, M. S. Kim, K. C. Kang, Y. W. Ku; H. J. Choi, S. M. Park, and K. R. Lee (2003). "An Experimental Coin-Sized Radio for Extremely Low-Power WPAN Application at 2.4GHz," IEEE J. Solid-State Circuits, 12, 2258-2268, December.

S.F.R. Chang, Wen-Lin Chen, Shuen-Chien Chang, Chi-Kang Tu, Chang-Lin Wei, Chih-Hung Chien, Cheng-Hua Tsai, J. Chen, and A. Chen (2005), A Dual-Band RF Transceiver for Multistandard WLAN Applications. IEEE Transaction on Microwave Theory and Techniques, 53, 1040-1055, March.

S. Sarkar, P. Sen, A. Raghavan, S. Chakarborty, and J. Laskar (2003). Development of 2.4 GHz RF Transceiver Front-end Chipset in 0.25μm CMOS, Proceedings of the 16th International Conference on VLSI Design.

Walter Schuchter, Guenter Krasser, and Guenter Hofer (2001). A Single Chip FSK/ASK 900MHz Transceiver in a Standard 0.25um CMOS Technology, IEEE RFIC Symposium.

W. Hioe, K. Maio, T. Oshima, Y. Shibahara, T. Doi, K. Ozaki, and S. Arayashiki, "0.18-um CMOS Bluetooth Analog Receiver With 88-dBm Sensitivity (2004). IEEE J. Solid-State Circuits, 39, 374-377, February.

Y. J. Jung, H. S. Jeong, E. S. Song, J. H. Lee, S. W. Lee, D. Y. Seo, I. H. Song, S. H. Jung, J. B. Park, D. K. Jeong, S. I. Chae, and W. Kim (2004). A 2.4-GHz 0.25um CMOS dual-mode direct-conversion transceiver for bluetooth and 802.11b, IEEE Journal of solid-state circuits, 39, July..

Y. K. Park, H. M. Seo, Y. K. Moon, K. H. Won, and S. D. Kim (2005). Low Power Radio Receiver Specifications of Ubiquitous System for Coexistence with Various Wireless Devices in 2.4GHz ISM-band, The 20th International Technical Conference on Circuits/System, Computers and Communications, July.

Y. Palaskas, Y. Tsividis, V. Prodanov, and V. Boccuzzi (2004). A Divide and Conquer Technique for Implementing Wide Dynamic Range Continuous-Time Filters, IEEE J. Solid-State Circuits, 39, 297-307, February.

15

Wireless Sensor Networks and Their Applications to the Healthcare and Precision Agriculture

Jzau-Sheng Lin[*], Yi-Ying Chang[*], Chun-Zu Liu[**] and Kuo-Wen Pan[**]
[*]Department of Computer Science and Information Engineering,
[**]Institute of Electronics Engineering
National Chin-Yi University of Technology, Taichung, Taiwan, R.O.C.

Abstract

Wireless connection based smart sensors network can combine sensing, computation, and communication into a single, small device. Because sensor carries its own wireless data transceiver, the time and the cost for construction, maintenance, the size and weight of whole system have been reduced. Information collected from these sensor nodes is routed to a sink node via different types of wireless communication approaches.

Healthcare systems have restricted the activity area of patients to be within the medical health care center or residence area. To provide more a feasible situation for patients, it is necessary to embed wireless communication technology into healthcare systems. The physiological signals are then immediately transmitted to a remote management center for analysis using wireless local area network. Healthcare service has been further extended to become mobile care service due to the ubiquity of global systems for mobile communications and general packet radio service.

It is important that using sensors to detect field-environment signals in agriculture is understood since a long time ago. Precision agriculture is a technique of management of large fields in order to consider the spatial and temporal variability. To use more sophisticated sensor devices with capabilities of chemical and biological sensing not only aids the personnel in the field maintenance procedure but also significantly increases the quality of the agricultural product.

In this chapter, we examine the fields in healthcare and precision agriculture based on wireless sensor networks. In the application of healthcare systems, a System on a Chip (SoC) platform and Bluetooth wireless network technologies were combined to construct a wireless network physiological signal monitoring system. In the application of precision agriculture, an SoC platform was also used combining the ZigBee technology to consist a field signals monitoring system. In addition to the two applications, the fault tolerance in wireless sensor networks is also discussed in this chapter.

Keywords: wireless sensor networks; healthcare; precision agriculture; Bluetooth; ZigBee.

1. Introduction to the wireless sensor networks

Owing to the rapid development of new medicines and medical technologies, the aged population have been resulted in a speed-up increase. Thus, more rehabilitation centers are created for the requirements of homecare as well as more medical personnel is needed to offer medical treatments and to prevent accidents for aged patients. To provide a more humane environment for these aged patients' physical and physiological heath care, monitoring and recording of their physiological status is very important [1-16]. It occupies a large portion of center's human resources to regularly observe and record the physiological status of patients. It still cannot guarantee to obtain the necessary patients' status information on time and to prevent accidents from happening even if we have sufficient professional nursing staff who works very carefully. In order to reduce the nursing staff's loading and prevent sudden situations that cause accidents, a physiological signal acquiring and monitoring system for the staff to collect the physiological status information of patients to the nursing center with physiological sensors module is essential.

Several technologies were used in the precision agriculture such as remote sensing, global positioning system (GPS), geographic information system (GIS), microelectronics and wireless communications [17, 18]. Most GPS and GIS with satellite systems provide images of great areas. Alternatively wireless sensor networks (WSNs), used for precision agriculture, give better spatial and temporal variability than satellites, in addition to permit collection of others soil and plant data, as temperature, moisture, pH, and soil electrical conductivity [19, 20].

Currently three main wireless standards are used namely WiFi, Bluetooth and ZigBee, respectively. Wi-Fi networks, a standard named IEEE 802.11, is a radio technology to provide reliable, secure, fast wireless connectivity. A Wi-Fi network can be used to connect computers to each other, to the Internet, and to wire networks. Wi-Fi networks work in the unlicensed 2.4 GHz and 5 GHz radio bands, with a data rate of 11 Mbit/s or 54 Mbit/s. They can provide real-world performance similar to that of the basic 10BASE-T wired Ethernet networks. Unlike a wired Ethernet, Wi-Fi cannot detect collisions, and instead uses an acknowledgment packet for every data packet sent.

Bluetooth is a protocol for the use of low-power radio communications over short distance to wirelessly link phones, computers and other network devices. Bluetooth technology was designed to support simple wireless networking of personal consumer devices and peripherals, including PDAs, cell phones, and wireless headsets. Wireless signals transmitted with Bluetooth cover short distances, typically up to 10 meters. Bluetooth devices generally communicate at less than 1 M bps in data transmission. The wireless Bluetooth technology is popularly used in several technique fields. Many researchers have used Bluetooth technology to their monitoring system [12]. Wireless mobile monitoring systems for physiological signal not only increase the mobility of uses but also improve the quality of health care [13].

ZigBee is a low-power, low-cost, wireless mesh networking standard. The low power allows longer life with smaller batteries, the low cost allows the technology to be widely developed in wireless control and monitoring applications and the mesh networking provides high dependability and larger range. ZigBee operates in the industrial, scientific and medical radio bands with 868 MHz, 915 MHz, and 2.4 GHz in different countries. The technology is intended to be simpler and less expensive than other WPANs such as Bluetooth.

Of those, ZigBee is the most promising standard owing to its low power consumption and simple networking configuration. The prospective benefits of using the WSN technologies in agriculture resulted in the appearance of a large number of R&D projects in this application domain. The job of the sensor network in this Chapter is to provide constant monitoring of field-environment factors in an automatic manner and dynamic transmitting the measured data to the farmer or researchers with WSN based on Zigbee and Internet. The real time information from the fields will provide a solid base for farmers to adjust strategies at any time.

Beside to develop a low cost, high performance and flexible distributed monitoring system with an increased functionality, the main goal of this chapter is to use a fault detection algorithm to detect fault sensing nodes in the region of fields. In the proposed strategy, wireless sensors send data via a Microprocessor Control Unit (MCU) and a wireless-based transmitter. The receiver unit receives data from a receiver and an SoC platform. And, these data are transmitted to the Internet through the RJ-45 connector. A remote data server stores the data. Any web browser, smart phone or PC terminal with access permission can view the data and remotely control the wireless network.

The rest of this chapter is organized as follows. Section 2 introduces the application to the healthcare technology, in which the system architecture of the monitoring system for the physiological signals including wireless-network acquiring unit and receiver unit with an SOC platform are discussed; The detail circuit of wireless-network acquiring unit and receiver unit for the application to the precision agriculture are mentioned in Section 3; The application scenario for the ZigBee based networks were demonstrated in Section 4; Section 5 describes the fault tolerance in WSN to detect the fault sensing nodes; Finally, the conclusions and the future work are indicated in Chapter 6.

2. The Application to the Healthcare technology

This Section proposed a wireless network physiological signal monitoring system which integrates an SoC platform and Bluetooth wireless network technologies in homecare technology. The system is constituted by three parts which include mobile sensing unit, Bluetooth module and web-site monitor unit. Firstly we use acquisition sensors for physiological signals, an MCU as the front-end processing device, and several filter and amplifier circuits to process and convert signals of electrocardiogram (ECG), body temperature and heart rate into digital data. Secondly, Bluetooth module was used to transmit digital data to the SoC platform with wireless manner. Finally, an SoC platform, as a Web server additionally, to calculate the value of ECG, the values of body temperature and the heart rate. Then, we created a system in which physiological signal values are displayed on Web page or collected into nursing center in real-time through RJ-45 of an SoC platform. The results show our proposed wireless network physiological signal monitoring system is very feasible for future applications in homecare technology.

Because of the fast development and wide application of Internet, homecare applications to provide health monitoring and care by sending personal physiological signals to Internet have become highly feasible. However, the health care systems have restricted the activity area of patients to be within medical health care center or within residence area. To provide more feasible manner for patients, it is necessary to embed wireless communication technology into healthcare systems. The physiological signals are then immediately transmitted to a remote management center for analysis by using wireless local area network. Homecare service has been further extended to become mobile care service due to the ubiquity of global system for mobile communications and general packet radio service. There are many researchers have used personal digital assistant (PDA) to monitor the patient's status remotely and accurately [14]. In 2006, Lin *et al.* [15] proposed a wireless physiological monitoring system named RTWPMS to monitor the physiological signals of aged patients via wireless communication channel and wired local area network. Body temperature, blood pressure, and heart rate signals are collected and then stored in the computer of a network management center in Lin's system. A wireless patch-type physiological monitoring microsystem was proposed by Ke and Yang [16] in which the skin temperature, ECG signals, and respiration rate are measured and shown by computer information center. In this section, we propose a wireless physiological signal monitoring system which integrates an SoC platform, Bluetooth wireless, and Internet technologies to home-care application to collect the heart rate, ECG, and body temperature into nursing center respectively. In the proposed monitoring system, we used an SoC platform to create a Web server that can reduce the device size significantly. In the proposed physiological monitoring system, we designed and implemented all of the application programs and hardware modules.

2.1 System architecture

Fig. 1 shows the architecture of the proposed wireless-network physiological signal monitoring system that includes mobile sensor units, Bluetooth transceiver module and Web server monitor system. The Bluetooth module is integrated into mobile unit as a transmitter as well as the SoC platform in monitor system worked as a receiver for physiological signals with a wireless manner. In order to get stable physiological signals, some amplifiers and filters are added into acquiring circuits. Finally, the physiological signal values can be displayed on Web page or collected into nursing center through RJ-45 of the SoC platform. According to the proposed architecture, a wireless network physiological signal monitoring system is implemented.

2.2 Mobile Physiological Signal Acquisition Unit

The main parts of this unit are mainly including the sensors of thermistor, ECG electrodes; acquiring circuit of heart rate, ECG, and body temperature; and MCU circuit respectively. In order to remove noise and amplify the physiological signals, filter and amplifier circuits are also added into the mobile unit. For the purpose of processing the heart rate, ECG, and body temperature signals and transferring them to Bluetooth module, an MCU named PIC16F877 is used.

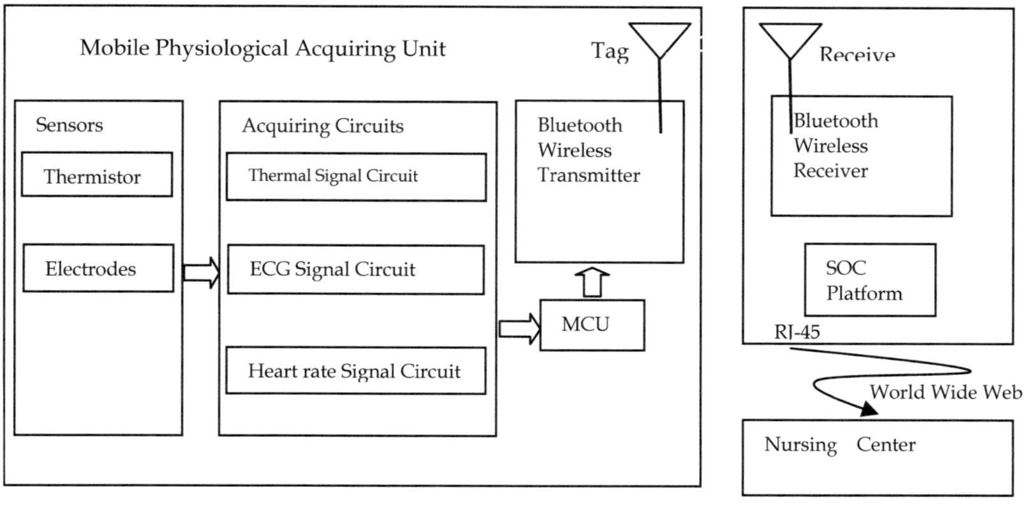

Fig. 1. The proposed architecture of wireless physiological signal monitoring system.

The body temperature is converted by an AD590 temperature sensor. The AD590 is a two terminal device that acted as a constant current element passing a current of 1 mA/°C. AD590 is particularly useful in remote sensing applications. The nominal current output of AD590 is 298.2μA at +25°C (298.2°K) and temperature coefficient is +1 μA/°K. After converting the output current of AD590 into a voltage signal, we change the temperature coefficient to +100 mV/°K by using an amplifier circuit and then send the signal to the ADC of MCU. The block diagram and circuit for body temperature acquisition system are shown as in Fig. 2.

In the proposed acquisition system, an instrument amplifier cooperates with AD590 and converts temperature signal into voltage. This instrument amplifier provides an extremely high input impedance, low output impedance, and higher common-mode rejection ratio (CMRR) to reject common-mode noise. In the front buffers, the lower OP amplifier got an aligned voltage from input port as well as the upper one transferred the temperature current to a voltage value. Because the HA17324 occupies four OP amplifiers (uA 741), we organized these three OP amplifiers in Fig.2 with an HA17324.

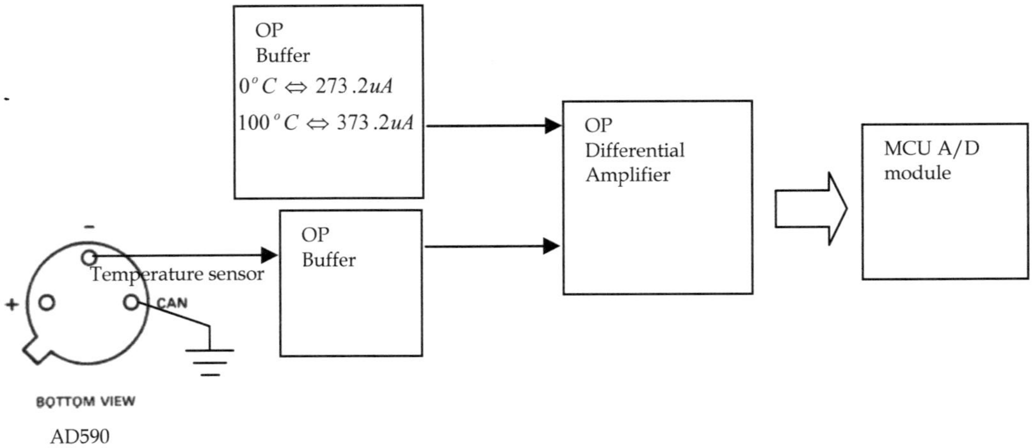

Fig. 2. The block diagram of body temperature signal acquisition system.

Electrocardiogram (ECG) is an electrical recording of the heart and is used in the investigation of heart disease. With each heart beat, an electrical impulse travels through the heart. Therefore, we can also calculate the number of heart beat with an interval to derive from the heart rate. This impulse causes the heart muscle to squeeze and pump blood from the heart. The electrical potential is an analog signal with bandwidth of 0.05 Hz to 100 Hz. It is generally around 1-mV peak-to-peak. Some of the noise can be cancelled with a high-input-impedance instrumentation amplifier (INA). Because of CMRR will result in greater rejection, we use AD620 as an INA in our signal acquisition circuit, which removes the AC line noise and amplifies the remaining unequal signals present on the inputs. In order to make signal lie in 0.05-100 Hz, we used a high-pass filter and a low-pass filter with the cut-off frequencies 0.0482 Hz and 106.103 Hz respectively. For the DC electrode, we used a high-pass filter to solve DC offset problem in which the cut-off frequency is 0.723 Hz. For the purpose of sending the analog signal to the A/D converter module in MCU, a clamping circuit was used to remain signals lie in 0 to5 volts. The block diagram and circuit for ECG signal acquisition module are shown as in Fig. 3.

The final part of the mobile physiological acquiring unit is the MCU in which the MicroChip PIC16F877 is used. The PIC16F877 features 256 bytes of EEPROM data memory, self programming, an In Circuit Debug (ICD), 2 comparators, 8 channels of 10-bit Analog-to-Digital (A/D) converter, and 2 capture/compare /PWM functions. The synchronous serial port can be configured as either 3-wire Serial Peripheral Interface (SPI™) or the 2-wire Inter-Integrated Circuit (I^2C™) bus and a Universal Asynchronous Receiver Transmitter (USART). To integrate Bluetooth communications module directly from the USART pins of the PIC microcontroller, the details of the complex Bluetooth protocol were not needed.

Fig. 4 displays the input and output interfaces of the MCU. In the MCU PIC 16F877, we used analog input ports RA0/AN0 and RA0/AN0 to extract the ECG and body temperature signals as well as a 4-MHz crystal was mounted on pins of oscillator1 (OSC1) and oscillator 2 (OSC2) as the system clock of the MCU. Then, the digital signals of ECG and body temperature are forward sent to the Bluetooth transmitter through data output (TX) on MCU.

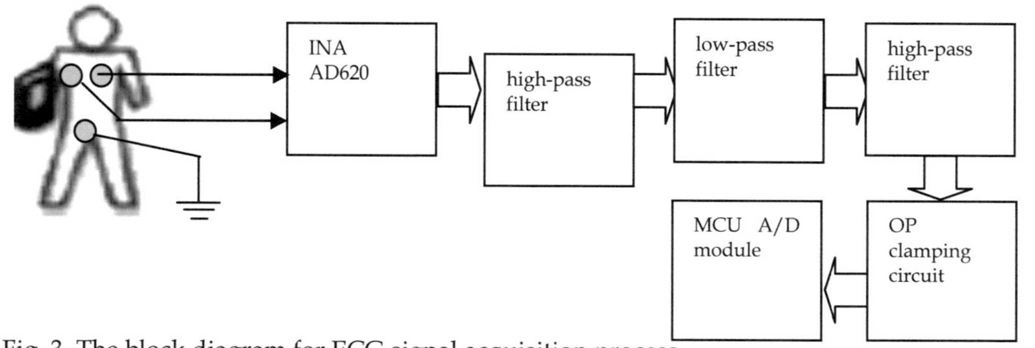

Fig. 3. The block diagram for ECG signal acquisition process.

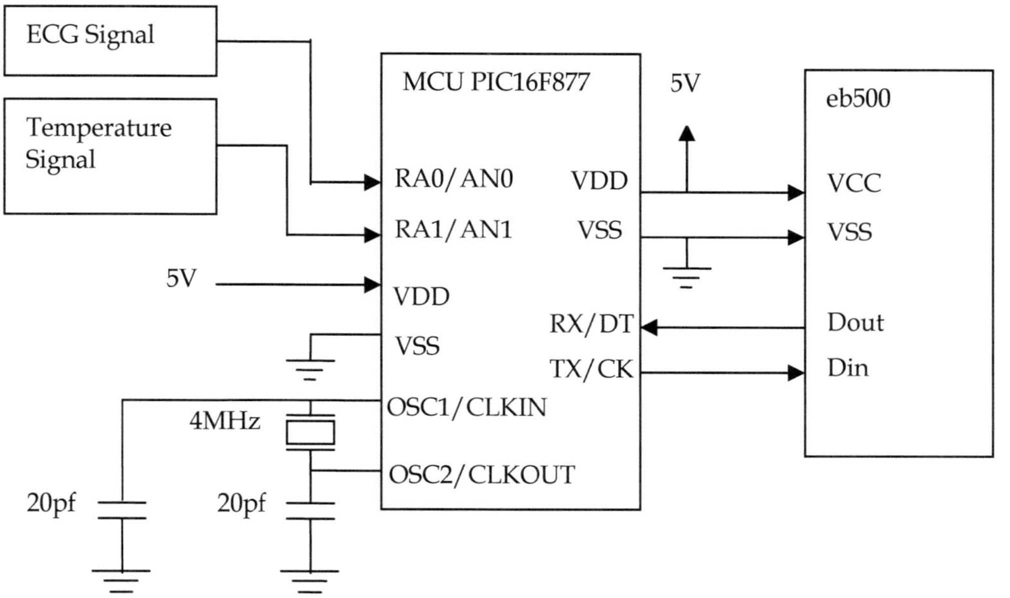

Fig. 4. The diagram of input and output signals on MCU PIC 16F877.

Fig. 5 shows the picture of the designed mobile physiological acquiring unit. In order to implement the trend of commercializing, we finished the layout of our mobile device that reduces its volume significantly. The heart rate, ECG, and body temperature signals can be acquired by physiological signal sensors. The signals were processed by amplifier, filter, and comparator circuits, and sent them out through eb500 module. In order to acquire physiological signals efficiently, we also use general battery to offer 5v for DC-DC regulator as the power supply for the mobile unit.

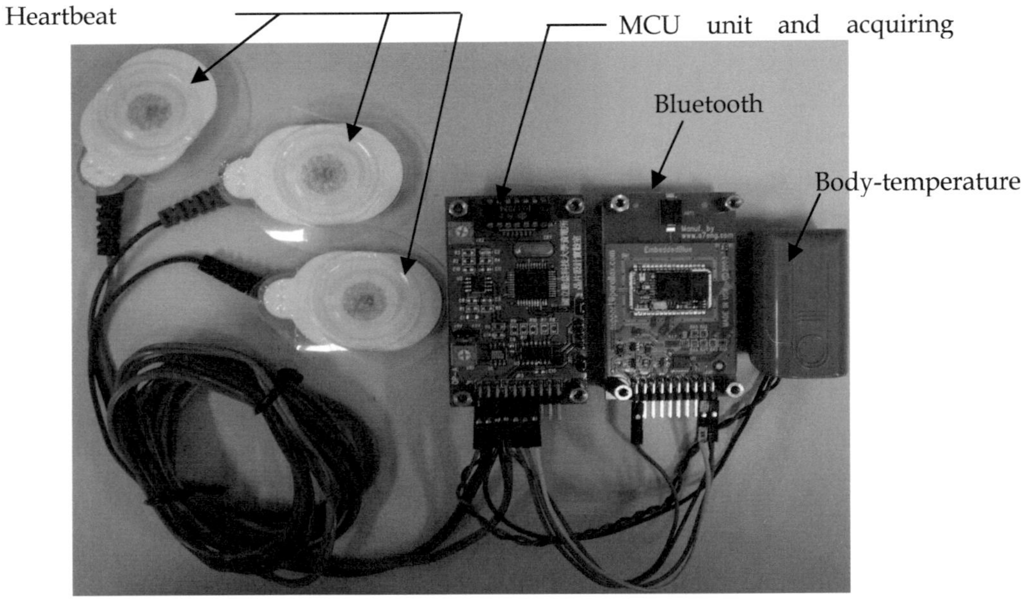

Fig. 5. The diagram of mobile unit with a Bluetooth transmitter

2.3 Bluetooth module

The used Bluetooth module in the proposed system is EmbeddedBlue 500 (eb500). EmbeddedBlue is a trademark of A7 Engineering. The eb500 module provides a point to point connection much like a standard serial cable. Connections are made dynamically and can be established between two eb500 modules or an eb500 module and a standard Bluetooth v1.1 or v1.2 device. Bluetooth utilizes frequency hopping in the 2.4GHz radio band and hops at a relatively fast pace with a raw data rate of about 1 Mbps. This translates to about 700 kbps of actual useful data transfer. The eb500 module supports a maximum sustained bidirectional data rate of 230.4 kbps.

In order to let two Bluetooth devices communicate each other, they must share at least one common profile. If a pocket PC is used to communicate with an EmbeddedBlue radio, it needs to make sure that they both support the same profile. The eb500 devices support the Serial Port Profile (SPP) which is one of the earliest and most widely supported profiles.

The eb500 module implements the SPP profile which enables it to appear like a traditional serial port. This virtually eliminates the need for the user to have specific Bluetooth knowledge and allows the radios to be integrated into applications very quickly. The eb500 module is a Class 2 intelligent Bluetooth module which communicates up to 10-meters that can make use of effectively at home environment. The eb500 supports two operating modes including command mode and data mode. Upon power up, the eb500 enters command mode and is ready to accept serial commands for modifying the baud rate and flow control settings. In command mode there are many commands that can be sent to change the baud rate, locate other devices, check the firmware version, etc.

2.4 Web Server Unit

Owing to the wide application of Internet, to access physical signals by using Internet through an embedded system is popular more and more. Using an embedded system not only can realize the equipment remote control, but also the system size can significantly be reduced. An external interface is essential to carry on the monitoring through the network. The users can manage and monitor the far-end system through Web browser which can simplify the design of human-machine interface.

We used an SOC platform built in XILINX SPARTAN-3 (SP3) [21] as a Web server and digital signal processing (DSP) unit which was implemented by using C language in order to transmit the physiological information to Web page or nursing center through the TCP/IP with a cable RJ-45. The SP3 FPGA uses eight independent I/O banks to support 24 different single-ended and differential I/O standards and allows you to easily migrate different densities across multiple packages. In the SP3 SOC platform, a built 10 base-T/100base-TX/FX IEEE 802.3u fast Ethernet transceiver named BCM 5221 is used as Ethernet PHY to transmit data to the Internet through RJ-45. The BCM 5221, designed by Broadcom Company, builds on a DSP PHY and full custom circuit design techniques to create a highly integrated and well-define physical layer solution. This development platform integrates many IP (Silicon Intellectual Property) modules including RS-232, RJ-45, USB, expand I/O pin etc. The Web server in SP3 SOC platform was developed by the Xilinx Embedded Development Kit (EDK), in which the Platform Studio (XPS) and IP cores (including a 32-bit soft- RISC-CPU MicroBlaze) are supported. The physiological data, received form Bluetooth receiver, are sent to and processed by the CPU (Microblaze) through a General Purpose Interface (GPI) IP. In addition, we organized off-chip memory with a 16Mega-byte SDARM as well as a hyper terminal through an internal IP named UART Lite. The architecture of the Web server and DSP unit constructed by an SOC with SP3 is shown as in Fig. 6. In the development platform, we use C language in the Xilinx's development platform and EDK version 8.1 to implement the Web server and DSP unit. Finally, SP3 platform combines eb500 to receive digital signal from mobile physiological acquiring device and calculate the heart rate, ECG, and body temperature values in the platform to transmit them to the Web page or nursing center.

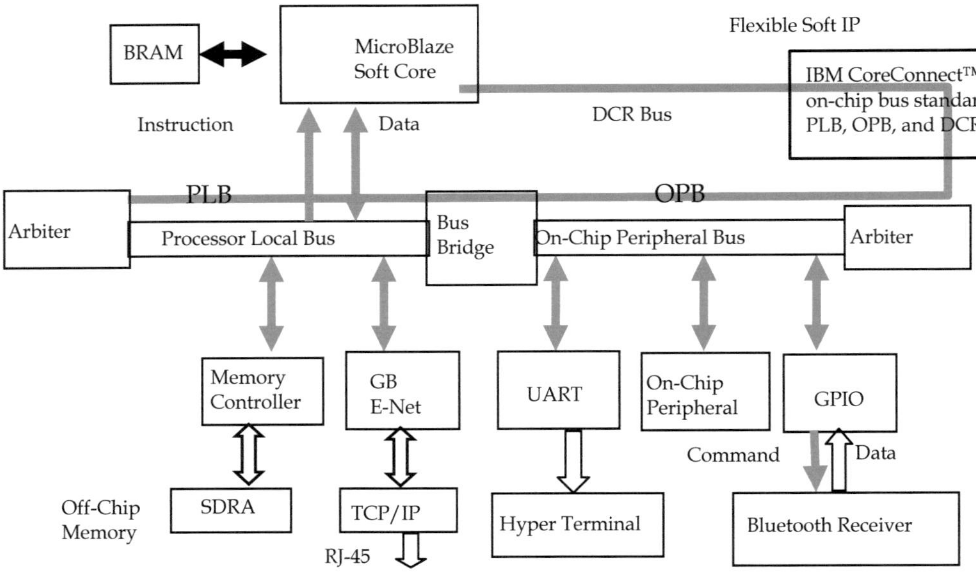

Fig. 6. The architecture of the Web Server Unit

The ECG signal was acquired through the AD620 with several millivolts. Then, the weak ECG signal was sent to the high-pass and low-pass filters, in which some noise signals are removed. In the real circuit, the drift problem for the base line of the ECG signal appeared in the frequencies between 0.0482 Hz and 106.103 Hz. Therefore, we added a clamping circuit to resolve the drift problem based line in the ECG signal. The clamped ECG signal can then be transmitted by the mobile unit and display on the Web page.

The heart rate signal can also be extracted from the intervals from a range of ECG signal. All of the above indicated hardware devices and application software have been integrated into a completed real-time wireless network physiological monitoring system. And, we use general battery solve the power supply problem of mobile device to simulate the mobile device. The front-end mobile monitoring device is light with a compact size.

The whole system has been successfully designed and tested. The physiological signals can then be accessed and stored into the physiological information database in information management system of the nursing center by a terminal or a computer in the Internet shown as in Fig. 7. Finally, the physiological signals can be displayed on the computer window through the Internet in nursing center like showing in Fig. 8. In Fig. 8, we extracted the physiological signals with a time interval about 20 ms. For the body temperature, we showed the recent 48 values with a curve manner and their average value (36.7 degrees). In addition, we also displayed the hard rate and the ECG curve about 270 points.

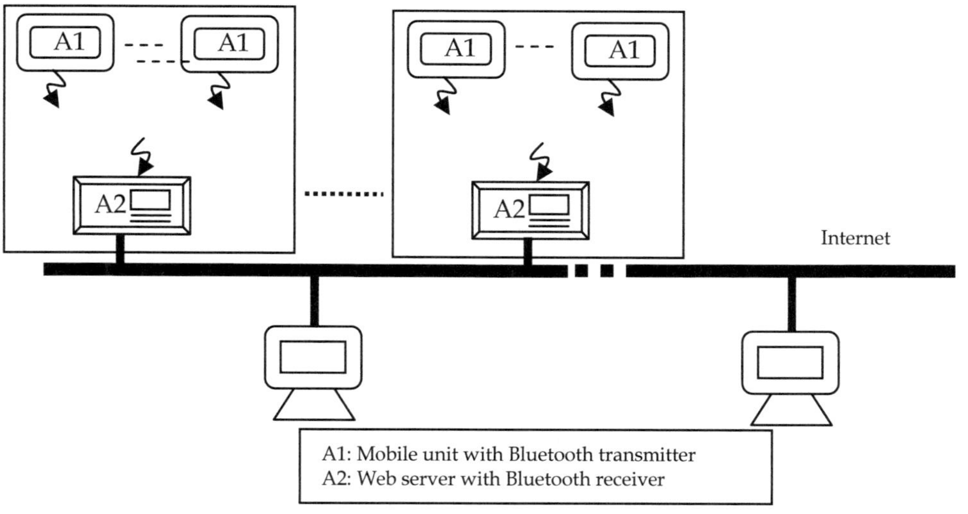

Fig. 7. Physiological signals monitoring system based on wireless and Internet Architecture.

3. The Application to the Precision Agriculture

This Section proposed a field signals monitoring system with wireless sensor network (WSN) which also integrated an SoC platform and Zigbee wireless network technologies in precision agriculture. The designed system is constituted by three parts which include field-environment signals sensing units, Zigbee transceiver module and web-site unit. Firstly we use acquisition sensors for field signals, an MCU as the front-end processing device, and several amplifier circuits to process and convert signals of field parameter into digital data. Secondly, Zigbee module was used to transmit digital data to the SoC platform with wireless manner. Finally, an SoC platform, as a Web server additionally, to process field signals. Then, we created a system in which field signal values are displayed on Web page or collected into control center in real-time through RJ-45 with the SoC platform. The experimental results show our proposed field-environment signals monitoring system is very feasible for future applications in precision agriculture.

In the initial effort, Alves-Serodio *et al.* [22] showed several concepts on technique for the supervision and control of agricultural systems such as greenhouse and animal live stocks-claim for the use of computer systems. The method basically focused on control of the environmental parameters in a low-cost way to generate the best agricultural product or animal living conditions. In [23] a web server based strategy was used where sensor nodes are setup with a web server to be accessed via the internet and make use of wireless LAN to provide a high speed transmission. The application of a web server assists to analyze distant agricultural fields over long periods of time whereby the whole dataset is accessible to general public. A pilot sensor network deployment in precision agriculture was proposed by Langendoen *et al.* [24]. In his algorithm the sensor nodes in the field measured relative humidity and temperature once per minute. They encoded ten samples in a single packet which was directly send or through multiple hops to a Wi-Fi gateway at the edge of the field. Wang *et al.* [25] discussed wireless sensors for agriculture and food industry in recent development and future perspective. In [26] an approach based on web server was used where sensor nodes were organized with a web server to be accessed via the internet and make use of wireless LAN to supply a high speed transmission.

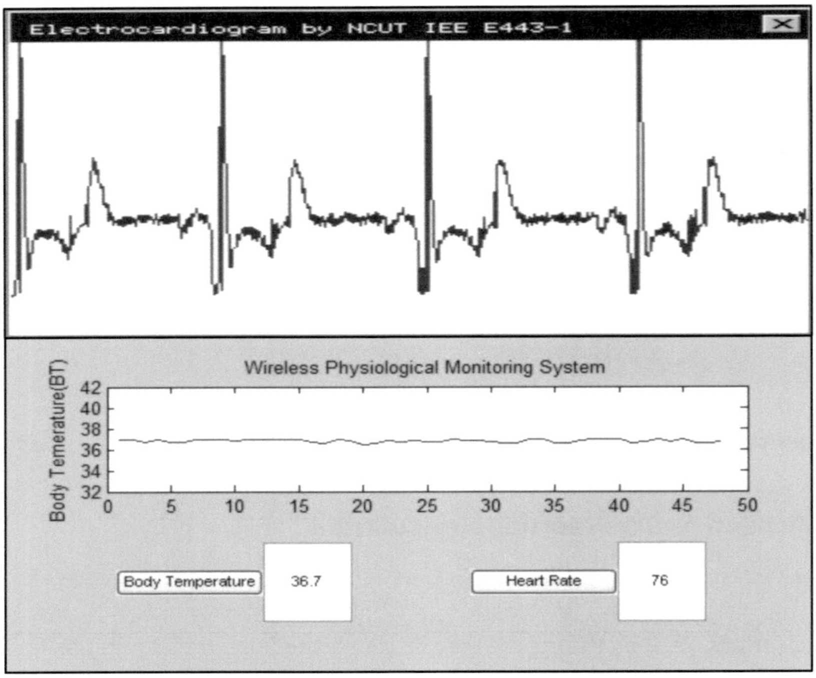

Fig. 8. Physiological-signal display window in nursing center

3.1 System Architecture

The advance of technology in wireless communications has developed small, low-power, and low-cost sensors. Sensor networks are developed to construct and control these sensor nodes, which have sensing, data processing, communication and control capabilities. Collecting information from these sensor nodes is routed to a sink node via different types of wireless communication approaches.

Fig. 9 shows the architecture of the proposed wireless-network monitoring system that includes sensors unit, Zigbee transceivers, an MCU, An SoC platform, and Web server. The MCU is a communicator and controller between sensors and Zigbee transmitter. The SoC platform in monitor system worked as a web server to receive the field-environment signals from a Zigbee receiver and transmit those signals to the Internet through RJ-45 interface. In order to get stable signals, some amplifiers are added into acquiring circuits. Finally, the field-environment signals can be displayed on Web page or collected into control center through RJ-45 on the SoC platform.

Acquisition Unit and Receiver Unit

The main part of the Wireless-Network Acquiring unit is mainly including the sensors of temperature and moisture in air and soil, CO_2, and illumination. In order to amplify the field-environment signals, amplifier circuits are also added into acquiring unit. For the purpose of processing these signals and transferring them to ZigBee wireless transmitter, an MCU named SPCE061A [27] is used. Firstly, the A/D converter bound on the MCU converts

the analog signal into digital manner. And, MCU calculates and organizes the data to desired format, and writes them to ZigBee wireless transmitter. Then, the ZigBee Transmitter sends these field-environment signals to the ZigBee receiver through a handshaking protocol. Finally, these signals are transmitted to the Receiver Unit.
The Receiver Unit is consisted of a ZigBee wireless receiver and an SoC platform. The field-environment signals, received by the ZigBee wireless receiver, were directly sent to the field information database on the Internet through a RJ-45 connecter and Web server built on the SoC platform.
WatchDog 3667 [28], products of Spectrum Technologies, Inc. including a 6 foot cable that is connected to an external port on a WatchDog Data Logger, was used as s sensor of soil temperature. Watermark 6450WD [29-31] (Spectrum Technologies, Inc.) was used to measure soil moisture. It consists of two concentric electrodes embedded in a reference matrix material, which is surrounded by a synthetic membrane for protection against deterioration. A stainless steel mesh and rubber outer jacket construct the sensor more durable than a gypsum block. The measured temperature range is -30 ~ $100^{\circ}C \pm 1^{\circ}C$ for the WatchDog 3667 while the detected moisture range is 0 ~ 200 cbars for the Watermark 6450WD.

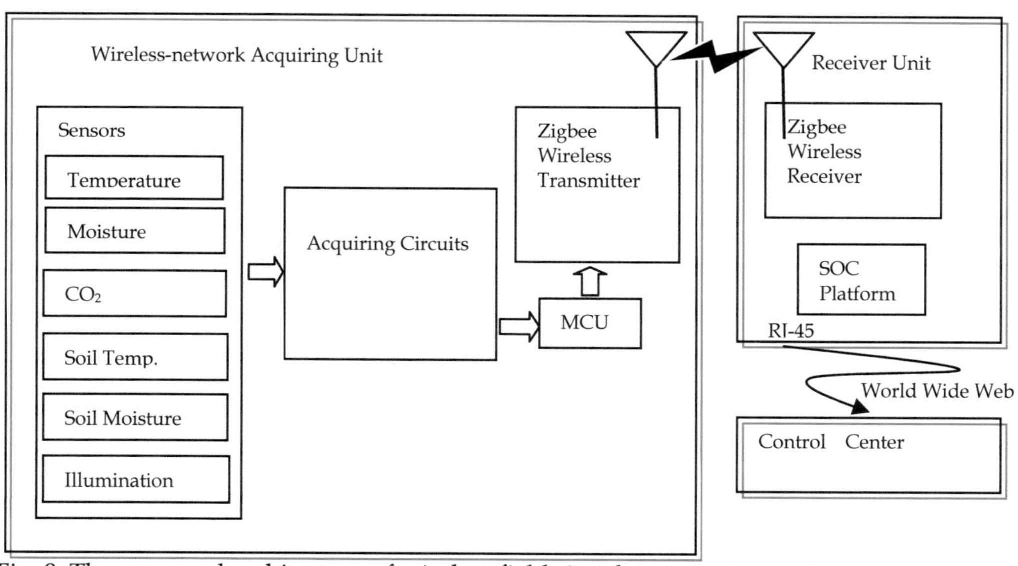

Fig. 9. The proposed architecture of wireless field signals monitoring system.

The module RHU-300M, products of Decagon Devices, Inc., was used in order to measure the temperature and moisture in the air. The range of measured temperature is 0 ~ $60^{\circ}C \pm 1^{\circ}C$ while the range of detected moisture is 10 ~ 95 %RH. For the purpose of detecting CO_2, the sensor REHS-135 [32], in which the operating humidity range is less than 95% Rh. And, the illumination was measured by using of the CDS photo-resister [33]. The completed hardware diagram of the acquiring system for these sensors to measure signals in the field-environment is shown as in Fig. 10.
The final part of the wireless-network acquiring system is the MCU in which the Sunplus SPCE061A is used. The SPCE061A features 2K words of SRAM and 32K words of Flash ROM data memory, 32 programmable input/output ports, 2 ports of 16-bit timer/counter, 7

channels of 10-bit Analog-to-Digital (A/D), 2 channels of 10-bit Digital-to-Analog (D/A) converters, and an In-Circuit-Emulation (ICE) port. Fig. 10 also displays the input and output interfaces of the MCU. In the MCU SPCE61A, we used analog input ports I/O A0 ~ A5 to extract the moisture, temperature and CO_2 in the air, soil temperature and moisture, and illumination. A crystal is mounted on pins of oscillator 1 (XI/R) and oscillator 2 (XO) as the system clock of the MCU. Then, the digital signals of field-environment are forward sent to the ZigBee transmitter through programmable I/O ports outputs I/O B7 and B10 respectively. To further improve communication, the nodes are enclosed in a small box while the sensors are also installed at a box with a height of 20, 40 or 60cm or embedded into soil for the soil temperature and moisture.

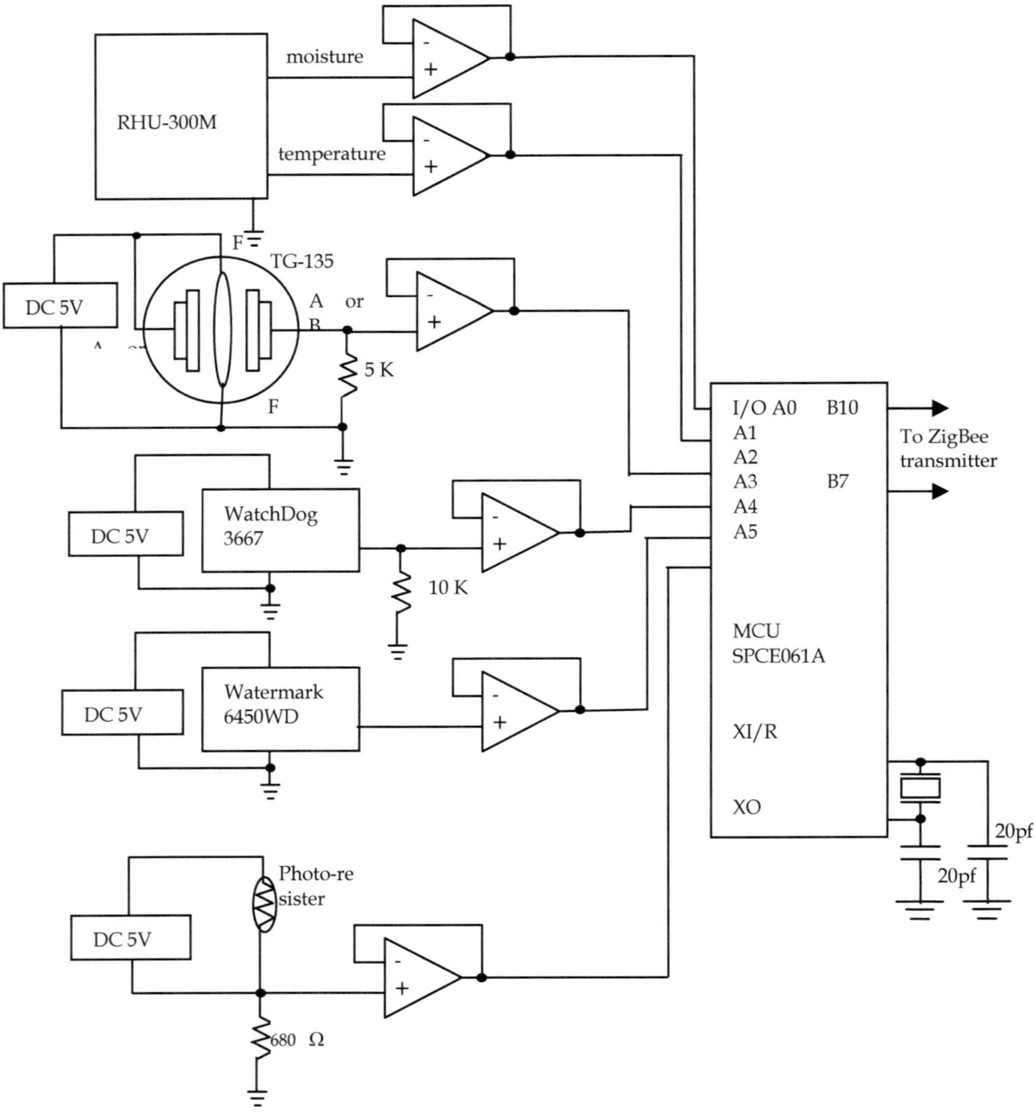

Fig. 10. The hardware diagram to measure signals in the field.

3.3 ZigBee Module

The used ZigBee transceiver module in the proposed system is module 3160 produced by Ready International Inc [34]. The 3160 modules provides a point to point connection much like a standard serial cable. Connections are made dynamically and can be established between server 3160 module and sensor module or between several sensor modules and a server module. ZigBee utilizes frequency hopping in the radio band and hops at a relatively pace with a raw data rate of about 250Kbps and a transmitting distance of about 200 m.

3.4 Web Server Unit

Owing to the wide application of Internet, to access field-environment signals by using Internet through an embedded system is popular more and more. In the web server unit, we also used an SOC platform built-in XILINX SPARTAN-3 (SP3) like the architecture shown as in Fig. 6, but Bluetooth receiver was changed as ZigBee receiver.

4. Application Scenario

ZigBee technology based wireless sensor can be used in a diverse, high volume sensor system. It can significantly save space and improve the reliability. Fig. 11 shows an application scenario in precision agriculture. As we know, to monitoring the real-time status of a wide field needs high-density sensors. As shown in the figure, each ZigBee receiver has quite mounts of sensors installed. MCU can poll each sensor quickly to get the sensing data. Since every sensor has a unique identification number, MCU can easy know the sensing data comes from which sensor and do respective operation.

The whole system has been successfully designed and tested. The field signals can then be accessed and stored into the field information database in the information management system of the control center by a terminal or a computer in the Internet like shown as in Fig. 11.

5. Fault Tolerance in Wireless Sensor Networks

5.1 Introductions to the Fault Tolerance in WSNs

WSNs have become a new data collection and monitoring system for different applications. The impressive advances in wireless communication have enabled the development of low power, low cost, and multifunctional wireless sensor nodes which consist of sensing, data processing, and communication components. These tiny sensor nodes can easily be deployed on a large-scale area to extract useful information from harsh or hostile environments, such as fire or rain etc. However, the character of these applications and network operational environment has also put strong impact on sensor network systems to maintain high service quality. In order to guarantee the network quality of service, it is necessary for the WSNs to be able to detect the faults and take actions to avoid further degradation of the service. Fault detection is an identifying scheme, in which an unexpected failure should be properly recognized by the network system. The fault detection approaches in WSNs can then be divided into centralized and distributed approaches [35].

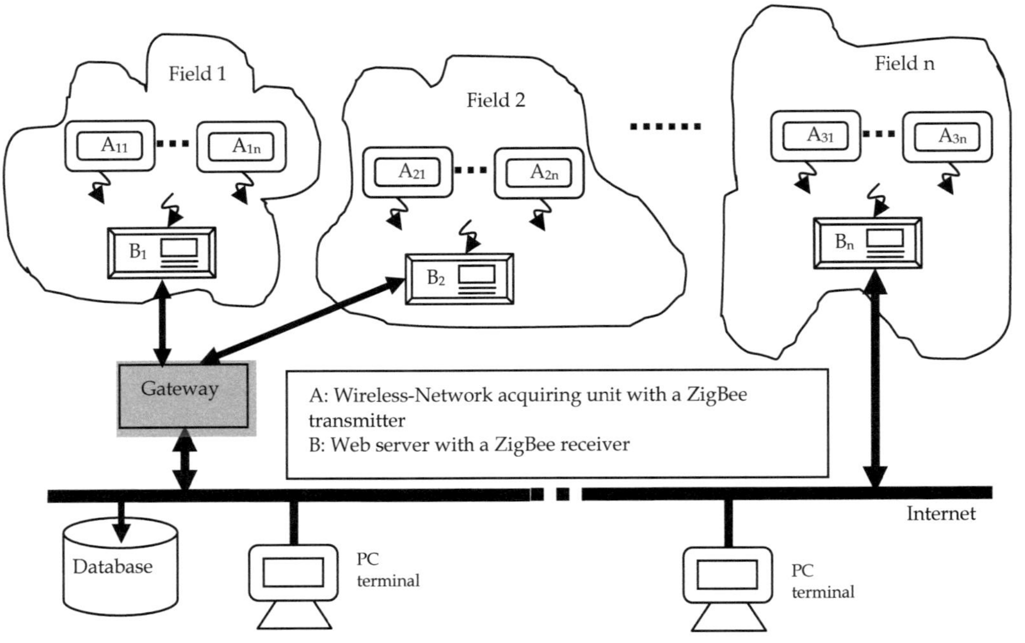

Fig. 11. The architecture of Field signals monitoring system in the precision agriculture based on wireless network and Internet.

In the centralized approaches, a geographically or logically centralized sensor node including a processing unit (PU) was used to take responsibility for monitoring and tracing failed or misbehavior node in a WSN. In several applications, a base station was used as a PU. In [36], the base station used marked packets which contained geographical information of source and destination locations to investigate sensors. It depends on nodes' response to identify and isolate the apprehensive nodes on the routing paths when an unnecessary packet drops or compromised data has been detected. Although the centralized approach is efficient and accurate to identify the network faults, resource-constrained sensor networks can not always periodically collect all the sensor measurements and states in a centralized manner. Additionally, this approach is not only extremely inefficient and expensive in consideration of a large-area sensor network, but multi-hops communication manner will also increase the response delay from the base station to faults occurred in the network. It is very expensive for the base station to collect information from every sensor and identify faulty nodes in a centralized approach. Therefore, a distributed strategy is highly preferred in WSNs.

Distributed approach emphases the local decision-making concept to allow a local node making certain decision before communicating with the central controller. The central controller should not be informed unless there is a fault occurred in the network in order to save delivering time. Harte *et al.* [37] proposed a node self-detection model to monitor the malfunction of the physical components of a sensor node through both hardware and software interface.

Clustering approach [38] has become an emerging strategy for constructing scalable and energy-balanced applications for WSNs. Tai *et al.* [39] built a cluster-based communication hierarch to split the entire network into different clusters and subsequently fault distribute manager into each region.

Distributed detection algorithm is used to have each node make a decision on faults. Clouqueur *et al.* [40] used fusion sensors to coordinate with each other to assure that they get the same global information about the network before making decision.
Fault detection through neighbor coordination is another strategy of fault management distribution, in which the network faults are detected and identified by nodes coordinate with their neighbors. Ding *et al.* [41] proposed a localized algorithm to identify doubtful node whose sensor readings have large difference against the reading value from the neighbors. Chen *et al.* [42] improved such localized algorithm to remove the node physical position.

5.2 The used Fault Detection algorithm in these two applications

In this chapter, Chen's algorithm [42] was applied in the WSN to monitor field signals for precision agriculture. Like shown as in Fig. 11, every field was considered as an interested area for the localized algorithm in distributed fault detection. Chen's algorithm was simulated under different number of faulty sensors in an example area and showed the simulation results with 97% faulty sensor detection accuracy with 25% faulty sensors. In this chapter, we assume that all system software as well as application software are already fault tolerant. We just focus on the hardware faults. On an interesting area, each node sends its measured value to all its neighbors. In the algorithm, a test value c_{ij} is generated by sensor S_i based on its neighbors S_j's measurements using measurement difference between S_i and S_j during a time interval with two predefined thresholds. The fault status of a node was determined to be likely good or likely faulty by using test value from its neighboring sensors. Finally, the good sensors are indicated in accordance with constrains in this approach.
In the simulation scenarios, we constructed two 13ode sensor arrays shown as in Fig. 12 to detect moisture and temperature in the air and soil, illumination, and CO_2 on $70 \times 20m^2$ and $72 \times 18m^2$ fields. The threshold values for these six parameters are predefined.
The experimental results proved that the localized fault detection algorithm can achieve high detection accuracy and low false alarm rate [39]. And, in the experimental environment we can easily detect the faulty sensor nodes.

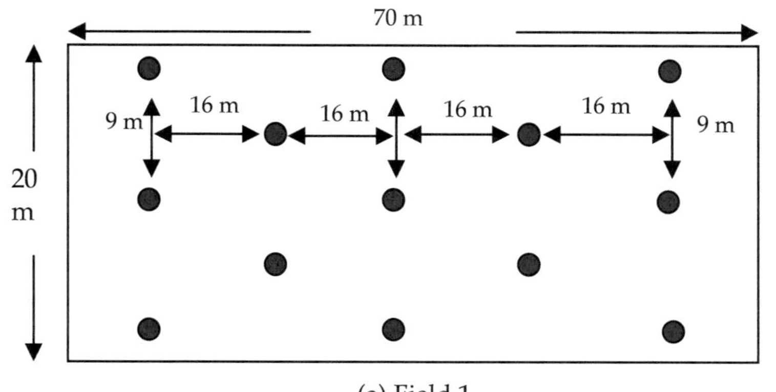

(a) Field 1

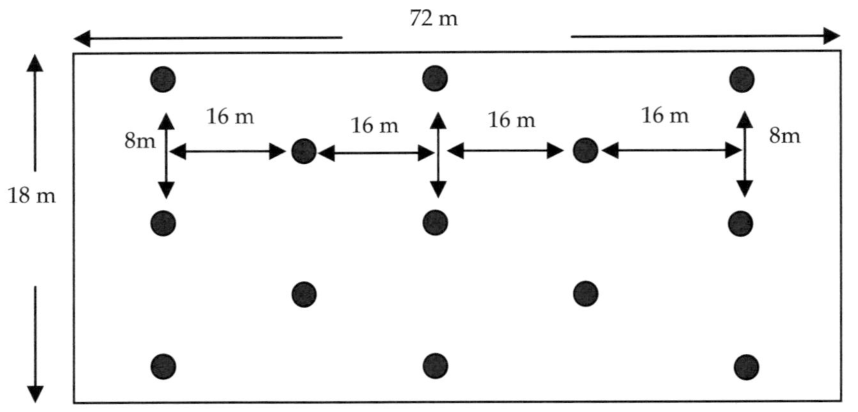

(b) Field 2

Fig. 12. Two simulation scenarios by using sensor node arrays with different distances.

6. Conclusions

A wireless network physiological signal and field signal monitoring systems in homecare technology and precision agriculture were proposed in this chapter. We have finished monitoring physiological signals such as heart rate, ECG, and body temperature as well as temperature and moisture in air and soil, CO_2, and illumination signals in the field. We used Bluetooth technique to solve wireless transmission problem and to finish physiological signals transceiver between mobile unit and Web server that might be useful in replacing cables of physiological signal monitoring system. Additionally, we also used ZigBee technique to finish field signals transceiver between acquiring unit and Web server that might be useful for field signal monitoring. Most of healthcare-monitoring and field-monitoring systems applications use mobile device and PC as main monitoring device in their system. We used an SOC platform as the Web server that can effectively to reduce cost and the physical size significantly. Because of the popularization of the internet that displays the physiological and field signal values on the Web page in real-time through RJ-45 of SP3 platform, the doctors or patient's family can easily take care of the patient's health status while the researchers or farmers can easily look out of the product's status in the precision agriculture anytime and any place through the Web page. Additionally, we also embedded the faulty sensor detection algorithm into sensor nodes on the two simulation fields and obtain feasible faulty sensor detection accuracy.

Although the fault detection algorithm can be implemented in the wireless sensor networks on the field to detect the faulty sensor nodes, we are still persecuted by the power supply with batteries for the sensor nodes. Low power consumption is one of the advantages of the Zigbee networks, but we must change batteries when the power were exhausted. Owing to the sunlight being sufficient on the field, the solar cell will be used to support the power for sensor nodes in the future.

7. Reference

P. Varady, Z. Benyo, and B. Benyo, "An open architecture patient monitoring system using standard technologies," *IEEE Trans. Inf. Technol. Biomed.*, vol. 6, no. 1, pp. 95–98, Mar. 2002.

J. Bai et al., "The design and preliminary evaluation of a home electrocardiography and blood pressure monitoring network," *J. Telmed. Telecare*, vol. 2, no. 2, pp. 100-06, 1996.

G. Williams, P. J. King, A. M. Capper, and K. Doughty, "The electronic doctor (TED)—A home telecare system," *in Proc. 18th IEEE Annu. EMBS Int. Conf., Amsterdam, The Netherlands*, Oct. 31-ov. 3, 1996, vol. 1, pp. 53-4.

P. Johnson and D. C. Andrews, "Remote continuous physiological monitoring in the home," *J. Telmed. Telecare*, vol. 2, no. 2, pp. 107-13, 1996.

K. Doughty, K. Cameron, and P. Garner, "Three generations of telecare of elderly," *J. Telmed. Telecare*, vol. 2, no. 2, pp. 71-0, 1996.

M. J. Rodriguez, M. T. Arredondo, F. del Pozo, E. J. Gomez, A. Martinez, and A. Dopico, "A home telecare management system," *J. Telmed. Telecare*, vol. 1, no. 2, pp. 86-4, 1995.

M. Rezazadeh and N. E. Evans, "Multichannel physiological monitor plus simultaneous full-duplex speech channel using a dial-up telephone line," *IEEE Transactions on Biomedical Engineering*, vol. 37, pp.:428 -32, 1990.

C. H. Ko, H. L. Chen, C. C. Kuo, G. Y. Yang, C. W. Yeh, B. C. Tsai, Y. T. Chiou, C. H. Chu, "Multi-sensor wireless physiological monitor module," *Proceedings of 56th Conf. of Electronic Components and Technology*, pp. 673-676, May 2006.

E. Jovanov, D. Raskovic, A. O. Lords, P. Cox, R. Adharni, and F. Andrasik, "Synchronized physiological monitoring using a distributed wireless intelligent sensor system," *The 25th Int. Conf. on IEEE Engineering in Medicine and Biology Society,* vol. 2, pp.1368-1371, Sept. 2003.

I. Pavlidis, "Continuous physiological monitoring," The 25th Int. Conf. on IEEE Engineering in Medicine and Biology Society, vol. 2, pp.1084-1087, Sept. 2003.

C. Baber, A. Schwirtz, J, Knight, H. Bristow, T. N. Arvanitis and F. Psomadellis, "Sensvest-on-body physiological monitoring system," *IEE Eurowearable*, pp. 93-98, Stevenage, Sept. 2003.

S.-N. Yu and J.-C. Cheng, "A wireless physiological signal monitoring system with integrated bluetooth and WiFi technologies," The 27th Int. Conf. on IEEE Engineering in Medicine and Biology Society, vol. 2, pp.2203-2206, Shanghai, China, Sept. 2005.

S. P. Nelwan, T. B. van Dam, P. Klootwijk, and S. H. Meij, "Ubiquitous mobiles access to real-time patient monitoring data," *Comput. Cardiology,* Rotterdam, the Netherlands, pp 557-560, September 2002.

Y.-H. Lin, I-C. Jan, P. C.-I. Ko, Y.-Y. Chen, J.-M. Wong, and G.-J. Jan, "A wireless pda-based physiological monitoring system for patient transport," *IEEE Trans. Biomed. Eng.*, pp 439-447, 2004.

B.-S. Lin, N.-K. Chou, F.-C. Chong, S.-J. Chen, "RTWPMS: A real-time wireless physiological monitoring system," *IEEE Transactions on Information Technology in Biomedicine*, vol. 10, pp. 647-656, 2006.

Ting-Chen Ke Kuo-Yu Yang, "A wireless patch-type physiological monitoring microsystem," *IEEE Sensors*, EXCO, Daegu, Korea, October 22-25, pp. 1143-1146, 2006.

N. Zhang M. Wang, and N. Wang; Precision Agriculture - a Worldwide Overview; Computers and Electronics in Agriculture, vol. 36, pp. 113-132, 2002.

C. T. Leon et al.; Utility of Remote Sensing in Predicting Crop and Soil Characteristics; Precision Agriculture, Kluwer Academic Publishers, vol. 4, pp. 359-384, 2003.

V. I. Adamchuk "On-the-go Soil Sensors for Precision Agriculture," *Computers and Electronics in Agriculture*, vol. 44, pp. 71-91, 2004.

R. Beckwith "Report from the Field: Results from an Agricultural Wireless Sensor Network," *Proceedings of the 29th Annual IEEE International Conference on Local Computer Networks* (LCN'04), pp. 471-478, 2004.

Xilinx International Co., XILINX SPARTAN-3, http://www.xilinx.com/products/ silicon_ solutions /fpgas/spartan_series/spartan3_fpgas/index.htm.

C.M.J Alves-Serodio, J. L. Monteiro, and C.A.C. Couto, "An integrated network for agricultural management applications," *IEEE International Symposium on Industrial Electronics*, Pretoria, South Africa, pp. 679.683, 1998.

T. Fukatsu and M. Hirafuji, "Field monitoring using sensor-nodes with a web server," *J. of Robotics and Mechatronics*, vol. 17, no. 2, pp. 164-172, 2005.

K. Langendoen, A. Baggio, and O.Visser,"Murphy loves potatoes: experiences from a pilot sensor network deployment in precision agriculture," *The 20th Int. Parallel and Distributed Processing Symposium*, pp. 25-29, April 2006.

N. Wang, N. Zhang, M. Wang, "Wireless sensors in agriculture and food industry—recent development and future perspective," *Comp. Electron. Agric.* vol. 50, no. 1, pp. 1-14 2006.

T. Fukatsu and M. Hirafuji, "Field monitoring using sensor-nodes with a web server," *J. of Robotics and Mechatronics*, vol. 17, no. 2, pp. 164-172, 2005.

Sunnorth, SPCE061A compiler user menu v1.0, www.sunnorth.com.cn.

Spectrum Technologies Inc., External temperature sensor category #3667, www.specmeters.com.

Spectrum Technologies Inc., Watermark soil moisture sensor category #6450WD, www.specmeters.com.

E. P. Eldredge, C. C. Shock, and T. D. Stieber, "Calibration of granular matrix sensors for irrigation management," *Agronomy Journal*, vol. 85, pp.1228-1232, 1993.

S. J. Thomson, T. Youmos, and K. Wood, "Evaluation of calibration equations and application methods for the Watermark granular matrix soil moisture sensor," *Appl. Eng. Agric.*, vol. 12, pp. 99-103, 1996.

Electronique-Diffusion Company Inc., REHS135, www.elecdif.com.

Allguy International Co., Ltd., CDS photoconductive cells, http://cn.commerce. com.tw/modules.php?modules=company&action=company_inside&ID= A0001712&s=h.

Ready International Inc., ZigBee 3160 module, http://www.ritii.com/ch/

M. Yu, H. Mokhtar, and M. Merabti, "A survey on Management in wireless sensor networks," www.cms.livjm.ac.uk/pgnet2007/Proceedings/Papers /2007-099.pdf.

Sapon Tanachaiwiwat, Pinalkumar Dave, Rohan Bhindwale, Ahmed Helmy, "Secure locations: routing on trust and isolating compromised sensors in location-aware sensor networks," *SenSys'03*, pp. 324-325, LA, CA, USA, Nov. 5-7, 2003.

S Harte, A Rahman, K M Razeeb, "Fault tolerance in sensor networks using self-diagnosing sensor Node," *The IEE Int. Workshop on Intelligent Environments*, pp. 7-12, 2005.

D. Estrin, R. Govindan, J. S. Heidemann, S. Kumar, "Next century challenges: scalable coordination in sensor networks," *In Mobile Computing and Networking*, pp. 263-270, 1999.

A. T. Tai, K. S. Tso, W. H. Sanders. "Cluster-based failure detection service for large-scale ad hoc wireless network applications in dependable systems and networks," *Proceedings of the 2004 Int. Conf. on Dependable Systems and Networks*, DSN ' 04. 2004.

T. Clouqueur, K. Saluja, P. Ramanathan, "Fault tolerance in collaborative sensor networks for target detection," *IEEE Transactions on Computers*, vol. 53, pp. 320-333, 2004.

M. Ding, D. Chen, K. Xing, X. Cheng. "Localized fault- tolerant event boundary detection in sensor networks," in Proceedings of INFOCOM, 2005.

J. Chen, S. Kher, and A. Somani, "Distributed fault detection of wireless sensor networks," DIWANS'06, Los Angeles, USA:ACM Pres, pp. 65-72, 2006.

16

On the Design and Analysis of Transport Protocols over Wireless Sensor Networks

Suman Kumar and Seung-Jong Park
Computer Science Department and Centre for Computation and Technology
Louisiana State University
USA

1. Introduction

Sensor networks are typically data driven where the whole network cooperates in communicating data from source sensors to sinks (typical repository/server). One of the main characteristics of a typical sensor node is the limited power supply it has (Kahn et al., 1999). Usually, it is battery operated which might last for some months to a year (depending on the type of application and other application specifications). Sensing nodes typically exhibit limited capabilities in terms of processing, communication, and especially, power (Pottie et al., 2000). Different application would have different constraints and priorities on how their sensor network must behave. Thus, energy conservation is of prime consideration in sensor network protocols in order to maximize the network's operational lifetime. Rather than sending individual data items from sensors to sinks, it is more energy efficient to send aggregated data. The net effect of this aggregation is, by transmitting less data units, considerable energy savings can be achieved which is the main idea behind in-network (Madden et al., 2002) aggregation and further distributed processing of the data.
Since enabling communication between sensors and sinks is the major role of sensor networks, many research works [Gopalsamy et al., 2002] have investigated energy-aware data delivery. However, sensor networks experience wireless errors and congestion more severely than other wireless networks because of the low capability to recover from losses and the high node-density. Therefore, robustness is also important to energy conservation since unreliable data delivery, which increases the probability of data retransmission under high loss rates, results in the consumption of a large amount of energy. Although the problem has been addressed by previous works [Heinzelman et al., 1999 & Ye et al., 2003] in the context of wireless ad-hoc networks, such approaches cannot be directly applied to the sensor environment. Because of the distinctive characteristics of multipoint-to-point communication vs. point-to-multipoint communication, the data delivery problem in sensor networks can be seen as consisting of two problems: downstream and upstream data delivery. Therefore, we address these problems as two separate ones. Firstly, a sink-to-sensors energy-aware data delivery scheme is proposed to solve the downstream problem while considering robustness simultaneously. Secondly, a sensors-to-sink energy-aware data delivery scheme is proposed to address the upstream problem.

Therefore, in this chapter, first we construct a probability model for existence of such redundancy among closely related sensor nodes. In the model, we assume sensor nodes are generated with two associated bi-variate Poisson distribution in a plane. We then propose a scalable framework for reliable data delivery. The proposed framework addresses and leverages the characteristics of the wireless sensor networks while achieving the reliability in an efficient manner. First, for downstream data delivery, we formulated the reliable data delivery problem theoretically using the minimum set cover problem and transformed it to the minimum dominating set (MDS) problem. For upstream data delivery, we formulate the perfectly correlated data aggregation problem using the Steiner minimum tree (SMT). We propose a decentralized aggregation method by integrating the shortest path tree and the minimum dominating set to approximate the optimal solution, the SMT. We evaluate the performance of the proposed approach with other previous schemes and we show that the proposed scheme performs substantially. With the help of proposed probability model for redundancy condition, we comment on the design of such schemes.

2. Condition for Data Redundancy between Sensing Nodes

In this section, we introduce a heuristic model for data redundancy in spatially distributed sensor network to characterize the amount of redundancy existing among near neighbour nodes. For the general scenario, although in our analysis we introduce two different kind of sensor nodes (further referred as A and B), it does not affect the general analysis for uniform sensor node scenario. However, it may lead to useful result considering that there are at least two kinds of sensor nodes that differ in some sense1 and still lead to a simplified analysis. We consider that whatever differences sensors have, they are distributed with the same master Poisson process. We recognise that the near neighbour distribution is the main factor contributing to the overlap of sensing regions among nodes that introduces data redundancy among sensor nodes. We give a probabilistic expression giving near node distribution and argue that for a given sensing range how many sensors can deliver partially redundant data.

2.1 System Model

Continuing our two node scenario and assuming data is uniformly distributed throughout the spatial region, the data collected by some node n_i in its sensing region A_i is proportional to the sensing area. Hence, data sensed in area $A_i = \gamma A_i$ Where, γ is some proportionality constant that depends on sensing ability of sensors. Hence, for sensing nodes A and B, the correlation factor is given by,

$$\frac{A \cap B}{A \cup B} \tag{1}$$

Assuming uniform node configuration of all the nodes, the sensing radius is r_s and transmission range is r_t. the sensing area is given as $A = \pi r_s^2$.
For a particular node say s, all the other nodes in area $4r_s^2$, shares some degree of redundant information with s. In figure 1, two nodes A and B has position vectors r and r′ respectively and r_s is their sensing range, the condition that these two nodes share redundant information is given by,

$$|r - r'| < 2r_s \quad (2)$$

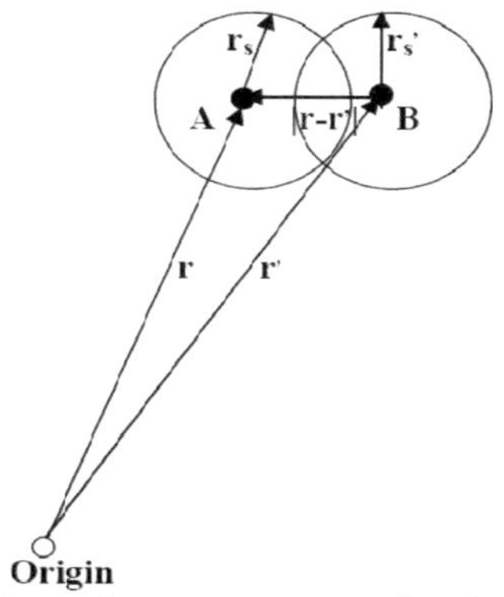

Fig. 1. Condition for Data Redundancy between two nodes A & B

Hence, to quantify the redundancy for all the neighbours around a sensor node we have to find out its near neighbour distribution in its own sensing range. Next section presents an analysis, assuming sensor nodes follows a spatial bi-variate distribution for sensor nodes, A and B. Here, we consider nodes A and B which are different in terms of sensing rate or some other figure of merit, say, sensing capability factor or can be totally different sensors.

2.2 Nearest Neighbour Distribution

Maritz (Maritz, 1952) obtained the probability generating function for the bivariate poison assuming that, in any interval of length dt, the combinations ($A\bar{B}, \bar{A}B, AB$ and $\overline{AB}$) of the two events A and B, occur with probabilities λdt, νdt, μdt and 1 - (λ+ν+μ)dt. Since, this analysis involves time bivariate distribution, we write the spatial bivariate distribution by following the same line of analysis by assuming event A represents the sensor type A and B represents sensor type B.

The distribution of the distance between two adjacent points, the nearest neighbour distribution considering marginal distributions are Poisson, we get the following relationship,

prob(X_{BB}(distance from a point B to next nearest point B)<r)=1-$e^{-(\mu+v)\pi r^2}$ (3)

and similarly for A. The distribution of the distance from a point A to a nearest point B may be derived as follows:

prob (X_{AB}(distance from a point A to nearest point B) > r)

= prob (A single) prob (distance from A to nearest B > r | A single) + prob (A double) prob (distance from A to nearest B > r | A double)

$$= \frac{\lambda}{\lambda+v} e^{-(\mu+v)\pi r^2}(1 + \int_{|x|>r} h(x,0)dx \quad (4)$$

Hence, prob (X_{AB} < r)=

$$1 - e^{-(\mu+v)\pi r^2}(\lambda + v\frac{\int_{|x|>r} h(x,0)dx}{\lambda+v}) \quad (5)$$

When A and B are independent, i.e. when $v = 0$, 5 reduces to the distribution of the distance from a random point to the nearest point B which is the same distribution as given in equation 3. For the sensing range $2r_s$ equation 4 gives the condition for two sensors sharing redundant data as below:

$$1 - e^{-(\mu+v)\pi r_s^2}(\lambda + v\frac{\int_{|x|>2r_s} h(x,0)dx}{\lambda+v}) \quad (6)$$

3. Down Stream Reliable Data Delivery over Sensor Network

In this section, we consider the problem of reliable downstream point-to-multipoint data delivery, from the sink to the sensors, in wireless sensor networks (WSNs). The need (or lack thereof) for reliability in a sensor network is clearly dependent upon the specific application the sensor network is used for. Consider a security application where image sensors are required to detect and identify the presence of critical targets. Given the critical nature of the application, it can be argued that any message from the sink has to reach the sensors reliably. The problem of reliable data delivery in multi-hop wireless networks is by itself not new, and has been addressed by several existing works in the context of wireless ad-hoc networks (Tang & Gerla, 2001). However, such approaches do not directly apply to a sensor environment because of three unique challenges imposed by the following considerations:
The issue of reliability is addressed in following context:
Downstream Reliability: We restrict the scope of this work to downstream reliability.
Communication and Node failures: A scheme that addresses reliability in a sensor network environment, has to deal with communication failures and node failures. The proposed algorithm will handle both communication and node failures.
Message size: We assume that the message size to be sent by the sink consists of one or more packets.
Metrics: We consider latency and energy consumption as the metrics of interest for comparison with other existing approaches. The goals is to minimize these metrics.
Network Model: We assume that both the sink and the sensors in the network remain static. We also assume that there is exactly one sink coordinating the sensors in the field. Further, since sensor networks have a large number of sensor nodes, the proposed approach must be scalable to the number of nodes in the network.

3.1 Design Choices and Challenges

We have following basic design choices:
1. A NACK based loss recovery scheme is preferable to an ACK based scheme as the latter suffers from the ACK implosion problem.
2. Local and dynamically assigned designated servers are essential to minimize the retransmission data overhead.
3. Out-of-sequence forwarding should be preferred to maximize the spatial reuse in the network.

We outline following challenges that need to be addressed to provide effective downstream data delivery:
1. Environment Constraints: It is evident that sensor network have two main constraints. First, Bandwidth and energy constraint and second frequent node failure problem.
2. ACK/NACK Paradox: This challenge stems out from the constraints imposed by typical message types that can be expected to use the downstream reliability. While the query-data and control code can be expected to be of non-trivial message size, queries pose a unique problem because of their short message sizes. While an ACK based recovery scheme would address the problems, its other deficiencies (in terms of ACK implosion) however clearly prohibit it from being used. Whereas, NACKs cannot handle the unique case of all packets in a message being lost at a particular node in the network. Since the node is not aware that a message is expected, it cannot possibly advertise a NACK to request retransmissions. NACK based scheme require in-sequence forwarding of data by nodes in the network to prevent a NACK implosion (Wan et al., 2002). This will clearly limit the spatial re-use achieved in the network.

3.2 Ideal Solution: Minimum Set Cover Problem

To solve the reliability problem at wireless sensor networks, it is necessary to formulate the problem into an optimization problem which has been known as a common and typical problem and investigated for optimal solutions. Assuming that the lost packet can be retransmitted and recovered by one of neighbours which received the lost packet before, a solution tries to designate a set of nodes, called recovery servers, which retransmit the lost packet in an optimal fashion. We will call this problem as loss recovery server designation problem. By the nature of local broadcasting of wireless communication, one recovery server can recover the lost packet of all neighbours around it. Therefore, it is optimal to minimize a size of the set of recovery servers covering all nodes which did not receive the packet. And it is necessary to find the optimal recovery sets for different loss patterns of each packet. The above loss recovery server designation problem can be defined as a set cover problem in the graph theory, the problem of covering a base set (nodes which did received a packet successfully) with as few elements of a given subset system (a set of recovery servers) as possible. However, Karp (Karp, 1972) showed that the decision version of the minimum set cover (MSC) is NP-complete. A common approach of coping with NP-hard problems is approximation algorithms that run in polynomial time and deliver solutions that are close to the optimal solution.
Therefore, we address the loss recovery server designation problem with an alternative which has similar complexity and advantages to solve the problem in decentralized fashion. In a graph, a dominating set is a subset of nodes such that for every node v in a graph, either a) v is in the dominating set or b) a direct neighbour of v is in the dominating set. The minimum dominating set problem asks for a dominating set of minimum size. The reason to choose MDS is considering the fact that MSC is equivalent to the MDS problem under L-reduction closely related to each other and have been shown to be NP-hard (Garey & Johnson, 1979). Although the MDS problem has different instances reduced from different instances of MSC problem, an instance for MDS problem can include a whole network by covering a set of nodes and edges which are not adjacent to a given set S. Therefore, we can handle the MDS problem without concerning the loss pattern S although there are trade-offs: the advantage of MDS is that we can solve MDS problem without considering different

instances for different loss patterns; and the disadvantage of MDS is that the cost of optimal solution for an instance of MDS is larger than that of optimal solution for an instance of MSC for given loss pattern S. we can use the approximated solution of MDS to solve the MSC which is the optimal solution of the loss recovery server designation problem

3.3 A Framework for Down Stream Data Delivery Scheme

The centerpiece of proposed design is an instantaneously constructible loss recovery infrastructure called the core. The core is an approximation of the minimum dominating set (MDS) of the network sub-graph to which reliable message delivery is desired. While using the notion of a MDS to solve networking problems is not new (Sivakumar et al., 1999), the contributions of this work lie in establishing the following for the specific target environment: the relative optimality of the core for the loss recovery process, how the core is constructed, how the core is used for the loss recovery, and how the core is made to scalably support multiple reliable semantics.

3.3.1 Core Construction

We assume that the first packet is reliably delivered for the initial discussions. The core forms the set of local designated loss recovery servers that help in the loss recovery process. The core is constructed using the first packet delivery. The reliable delivery of the first packet determines the hop count of the node in the network, which is the distance of the node from the sink. A node, which has a hop count that is a multiple of three, elects itself as a core if it has not heard from any other core node. In this fashion, the core selection procedure approximates the MDS structure in a distributed fashion (Figure 3). The uniqueness of the core design in this approach lies in the following characteristics: (i) the core is constructed using a single packet flood, more specifically during the flood of the first packet; and (ii) the structure of the sensor network topology (with sensors placed at fixed distances from the sink) is leveraged for more efficient, and fair core construction.

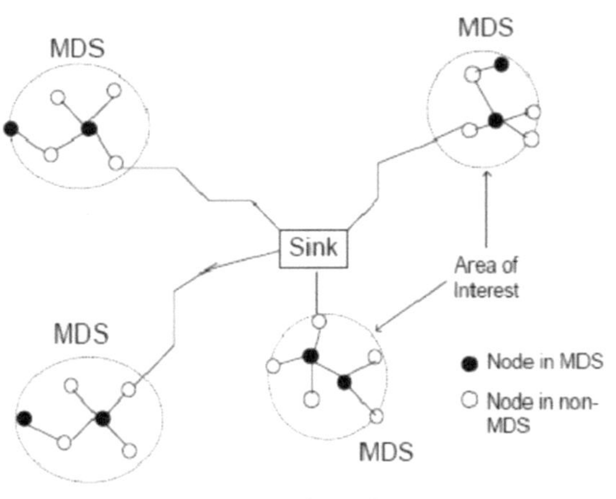

Fig. 3. Core Construction as an approximation of MDS

The core construction uses following algorithm:
Sink: When the sink sends the first packet, it stamps the packet with a "band-id" (bId) of 0.

When a sensor receives the first packet successfully, it increments its bId by one, and sets the resulting value as its own band-id. The band-id is representative of the approximate number of hops from the sink to the sensor.

Nodes in 3i bands: Only sensors with band-ids of the form 3i, where i is a positive integer, are allowed to elect themselves as core nodes. When a sensor S_0 with a band-id of the form 3i forwards the packet (after a random waiting delay from the time it received the packet), it chooses itself as a core node if it had not heard from any other core node in the same band. Once a node chooses itself as a core node, all packet transmissions (including the first) carry information indicating the same. If any node in the core band that has not selected itself to be a core receives a core solicitation message explicitly, it chooses itself as a core node at that stage. Every core node S_3 in the 3(i+1) band should also know of at least one core in the 3i band. If it receives the first packet through a core in the 3i band, it can determine this information implicitly as every packet carries the previously visited core node's identifier, bId, and Amap. However, to tackle a condition where this does not happen, S_3 maintains information about the node (S_2) it received the first packet from, and the S_2 node maintains information from the node (S_1) it received the first packet from. After a duration equal to the core election timer, S_3 sends an explicit upstream core solicitation message to S_2, which in turn forwards the message to S_1. Note that by this time, S_1 will already have chosen a core node, and hence it responds with the relevant information.

Nodes in 3i+1 bands: When a sensor S_1 with a band-id of the form 3i+1 receives the rst packet, it checks to see if the packet arrived from a core node or from a non-core node. If the source S_0 was a core node, S_1 sets its core node as S_0. Otherwise, it sets S_0 as a candidate core node, and starts a core election timer. If S_1 hears from a core node S_0 before the core election timer expires, it sets its core node to S_0 . However, if the core election timer expires before hearing from any other core node, it sets S_0 as its core node, and sends a unicast message to S_0 informing it of the decision.

Nodes in 3i+2 bands: When a sensor S_2 with a band-id of the form 3i+2 receives the first packet, it cannot (at that point) know of any 3(i+1) sensor. Hence, it forwards the packet without choosing its core node, but starts its core election timer. If it hears from a core node in the 3(i+1) band before the timer expires, it chooses the node as its core node. Otherwise, it arbitrarily picks any of the sensors that it heard from in the 3(i+1) band as its core node and informs the node of its decision through a unicast message. If it so happens that S2 does not hear from any of the nodes in the 3(i+1) band (possible, but unlikely), it sends an anycast core solicitation message with only the target band-id set to 3(i+1). Any node in the 3(i+1) band that receives the anycast message is allowed to respond after a random waiting delay. The delay is set to a smaller value for core nodes to facilitate re-use of an already elected core node. A boundary condition that arises when a sensor with a band-id of 3i+2 is right at the edge of the network, is handled by making the band act just as a candidate core band (3i). Such a condition can be detected when nodes in that band do not receive any response for the anycast core solicitation message. Thus, at the end of the first packet delivery phase, each node knows its bId, whether it is a core node or not, and in the latter case its core node information. In addition, every core node in the 3(i+1) band knows of at least one core node in the 3i band.

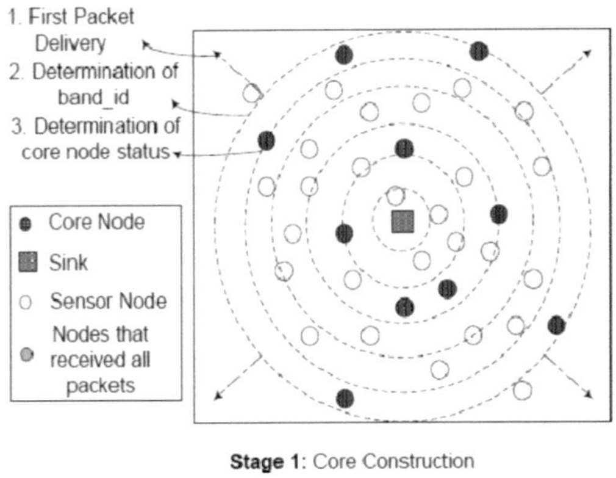

Fig. 4. Core Construction

3.3.2 Loss Recovery Process

Once the core is constructed, the framework employs a two-stage recovery process that first involves the core nodes recovering from all lost packets, and then the recovery of lost packets at the non-core nodes. The reasons for using two-stage recovery are threefold: (i) the number of non-core nodes will be a substantial portion of the total number of nodes in the network, and hence precluding any contention from them is desirable; (ii) when the core nodes perform retransmissions for other core nodes, holes corresponding to a single packet among a core node's neighbours would also be filled with a single retransmission; and (iii) when only the core nodes are performing retransmissions during the second phase, due to the nature of the core (ideally, no two core nodes are within two hops of each other), the chances for collisions between retransmissions from different core nodes are minimized. The recovery process for the core nodes is performed in parallel with the underlying default message-forwarding (Figure 5). This parallel recovery process for the core nodes does not increase the contention in the network significantly because the fraction of core nodes is very small compared to the total number of nodes in the network, and all requests and retransmissions are performed as unicast transmissions to the nearest upstream core that has a copy of the lost packet.

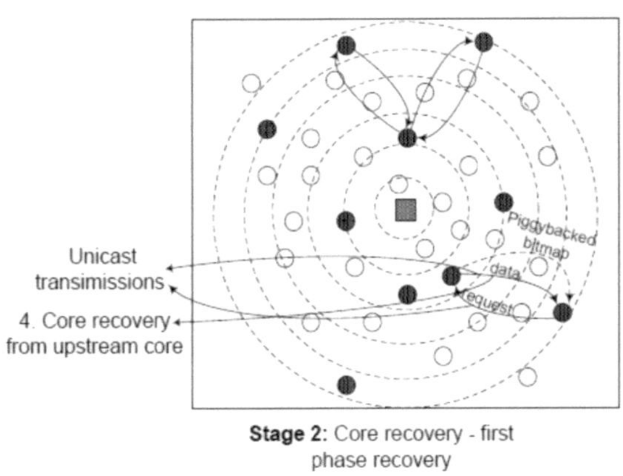

Fig. 5. Loss recovery for Core Nodes

The second phase of the loss recovery starts only when a non-core node overhears an A-map from the core node indicating that the core node has received all the packets in a message. Hence, the second phase of the loss recovery does not overlap with that of the first phase in each local area, preventing any contention with the basic flooding mechanism, and with the first phase recovery. To inhibit unnecessary retransmission requests, proposed scheme uses a scalable A-map (Availability Map) exchange between core nodes that conveys meta-level information representing availability of packets with bits set. Any downstream core node initiates a request for a missing packet only if it receives an A-map from an upstream core node with the corresponding bit set. The core recovery phase is highly efficient as the core nodes initiate requests only when they are sure of an upstream core node having a particular packet.

3.3.3 Role of WFP Pulse Transmission

Reliable single packet delivery is leveraged for the instantaneous core construction. To achieve that, we use WFP pulse transmission. WFP Pulse can be regarded as a short period signal which does not include any information, the transmission period of the WFP pulse is significantly smaller when compared to the transmission time T_D required for a regular data packet. Also, twice the regular transmission power is used to transmit the pulses to achieve relative amplitude of 3dB at the receiver. To increase the robustness of the pulse detection, every set of pulse transmission includes p pulses transmitted consecutively within a period T_P ($T_P << T_D$). Figure 6 shows the transmission scheme for the WFP pulse. Hence, receivers infer an incoming WFP signal only after detecting p pulses. As shown in figure, the WFP pulse is forced in this design.

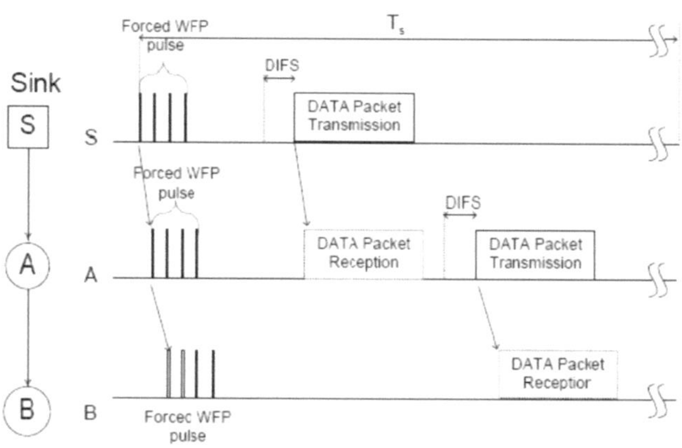

Fig. 6. Example for Single or First Packet Delivery

Figure 6 shows the basic procedure of the single or the first packet delivery with a simple topology. When a sink wants to initiate a reliable single first packet delivery, it sends a set of forced WFP pulses without sensing the wireless channel. When neighbouring sensors hear WFP pulses, they send a set of forced WFP pulses immediately. After a deterministic period that is set based on the diameter of the network, the sink transmits the single first data packet subject to the medium access scheme, e.g., CSMA. If the node A receives the single/first packet, it changes its operation from the advertisement mode to the delivery mode by halting the WFP pulses, and by sending the single/first data packet after carrier-sensing. However, if the single/first packet is lost, nodes will continue to transmit the WFP pulses, which in turn trigger retransmissions. Figure 7 shows the case of retransmission. Since the forced WFP pulses sent every Ts period play the role of a NACK signal, node B will wait for a duration of at least Ts to send next set of forced WFP pulses. Therefore, the latency for the single/first packet delivery is directly dependent upon T_s.

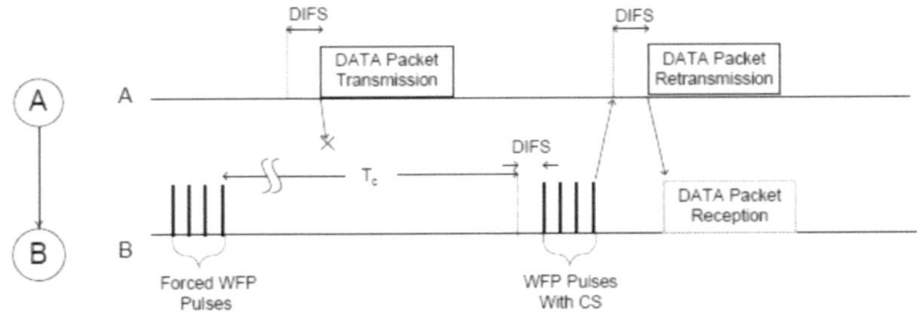

Fig. 7. Loss Recovery using WFP Pulse Transmission

To reduce the latency, it uses another kind of WFP pulse which a node sends after a regular carrier sensing operation. Node B sends p number of WFP pulses after carrier-sensing (WFP_{cs}) opportunistically (unless it has received the single/first packet) with a period T_c which is smaller than T_s. The period T_c should be proportional to the hop distance of the node B from the sink because a node should wait until the upstream nodes between the node and the sink receives the single/first packet. Since a node senses the state of channel before transmitting WFP_{cs} pulses, the WFP_{cs} pulses have a lesser probability of colliding with data packets than WFP pulses. When a node gets to transmit WFP_{cs} pulses, it resets the timer corresponding to the Ts time period for forced WFP pulses.

4. A Framework for Energy Efficient Upstream Data Delivery

In Section 2, probability condition (Equation 6) is derived based on near neighbour distribution for spatial correlation among data between neighbouring nodes. In this Section, we consider the problem of data aggregation in environments where the data from the different sensors are spatially correlated to each other. To do that, we present a simple, scalable, and distributed approach for approximating the Steiner minimum tree, and thereby achieve the potential cost benefits introduced earlier. Moreover, we can solve the upstream data delivery problem without any overhead because the proposed approach uses the same minimum dominating set structure, the core, which already has been constructed through the query delivery. To aggregate perfectly correlated data in an energy-efficient way, we use two structures that have been constructed during downstream data delivery: (i) the minimum dominating set (MDS) which is same to the core structure proposed in Section 4.6 and (ii) the shortest path tree which is constructed through a basic flooding. The purpose of the MDS structure is to aggregate correlated data from neighbouring sources; that of SPT is to gather aggregated data among core nodes in the MDS. The correlation factor depends on the degree of correlation i.e., the probability of finding a near neighbour node to a particular node. The probabilistic model is helping to design such an Up-stream data delivery mechanism however, it is not limited to any particular case of distribution and hence provides a generalized approach.

Although there have been many previous works in (Hwang et al., 1992) on the approximation of the SMT, those schemes still require computational and communication overheads that WSNs cannot support. In this section, we design an aggregation structure that approximate the optimal solution in a distributed fashion with less amount of overhead than distributed approximation of the SMT. From the definition of the Steiner minimum tree (SMT), we need to find an additional set of nodes that are not sources and inserted into the SMT in order to achieve the shortest connectivity. In graph theory, this set is called "Steiner points. Therefore, one of the above heuristics also tries to find these Steiner points. However, since these Steiner points depend on the locations of sources, we need to find the optimal set of Steiner points after we know the exact locations of sources. Instead of solving the SMT problem of which optimal solutions are different to each other based on given set of sources, we address it with the minimum dominating set (MDS) problem of which optimal solution is not changed irrespective of given set of sources. Assuming perfect correlation among all data, it is well known that the early aggregation around sources is to reduce redundant data in tree structures. And, we can utilize the above heuristic using the MDS approach. Each node in MDS can work as a Steiner point if it has any neighbouring sources around it.

After a query flooding constructs the core structure, data aggregation can use the core to find the set of Steiner points which aggregate data from neighboring sources. Then the data at some core nodes can be forwarded to its upstream core locating at inside core band since the core structure has the shortest path information toward a sink. Eventually, all data from core nodes will reach a sink through the shortest path that was constructed while a query was flooded. Although there is a gap between the optimal solution of the Steiner minimum tree and the approximated solution using the minimum dominating set, the proposed MDS approach can obtain a promising result compared to other approximations that assume centralized coordination and high computational complexity.

The following are the key goals that the design of proposed data aggregation strategy is based on following:

Perfect Correlation: Since our focus is on the aggregation problem, assuming all data from sensors are perfectly correlated, the amount of aggregated data is equal to the amount of original data before aggregation.

Efficiency: Since the energy conservation is the critical issue in WSNs, the goal of design is to minimize the energy consumption at data aggregation. To minimize the energy consumption, it is better to reduce redundancy among data while data are delivered. Therefore, the proposed scheme will aggregate correlated data as soon as and as much as possible to reduce redundancy.

Scalability: In general, WSNs might have more than tens of thousands sensors. The proposed scheme should be operated efficiently with reasonable amount of overhead linearly increasing to the scale of WSNs.

Decentralization: Since using global information in a distributed environment such as a sensor network can incur high overheads, the proposed scheme should use purely local information in its approach. Then it will be operated in a decentralized fashion over large scale of WSNs.

Loose Synchronization: To minimize the cost of aggregation, most of theoretical solutions use tree structures, e.g., the shortest path tree, the minimum spanning tree and the Steiner

minimum tree. Although these tree structures reduce the redundancy among data, they also requires synchronization among nodes that transmit, aggregate or forward data. However, since the synchronization is also one of hard problems in WSNs, the proposed scheme will relax the degree of synchronization so that it can be operated without assumption of other synchronization algorithms.
Mobility and Node Failures: The dynamic change of network topology due to mobility and node failures makes aggregation schemes in WSNs inefficient and even more out of service. Therefore, the proposed scheme will address this problem by constructing an aggregation structure, dynamically and instantaneously.

4.1 Core Construction

Same core construction mechanism is used as presented in section 3. Based on this core structure, a node in a network should be one of core nodes, non-core nodes, or leaf nodes.
A core node is a node at a core band of which band-id2is 3i. Two core nodes in the same core band should have at least two-hop distance between each other to reduce the total number of core nodes. A core node also keeps the information of a precedent in the shortest path tree root at a sink, so that the core at 3i band can transmit the data to another core node at inner core 3(i-1) band, eventually.

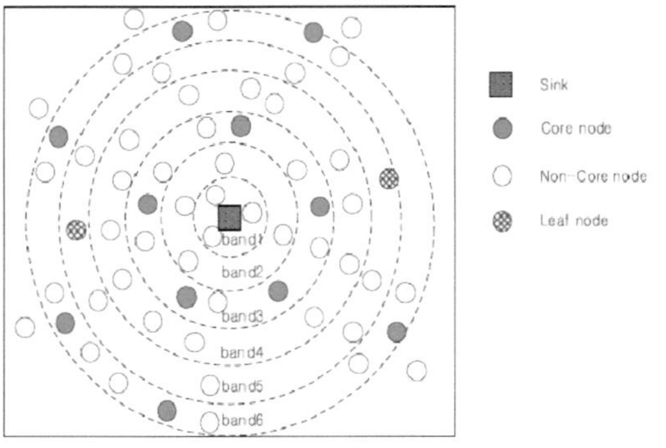

Fig. 8. Instantaneous Core Construction in Up-Stream Data Delivery Scheme

All nodes at non-core bands 3i+1 or 3i-1 should be a non-core node. And some nodes at core band 3i might become a non-core node based on the core construction procedure. All non-core nodes should access two nodes: its core node at 3i band and its precedent in the SPT, of which band-id is less than its band-id. Some non-core nodes at 3i+1 or 3i-1 band cannot have a neighboring core node at 3i band. In this case, they can still access a core node at 3i band through its neighboring non-core node at 3i band indirectly. For exceptional cases, some non-core nodes of which band-id is 3i+2 cannot have any neighboring nodes located at core band 3i+3. These non-core nodes declare themselves as a leaf node. Then they always transmit data to a precedent that is a non-core node at inner band 3i+1. Figure 8 shows the instant result for core construction by disseminating a query through a network.

4.2. Two Stages of Data Aggregation

Stage 1: Original Data Transmission
We assume that all nodes know the start time of data transmission for each query.

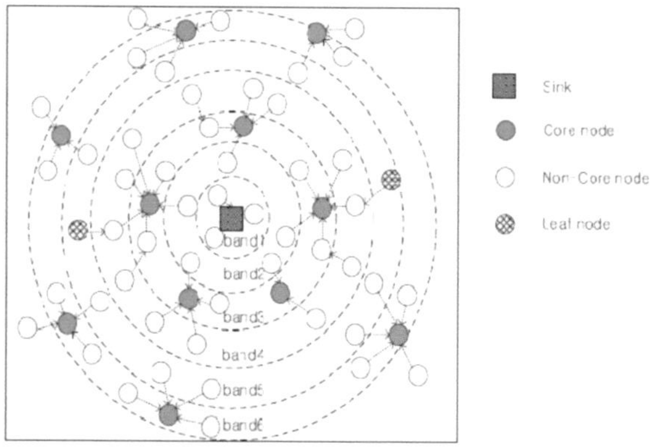

Fig. 9. Stage 1: Original Data Transmission

If a non-core node at 3i-1 or 3i+1 band is a source node, it will transmit data to its core at core bands after some delay. If the receiving node at core band 3i does not declare itself as a core node, it will forward the data to its core node at the same core band 3i. We use a contention-free medium access control scheme to coordinate all non-core sources around a core node based on the number of non-core nodes around the core node. In Figure 9, all non-core nodes, white circles, send data to core nodes, gray circles. Between different groups around each core node, we don't need to consider scheduling because they are separated with each other at least two-hop distance.

If a leaf node at 3i+2 band is a source node, it will transmit data immediately to its neighbouring non-core node at 3i+1 bands so that the neighbouring non-core node can receive the data successfully before it sends its own data. In Figure 9, leaf nodes at band 5, checked circles, send data to transmit data successfully to a non-core node within that delay.

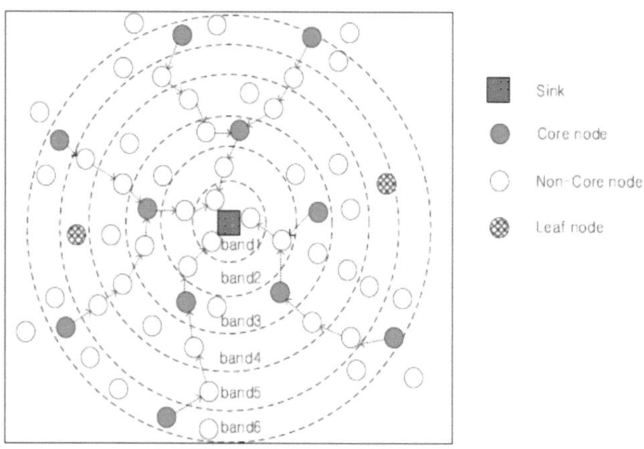

Fig. 10. Stage 2: Aggregated Data Transmission

If a core node at core bands is not a source node, it does not need to transmit data unless it receives any data from its non-core nodes or core nodes at outer core band. Although the core node has data to send, it will wait for some time. so that it can wait and aggregate its own data with incoming data from other core nodes that are located at outer bands.

Stage 2: Aggregated Data Transmission
After stage 1, we assume that all data from non-core nodes are received by core nodes and aggregated with other data. The remaining procedure is to deliver the aggregated data to a sink. To deliver these aggregated data, this scheme uses the shortest path tree that was constructed during the corresponding query flooding. Figure 10 shows delivery paths between core nodes at different core bands. Compared to the original shortest path tree, the paths have some differences. Instead of reaching a sink directly using the SPT, it is better to reach another core node at inner band since it can reduce redundancy among other aggregated data. Whenever a non-core node at core bands receives aggregated data from other core nodes at outer bands, it will forward them to its core node at the same core band.

5. Peformance Analysis and Discussion

This section is focussed on a formal analysis and performance evaluation of the protocol design proposed in this chapter.

5.1. Downstream Data Delivery

For easy reference we call our downstream data delivery scheme GARUDA. The NS2 simulator is used for all evaluations. For all experiments: (a) the rst 100 nodes are placed in a grid fashion within a 650m x 650m square area to ensure connectivity, while the remaining nodes are randomly deployed within that area, and the sink node is located at the centre of one of the edges of the square; (b) transmission range of each node is 65m ; (c) channel capacity is 1 Mbps; and (d) each message consists of 100 packets (except for the single packet delivery part); and the size of packet is 1 KB. CSMA/CA is used as the MAC protocol. We use basic flooding as the routing protocol. All the simulation results are shown after averaging the metrics over 20 randomly generated topologies and calculating 95% confidence intervals. We choose a fixed packet loss rate of 5% for wireless channel error, and vary the number of nodes in the network, which in turn increases the degree of contention in the network.

5.1.1 Latency

The latency involved in receiving a single packet reliably and multiple packet delivery with increasing number of sensors is presented in Figure 11(a) and (b) respectively for both the proposed framework and the ACK based scheme. The latency of the proposed scheme was significantly smaller because of the two radio approach, which used an implicit NACK scheme. This means that there was no explicit NACK sent to the sender of a packet if a packet was not received, thus not increasing the load in the network. Although, our core construction scheme used out-of-sequence delivery, we piggybacked the A-map of the core node along with the transmission of each packet which allows the non-core nodes to wait for the core to recover from all loses prior to any retransmission requests thus eliminating the NACK implosion problem.

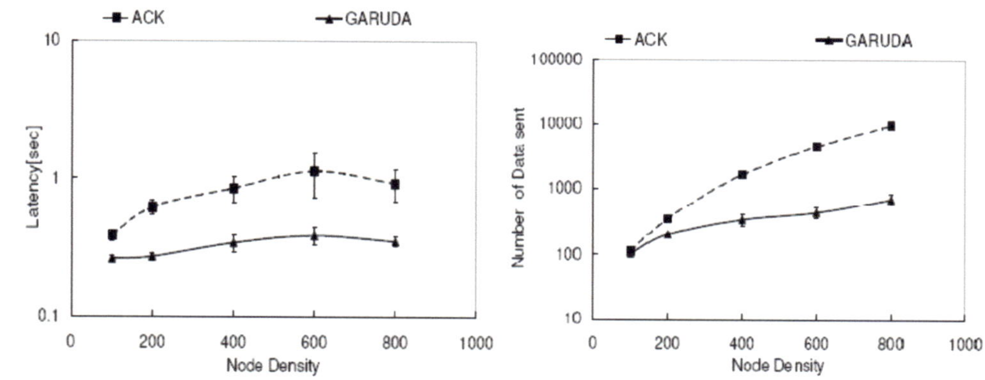

(a) (b)

Fig. 11. Latency Comparison between proposed Down Stream Data Delivery Scheme and Basic ACK Scheme for (a) First/Single Packet Delivery and (b) Multiple Packets Delivery

5.1.2 Number of Data Packet Sent

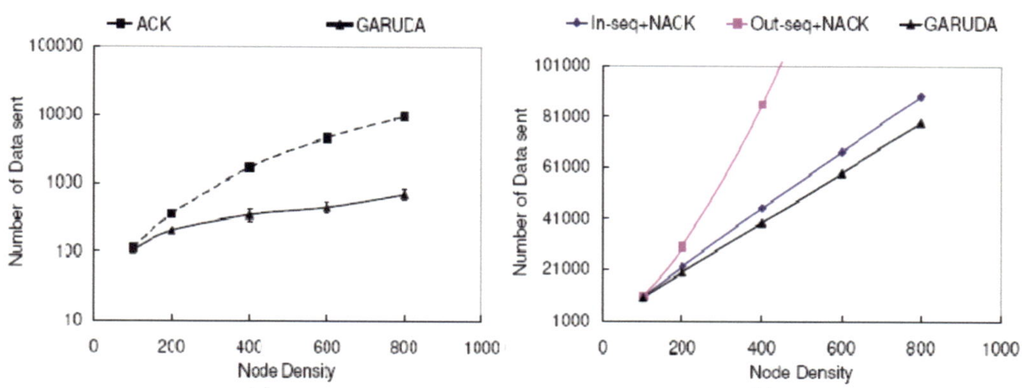

(a) (b)

Fig. 12. Number of Data Packets Sent among among the proposed approach and (a) Basic ACK Scheme for First/Single Packet Delivery (b) Alternatives for Multiple Packets Delivery

Figure 12(a) shows the number of data sent by the proposed framework and the ACK based scheme. It is interesting to note that in our proposed framework, the number of data sent increased more or less linearly (with a slope of 1 approximately) as the number of nodes increased. The implicit NACK scheme coupled with the inherent redundancy involved in the flooding process itself is the main reason for this trend. The implicit NACK scheme alleviates congestion related losses, while the inherent redundancy and the broadcast nature of the flooding process ensures that the packet is received successfully without any need for retransmission even in the presence of losses. For the ACK based scheme, the number of data packets sent was appreciably higher and showed a nonlinear increasing trend with increasing number of nodes in the network. This is again because of the increased load in the network due to the presence of ACK transmissions thus increasing the losses in the network. We observe in the case of multiple packet delivery (Figure 12(b)), that the proposed scheme outperforms alternative schemes.

5.1.3 Energy Efficiency

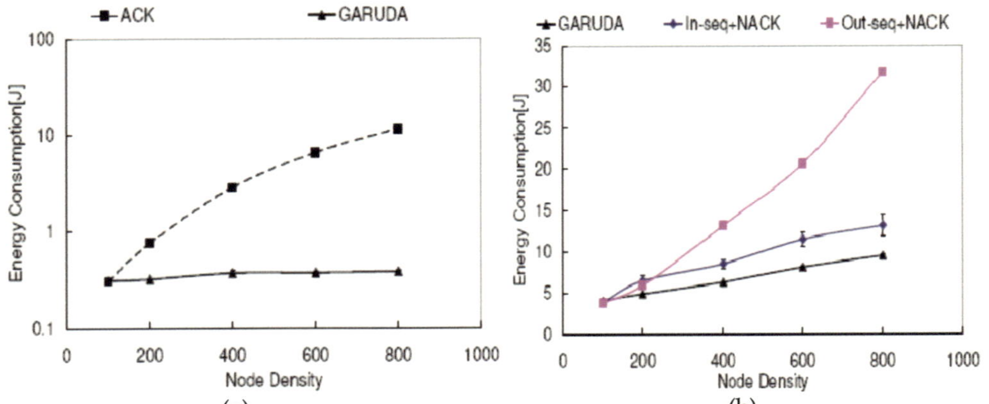

Fig. 13. Energy Consumption per node Comparison between proposed Down Stream Data Delivery Scheme and Basic ACK Scheme for (a) First/Single Packet Delivery and (b) Multiple Packets Delivery

Figure 13 shows energy consumption per node comparison between proposed scheme and other alternative schemes. The average energy consumed per node is significantly smaller for the our case when compared to the other two cases (Figure 13(b)). The average energy consumed for all three cases was directly proportional to the number of transmissions, which was the sum of the number of requests sent and the number of data sent per node. Hence, the reduction in energy consumption follows.

5.2 Up Stream Energy Efficient Data Delivery

Likewise, we refer GARUDA-UP in the graphs for easy reference in this section for upstream data delivery scheme proposed in this chapter. We assume a typical one-shot query-response model in sensor networks. In this model, a sink broadcasts a query to the entire network and sensors that have corresponding information will reply with one message.

In terms of message size, we assume that every source sends one message of the same size, but the specific length of the message does not matter. We use a discrete event simulator for all evaluations. The simulation topologies are largely similar to that used in general sensor networks: 2000 to 8000 nodes uniformly distributed within a circular field of radius 400m. The number of sources that generate messages for one specific query varies from 1/10 to 1/4 of the total number of nodes in the network. We compare GARUDA-UP with SPT since most of the current routing protocols in the context of WSNs such as Directed Diffusion and GPSR try to approximate the message complexity of SPT. We are interested in how GARUDA-UP performs better compared to the centralized algorithm. We also compare it with MST, which represents the optimal solution in the target environment. Ideally, we should have compared it with the Steiner minimum tree. But as we mentioned before, the computation overhead is very high, especially since we are considering thousands of nodes, and the time it takes to generate even one sample is prohibitive. For this reason, we use MST to approximate Steiner Tree performance which has the same message complexity order, that of Steiner minimum tree, but a much less computation cost. We generate SPT with Dijkstra's algorithm and MST with Prim's algorithm. We evaluate the GARUDA-UP approach using message complexity that is equal to the total cost of data aggregation. For message complexity, we measure the total number of transmissions required for all responses to reach the sink. To focus on the comparison of aggregation efficiency of different structures, we assume a perfect MAC layer that avoids collisions for all approaches. All the simulation results are derived after averaging results over 10 random seeds and are presented within 95% confidence intervals.

5.2.1 Node Densities

From Figure 14(a) and (b), we observe that proposed scheme outperforms the SPT scheme under all situations. Therefore, from the simulation results, we can say that GARUDA-UP is a good decentralized approximation to the MST. We can also see that the cost of the SPT increases faster than that of the proposed approach as the number of nodes increases. This is expected since more number of nodes reduces the efficiency of aggregation in the SPT as the paths chosen by different sources are less likely to overlap.

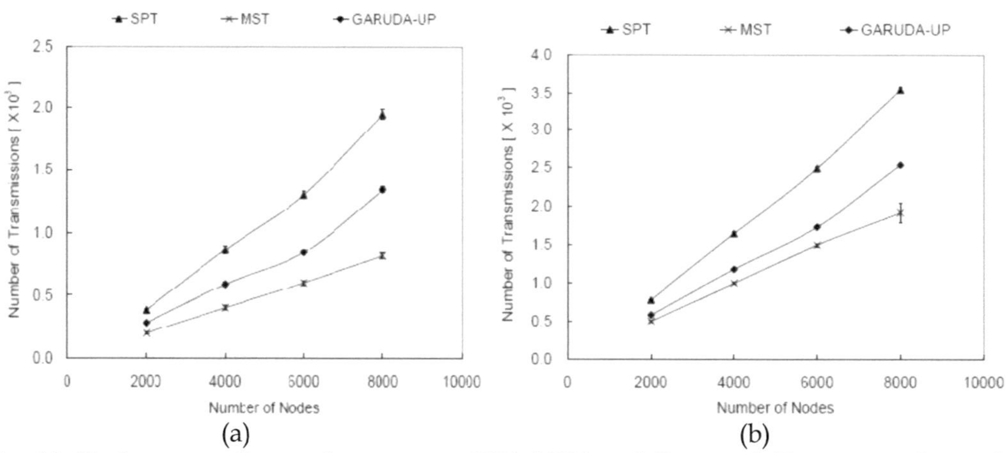

Fig. 14. Performance Comparison among SPT, MST, and Proposed Upstream scheme for Varying Number of Nodes and Fixing the Ratio of Number of Nodes to that of Sources to (a) 10 and (b) 4

Therefore, the proposed approach can be considered as a more scalable decentralized approach as the number of nodes increases. Furthermore, it is observed that the difference between two schemes increases as the ratio of the number of sources to the number of nodes, decreases because more number of sources increase the probability of aggregation for the SPT.

5.2.2. Role of Redundancy

We outline that upstream design is very much dependent on spatial distribution of sensor nodes in a plane. It is interesting to know with what probability we can find sensor nodes in the neighbourhood that support this kind of scheme. To illustrate this, we take an isotropic

Gaussian case and show how probability of detection of redundancy varies with distance. We integrate equation 6 over the region of the area $4\pi r_s^2$. We get the following equation:

$$p(X_{AB} < 2r_s) = 1 - e^{-(\mu+v)\pi r_s^2}\left(\lambda + v\frac{\frac{1}{4}\sqrt{\frac{\pi}{a}} - \sqrt{\frac{1}{4a}}\mathrm{erf}\,(2ar_s)}{\lambda+v}\right) \quad (7)$$

where,

$$a = \frac{1}{2(\sigma_1{}^2 + \sigma_2{}^2)}$$

and erf(x) is error function for each element of x.
Event type AB is related to the event that A and B both occur i.e., the event type AB. Hence, we are interested in the variation of $pr(X_{AB} < r)$ with radial distance r. In figure 4, we show the probability variation for three values. As we can see, when it is very low (0.001), the $pr(X_{AB} < r)$ is low and as we increase the value of the probability, its value increases. Those values play an important role in nearest neighbour probability. For a densely distributed sensor network, value is large and hence results into more redundant data collected by neighbourhood nodes.

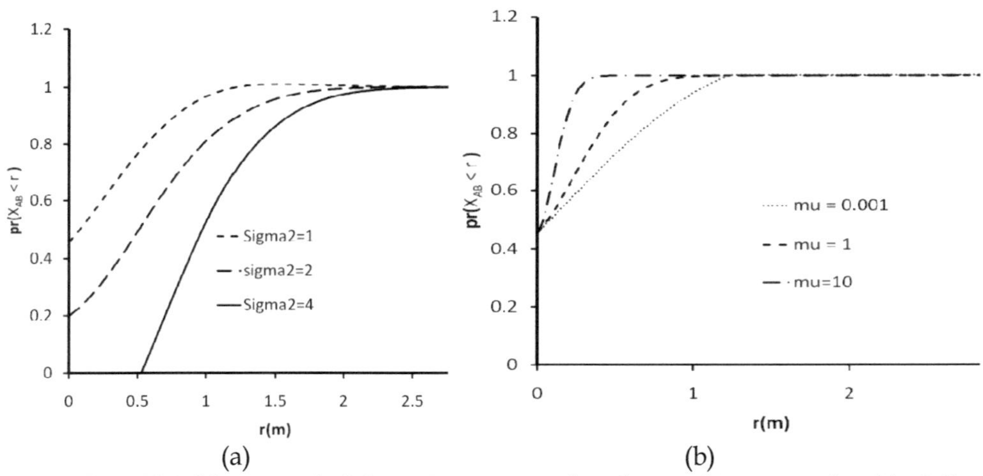

(a) (b)

Fig. 15. Near Neighbour probability Variation with relative distance r for (a) different σ_2(Sigma2) values for λ = 0:1, μv = 0:1, v= 0.1 and = σ_1 and (b) different v (mu) values for σ_2 =1 values for λ = 0.1, μ = 0.1, v= 0.1 and σ_1= 1

Figure 15(b) shows the variation of $pr(X_{AB} < r)$ with distance r for different variance values. As the value of variance increases, we expect less redundancy in the data values as shown in the figure. The figure demonstrates that variance is one critical design parameter. In both figures (15(a) and (b)), we see as distance increases, probability to find a nearest neighbour point increases and converge to 1. However, not necessarily they all satisfy the condition of redundancy. Only those sensors first the condition of redundancy that are separated by not more than sum of their sensing ranges. There are different sensing ranges for different sensor networks and under the given distribution, we can easily calculate the probability of two sensors overlapping each other's sensing ranges. We show the figures for only 3m range as after that the probability converge to 1 (however it may not be true for all the distributions) and remain constant for larger values of radial distance. Hence, the upstream protocol gives better results if nodes are closely located and can exploit the redundancy.

6. Conclusion

Dense deployment of sensor network results in better operation using collaborative nature of wireless sensor networks. This collaboration results in redundant data which proved as a unique characteristic of a typical sensor network. In this chapter, we introduced a redundancy model. In this model, we observed that redundant data occurs when the nearby sensor devices are separated by not more than twice their sensing radius. It is seen that when condition of redundancy meets near neighbour distribution, which is a very important factor which gives the degree of overlap among sensors in the near neighbour distribution. We proposed the reliable downstream data delivery. Reliable data delivery problem is formulated theoretically using the minimum set cover problem and transformed it to the minimum dominating set (MDS) problem for a practical and feasible standpoint. Proposed framework consist of (i) the core to approximate the MDS; (ii) WFP pulses to tackle a new challenge, lost-all-packet problem; (iii) two-stage recovery to reduce possibility of collision as well as utilize the broadcast nature of wireless networks; and (iv) A-map to prevent error propagation. Performance of this scheme is evaluated with other previous schemes; and showed that it outperforms other schemes in terms of latency and the number of retransmissions and per node energy efficiency.

Upstream energy efficient framework is formulated for the perfectly correlated data aggregation problem by using the Steiner minimum tree (SMT) and showed the upper bound for message complexity. We also compare the performance of this approach with the SPT and the minimum spanning tree (MST) through simulations and showed that it outperforms the SPT and closely approaches the SMT with less computational complexity and without global coordination. We believe in addition to the shortest path tree and the minimum dominating set, one can also exploit the characteristics of the minimum spanning tree or minimum set cover with small amount of overhead and distributed coordination.

The other way is to find an optimal solution for the upstream data aggregation problem assuming a correlation factor between 0 and 1; and then design an approximation solution for the general aggregation problem in a decentralized fashion so that one can implement it over wireless sensor networks. We also discuss the role of near neighbour distribution in design of such schemes.

7. References

GAREY, M. R. & JOHNSON, D. S.(1979), Computers and Intractability, A Guide to the Theory of NP-completeness. Freeman, 1979.

GOPALSAMY, T., SINGHAL, M., PANDA, D., and SADAYAPPAN, P.(2002), A reliable multicast algorithm for mobile ad hoc networks, in Proceedings of 22nd International Conference on Distributed Computing Systems, (Vienna, Austria), pp. 563-570, July 2002.

HEINZELMAN, W. R., KULIK, J., and BALAKRISHNAN, H.(1999), Adaptive protocols for information dissemination in wireless sensor networks, in MOBICOM, pp. 174-185, Aug 1999.

HWANG, F., RICHARDS, D., and WINTER, P.(1992), The Steiner Tree Problem. North-Holland,1992.

Kahn, J.; Katz, & Pister K. (1999). Next century challenges: Mobile networking for smart dust. In Fifth annual international conference on Mobile computing and networking (MobiCom 99), pages 263-270, Seattle, USA, August 1999.

KARP, R. M.(1972), Reducibility among combinatorial problems, Complexity of Computer Computations, pp. 85-103, May 1972.

Madden, Samuel ; Franklin, Michael J. ; Hellerstein, Joseph M.; & Hong, Wei(2002) . Tag: a tiny aggregation service for ad-hoc sensor networks. Proceedings of the 5th symposium on Operating systems design and implementation, 39:131-146, 2002.

Maritz, J. S.(1952). Note on a certain family of discrete distributions. Biometrika, 39:196-8, 1952.

Pottie, G.J. and Kaiser, W.J.(2000). Wireless integrated network sensors. Communications of the ACM, 43(5):51-58, May 2000.

SIVAKUMAR, R., SINHA, P., & BHARGHAVAN, V.(1999), CEDAR: a Core-Extraction Distributed Ad hoc Routing algorithm, IEEE Journal on Selected Areas in Communications (Special Issue on Ad-hoc Routing), vol. 17, pp. 1454-1465, Aug. 1999.

TANG, K. and GERLA, M.(2001), MAC reliable broadcast in ad hoc networks, in Proc. IEEE MILCOM, (Virginia, USA), pp. 1008-1013, Aug. 2001.

WAN, C.-Y., CAMPBELL, A., & KRISHNAMURTHY, L.(2002), PSFQ: A Reliable Transport Protocol for Wireless Sensor Networks, in Proc. ACM International Workshop on Sensor Networks and Architectures, (Atlanta, USA), pp. 1-11, Sept. 2002.

YE, Z., KRISHNAMURTHY, S., & TRIPATHI, S.(2003), A framework for reliable routing in mobile ad hoc networks, in Proceedings of IEEE INFOCOM, (San Francisco, USA), pp. 270-280, Mar. 2003.